ARTIFICIAL INTELLIGENCE IN OPTIMAL DESIGN AND MANUFACTURING

PRENTICE HALL SERIES ON ENVIRONMENTAL AND INTELLIGENT MANUFACTURING SYSTEMS

M. Jamshidi, Editor

Volume 1: ***ROBOTICS AND REMOTE SYSTEMS FOR HAZARDOUS ENVIRONMENTS,*** edited by M. Jamshidi and P.J. Eicker.

Volume 2: ***FUZZY LOGIC AND CONTROL: Software and Hardware Applications,*** edited by M. Jamshidi, N. Vadiee, and T. J. Ross.

Volume 3: ***ARTIFICIAL INTELLIGENCE IN OPTIMAL DESIGN AND MANUFACTURING,*** edited by Z. Dong.

ARTIFICIAL INTELLIGENCE IN OPTIMAL DESIGN AND MANUFACTURING

Zuomin Dong, Editor

P T R PRENTICE HALL
Englewood Cliffs, New Jersey 07632

Library of Congress Cataloging-in-Publication Data

Artificial intelligence in optimal design and manufacturing/Zuomin Dong (Ed.)
p. m.
Includes index.
ISBN 0-13-037540-3
1. CAD/CAM systems. 2. Production planning. I. Dong, Zuomin.
TS155.6.A77 1994
670'.285'63--dc20 93-552
CIP

Editorial production: *bookworks*
Acquisitions editior: *Michael Hays*
Manufacturing buyer: *Mary Elizabeth McCartney*

Printed in the United States of America

10 9 8 7 6 5 4 3 2 1

ISBN 0-13-037540-3

Prentice-Hall International (UK) Limited, *London*
Prentice-Hall Australia Pty. Limited, *Sydney*
Prentice-Hall Canada Inc., *Toronto*
Prentice-Hall Hispanoamericana, S.A., *Mexico*
Prentice-Hall of India Private Limited, *New Delhi*
Prentice-Hall of Japan, Inc., *Tokyo*
Simon & Schuster Asia Pte. Ltd., *Singapore*
Editora Prentice-Hall do Brasil, Ltda., *Rio de Janeiro*

Contents

Series Editor's Foreword

The manufacturing and environment constitute two of the most important economical and ecological activities of any nation. These activities which seem to be conflicting at first, can and should be closely related in the design for manufacturability of a given product or process. A field that can help close potential gap(s) between manufacturing and environment is artificial intelligence. With the advent of artificial intelligence (AI) techniques such as expert systems, fuzzy logic, and neural networks, one can try to make manufacturing processes more cost effective and at the same time environmentally compliant. The mission of this series of books is to bring the state-of-the-art in new and innovative technologies, techniques, and practices in "intelligent manufacturing," "environmentally-compliant manufacturing," and "intelligent environmental protection and monitoring" to the attention of the world community of practicing engineers. The environment is defined as all aspects of the world surrounding us. Examples are air, water, soil, energy resources, other earth resources, population, etc.

The object of this volume, superbly put together by its editor Professor Zuomin Dong of the University of Victoria in Canada, is to introduce recent progress in artificial intelligence and its applications in optimal design and manufacturing, with a focus on the combined use of AI, numerical optimization, and geometric modeling techniques. The book is intended to be of immediate use to practicing engineers in the industry who will be facing issues of the chapters of this book on a weekly basis. As such, as mentioned by Professor Dong elsewhere, it is not a text for AI or geometric modeling, but rather an AI-based set of tools and experiences for optimal production design and manufacturing. As scrics editor, I wish to thank Professor Dong for his superb selection of distinguished authors and practitioners in the field and for making this project a reality which, I am sure, will benefit many engineers for years to come.

Mo Jamshidi, Series Editor
Prentice Hall Series on:
Environmental and Intellignet Manufacturing Systems

Preface

In the last two decades, computer technology has been developing rapidly, penetrating into every phase of mechanical design and manufacturing. However, most computer applications in design and manufacturing have focused on specific tasks such as graphics display, numerical analysis, process control, and simulation. An integrated approach, leading to faster product development, higher productivity, lower costs, and better products, is still being developed.

Recent advances in artificial intelligence (AI) provide a new means for making engineering judgments that require knowledge deduction and symbolic reasoning. The combination of artificial intelligence and well-developed numerical optimization techniques, in particular, provides tools for making qualitative decisions and quantitative evaluations of those decisions for an optimal solution. The integration of artificial intelligence and numerical optimization with geometric modeling further enhances these tools for computer-automated design and manufacturing. The approach has been recognized as much more effective than stand-alone AI, optimization, and graphics systems in improving the overall productivity of design and manufacturing.

Recognition of the potential benefits introduced by combining these advanced technologies in engineering practice and the potential impact of this approach on the formation of an integrated and concurrent engineering environment has led to considerable growth of research in this area. However, given the complexity of design and manufacturing problems and the wide scope of artificial intelligence, numerical optimization, and geometric modeling techniques, it is a difficult task to identify the appropriate software tools and combine the software modules to suit a particular application. The selection of an appropriate AI tool and the proper integration of the selected AI tool with numerical analysis and optimization routines, as well as computer graphics environments, are common tasks faced by many engineers and researchers seeking solutions for applying artificial intelligence to design and manufacturing.

This book introduces recent progress in artificial intelligence and its applications in optimal design and manufacturing, with a focus on the combined use of artificial intelligence, numerical optimization, and geometric modeling techniques. Methods for applying this Quantitative Intelligent System approach to typical design, planning, and manufacturing problems are discussed in-depth and illustrated using practical examples. The book does not cover the artificial intelligence, numerical optimization, and geometric modeling theory which may be found in existing textbooks. Emphasis is placed on the appropriate selection of an AI tool and the creative combination of the selected AI tool with the numerical optimization method and geometric modeling environment for a given design or manufacturing application. A broad range of AI techniques, from the traditional predicate calculus-based search schemes and expert systems, to the emerging applications of fuzzy mathematical techniques and artificial neural network systems are covered in the book.

The book has four parts and thirteen chapters. Part I, Intelligent Design Systems consists of Chapters 1 to 4, and focuses on the application of artificial intelligence, numerical optimization, and geometric modeling, as well as their combined use in mechanical design. These applications include computer-aided machining fixture design, automated generation of sculptured surface models in a spatially constrained environment, an intelligent design system for transmissions, and an intelligent computer-aided design system. Part II, Product and Production Planning Using Intelligent Systems is formed by Chapters 5 to 7, and covers product planning, production sequence generation, and production scheduling using knowledge modeling, knowledge-based systems, numerical optimization, and computer automation. Part III, Intelligent Manufacturing Systems – Planning and Control consists of Chapters 8 to 10, and covers the automated process planning of sculptured part machining, intelligent rough machining of sculptured parts, and neural networks application to welding process control. Part IV, Motion Planning for Intelligent Robotic and Manufacturing Systems is composed of three chapters organized in tandem. Chapter 11 discusses the on-line environment perception and modeling from the 3-D range data of a machine vision system. Chapter 12 focuses on motion planning using nonlinear programming for a given geometric environment. Chapter 13 introduces on-line collision identification and distance calculation methods using artificial neural networks to speed up the nonlinear-programming-based motion planning. A combination of the techniques presented in these three chapters forms an innovative approach for the on-line motion planning of intelligent robotic and manufacturing systems.

The intended audience of the book consists primarily of practicing engineers wishing to update their knowledge of recent progress in computer-automated design and manufacturing. The book also serves as a reference for managers of industrial firms and researchers interested in industrial applications.

I hope that this book can make the rapidly advancing quantitative intelligent system methodology more accessible and provide readers with some references on the combined use of AI, optimization and geometric modeling in design and manufacturing. I also hope that the book can promote broader engineering applications and stimulate further research in this area. The further development of this methodology will inevitably lead to more promising tools for facilitating a fully computer-integrated and concurrent engineering environment.

Acknowledgements

I wish to thank all the chapter authors for their contributions to this book. Many of the authors are well-known scholars in design automation and advanced manufacturing. Some of them are respected instructors and colleagues of the editor. Their names are given in the list of contributors. My thanks also go to Michael Hays, executive editor and Lisa Garboski, production editor at Prentice Hall, and Mo Jamshidi, series editor of this book, for their help with the book's production. I am grateful to Suzanne Stevens, Deyi Xue and Ping Liu who helped prepare the manuscripts, and my wife Diana and my son Stephen for their love and support.

Zuomin Dong
University of Victoria
Victoria, B.C. Canada

List of Contributors

Xin Dong
Applicon Inc.
829 Middlesex Turnpike
Billerica, MA 01821

Zuomin Dong
Department of Mechanical Engineering
University of Victoria
P.O. Box 3055
Victoria, BC, Canada V8W 3P6

Gary W. Fischer
Department of Industrial and Management Engineering
College of Engineering, 4132 EB
Iowa City, IA 52242

Charles R. Gilman
Rensselaer Design Research Center
Rensselaer Polytechnic Institute
Troy, NY 12180-3590

H. H. Tom Huang
Department of Industrial and Management Engineering
College of Engineering, 4132 EB
Iowa City, IA 52242

Andrew Kusiak
Department of Industrial and Management Engineering
College of Engineering, 4132 EB
Iowa City, IA 52242

Yuan-Shin Lee
School of Industrial Engineering
Purdue University
West Lafayette, IN 47907

Hui Li
Department of Mechanical Engineering
University of Victoria
P.O. Box 3055
Victoria, BC, Canada V8W 3P6

Chung-Yuh Liu
Department of Mechanical Engineering
University of Michigan – Dearbon
Dearborn, MI 48128-1491

Roger W. Mayne
Department of Mechanical and Aerospace Engineering
State University of New York at Buffalo
Amherst, NY 14260

James H. Oliver
Department of Mechanical Engineering
Iowa State University
2026 H.M.Black Eng.Bldg.
Ames, IA 50011-2160

Ron Podhorodeski
Department of Mechanical Engineering
University of Victoria
P.O. Box 3055
Victoria, BC, Canada V8W 3P6

Biren Prasad
Electronic Data Systems
General Motors Corporation
CPC Headquarters
Warren, MI 48090

Ming Rao
Department of Chemical Engineering
University of Alberta
520A Chem. Min. Eng. Bldg.
Edmonton, AB, Canada T6G 2G6

Yury Stepanenko
Department of Mechanical Engineering
University of Victoria
P.O. Box 3055
Victoria, BC, Canada V8W 3P6

Michael J. Wozny
Rensselaer Design Research Center
Rensselaer Polytechnic Institute
Troy, NY 12180-3590

Philip Theruvakatti
Department of Mechanical Engineering
Iowa State University
2026 H.M. Black Eng. Bldg.
Ames, IA 50011-2160

Deyi Xue
Department of Mechanical Engineering
University of Victoria
P.O. Box 3055
Victoria, BC, Canada V8W 3P6

Tetsuo Tomiyama
Department of Precision Machinery Engineering
University of Tokyo
Hongo 7-3-1, Bunkyo-ku
Tokyo 113, Japan

Helai Yao
Department of Mechanical Engineering
University of Victoria
P.O. Box 3055
Victoria, BC, Canada V8W 3P6

Geoffrey W. Vickers
Department of Mechanical Engineering
University of Victoria
P.O. Box 3055
Victoria, BC, Canada V8W 3P6

Hiroyuki Yoshikawa
Department of Precision Machinery Engineering
University of Tokyo
Hongo 7-3-1, Bunkyo-ku
Tokyo 113, Japan

H. P. Ben Wang
Department of Industrial Engineering
FAMU/FSU College of Engineering
2525 Pottsdamer St, P.O. Box 2175
Tallahassee, FL 32316-2175

Jing Yuan
Department of Mechanical Engineering
University of Windsor
401 Sunset Ave.
Windsor, ON, Canada N9B 3P4

Juite Wang
Department of Industrial and Management Engineering
College of Engineering, 4132 EB
Iowa City, IA 52242

Ji Zhou
Department of Mechanical Engineering
Huazhong University of Science and Technology
Wuhan, 430074 China

Qun Wang
Department of Chemical Engineering
University of Alberta
520A Chem.Min.Eng. Bldg.
Edmonton, AB, Canada T6G 2G6

ARTIFICIAL INTELLIGENCE IN OPTIMAL DESIGN AND MANUFACTURING

Part I

Intelligent Design Systems

Chapters 1 to 4 of Part I, focus on the application of artificial intelligence, numerical optimization and geometric modeling, as well as their combined use in mechanical design. These applications include computer-aided machining fixture design, automated generation of sculptured surface models in a spatially constrained environment, an intelligent design system for transmissions, and an intelligent computer-aided design system.

Chapter 1, Feature-Based Fixture Design and Set-up Planning, by Xin Dong, Charles R. Gilman, and Michael J. Wozny, presents an intelligent computer-aided design system for designing machining fixtures. Utilizing recent developments in solid geometric modeling, graphical user interface, and knowledge-based engineering, the system assists a fixture designer in the selection of surfaces for locating, clamping, and support. Interactive part and assembly modeling capabilities are provided for defining the workpiece and the fixture elements. Kinematic analysis of deterministic positioning and total constraint for determining the clamping forces of the fixture are also supported. The system provides a valuable computer tool to fixture designers and reduces the lead time in fixture design and manufacturing.

Chapter 2, Automated Generation of Sculptured Surface Models, by James H. Oliver and Philip Theruvakatti, presents a novel approach to sculptured surface model generation that incorporates all available design and manufacturing constraints. Specifically, the chapter describes the latest research results toward the automated generation of sculptured surface models in a spatially constrained environment, subject to specified intrinsic shape properties. To address simultaneously the independent goals of global obstacle avoidance and local control of intrinsic shape properties, surface synthesis is formulated as a combinatorial optimization problem and solved via simulated annealing. Several examples are presented to demonstrate the optimal surface synthesis technique.

Chapter 3, An Intelligent Approach to Conceptual Design of Mechanical Systems, by Qun Wang, Ming Rao, and Ji Zhou, presents an integrated distributed intelligent system for transmission (IDIST). The system combines expert systems with conventional CAD techniques for designing transmissions according to customer's specifications. The difficulties and characteristics of conceptual design for mechanical systems are addressed first. A general problem-solving strategy, which consists of five elements – design task definition, effective conceptual design, parameter determination, analysis, and comprehensive evaluation – is presented. The associated control structure and a tool for developing the expert systems suitable for the conceptual design of a mechanical system are described. Several examples are illustrated.

Chapter 4, Developing an Integrated and Intelligent CAD System, by Tetsuo Tomiyama, Deyi Xue, and Hiroyuki Yoshikawa, presents the framework of a general intelligent CAD system – intelligent, integrated and interactive computer-aided design (IIICAD). The representation of design process information using a computable design process model and the organization of design object information using a metamodel mechanism are discussed. A new design knowledge representation language – integrated data description language (IDDL) – and its use in intelligent design are illustrated. Implementation of the IIICAD system using the object-oriented programming language Smalltalk-80 and a dedicated solid modeler is outlined. The chapter provides a good picture of future intelligent CAD systems.

Chapter 1

Feature-based Fixture Design and Set-up Planning

XIN DONG, CHARLES R. GILMAN, and MICHAEL J. WOZNY
Rensselaer Design Research Center, Rensselaer Polytechnic Institute

1.1 Introduction

Long- and short-term market demand shifts have forced major manufacturers to adopt flexible manufacturing systems. The three major physical elements that make up a flexible manufacturing system are workstations, material handling systems, and workholding tooling. Although considerable advances have been made in both workstations and material handling systems, workholding devices remain a weak link in achieving full system flexibility because they are extremely sensitive to product design and method of manufacture.

Fixtures are workholding devices used frequently in manufacturing operations to hold a workpiece in a prescribed manner of firmness and location. By using a fixture in machining, a part is presented to the cutting tool in a fixed relationship in order to achieve the prescribed geometric accuracy.

Fixture design is considered a complex task and requires a high degree of technical skill usually acquired through years of experience in machine shops. From a technical point of view, a fixture designer must take into consideration workpiece stability, locating accuracy, workpiece deflection under the clamping and cutting forces, ease of workpiece loading and unloading, and so on. Because of all the engineering knowledge involved, the design of fixtures has largely been done manually. It often takes days or weeks to come up with a sound fixture design.

Computer-aided fixture design tools are becoming a necessity. This is due to two reasons. First, experienced fixture designers are more difficult to find; it takes years for novice engineers to become proficient. Second, because of the ever increasing competition in the global market, the manufacturing industry is under pressure to reduce the lead time in new product development. Since fixture design has been a bottleneck in product development, it has become necessary to develop and use computer tools for this task.

Recent advances in solid geometric modeling, graphical user interfaces, and knowledge-based engineering have made computer-aided fixture design possible. This chapter describes an intelligent computer-aided design system for machining fixtures. This system assists a fixture designer in the selection of surfaces for locating, clamping, and support. Interactive part and assembly modeling capabilities are provided for defining the workpiece and the fixture elements. A set-up plan is automatically generated by the system. It includes the number of set-ups and the sequence of these set-ups. The plan generated satisfies both static and dynamic constraints and is optimal with respect to the criteria considered.

This system will provide a valuable computer tool to fixture designers and help reduce the lead time in product development.

1.2 Related Work

Kinematic principles for fixtures have been the subject of many research efforts. Asada and By described the concept and hardware implementation of automatically reconfigurable fixtures [1]. In their work, mathematical formulations are developed for conditions of deterministic positioning, strong and weak accessibility, and detachability. Mani and Wilson presented a synthesis procedure for the selection of locating and clamping points [12]. They used a rating scheme for selecting the faces of a workpiece for locating and clamping. A theory for determining the kinematic freedom of motion of the workpiece has been developed. Chou and colleagues derived analytical formulations for deterministic location, clamping stability, and total constraint [6]. They developed a synthesis procedure to select locating and clamping points and clamping forces so as to satisfy the requirement for clamping stability and total constraint. Menassa and DeVries developed methods for selecting locating and clamping points by applying kinematic theory and good engineering rules [14, 15]. They further formulated an optimization problem for minimizing the workpiece deflection through the positions of support points for a machined part.

Computer-aided design (CAD) is used by a number of systems to provide the part geometry and feature information and to model the fixture elements. Nee and colleagues reported on a knowledge-based program in which fixture elements are parametrically constructed with an interface program calling standard commands from a two-dimensional CAD package [16]. Nee later presented a framework for an object/rule-based automated fixture design system using a solid modeler [17]. Bidanda and Cohen developed a computer-aided fixture selection system for concentric, rotational parts using the geometry and tolerance data created in a CAD system [2]. Sakurai and Gossard used feature recognition to generate feature information from a geometric modeling system for fixture design and set-up planning [18]. Dong and colleagues presented the concept of feature state for capturing the in-process characteristics of machining features [8].

Several researchers have applied artificial intelligence and knowledge-based engineering techniques to fixture design. Bidanda and Cohen used a rule-based system for fixture selection for concentric rotational parts [2]. Markuse and colleagues used rules to configure a fixture from a set of fixture elements [13]. The system uses the branch and bound strategy to avoid exhaustive search. Jiang and colleagues described a computer-aided group technology fixture design system in which the agreement level of a given part to a part family is represented as a fuzzy number [11]. The decision logic of the system is modeled by decision tables and decision table links. The use of expert systems for fixture design was also reported by other researchers [5, 9, 10].

A set-up is a group of cuts that are all made while the part is clamped in one particular position. In machining, an important task is to determine the set-ups and the sequence of these set-ups. A valid set-up plan must satisfy static as well as dynamic constraints. Static constraints include process geometric constraints, good manufacturing practices, location constraints and reference constraints. Dynamic constraints, on the other hand, are due to locating and clamping requirements [5].

Ferreira and Liu developed a rule-based system for generating a set of orientations for machining a workpiece [9]. Rules are used for arranging the order of the objectives, generating candidate orientations, and trimming off the lines of reasoning which have low probability of success. Overall, it uses a generate and evaluate strategy, that is, it first generates some candidate orientations and then evaluates them based on the order of its set of objectives. The evaluation criteria include the availability of a support face, workpiece stability under gravity and cutting force, and locating accuracy. The paper did not address how to select a combination of orientations for a part and the sequence of the set-ups corresponding to these orientations.

Boerma and Kals reported on the development of a computer-aided planning system for the selection of set-ups and the design of fixtures in part manufacturing [3]. In their system, closely related features are machined in the same set-up, while less closely related features may be machined in different set-ups. The paper stresses that the set-up selection has to be preceded by an evaluation of the tolerances concerning the geometric relations between different features.

Chang described a strategy for generating set-up plans in QTC, an expert process planning system [5]. In QTC, static constraints are represented by a precedence diagram. To come up with a set-up plan, feature clusters are created, where each cluster is a group of features which have the same approach direction. These clusters are refined using the precedence constraints and may be regrouped when necessary. While the above strategy seems to be a sound approach, no details are provided in the reference on the refinement and regroup process.

Precedence constraints are often due to feature interactions. A feature interaction occurs when cutting one collection of features affects the way others can be made. Hayes and Wright explored the use of specialized rules to detect such interactions [10]. For example, to identify the interaction between a hole and an angle, a rule is used to check whether or not the angle is cut from the clamping surfaces for the hole in a vise. The planner spends a large amount of time in the early planning stages looking over the part specification for feature interactions and exploring the limitations that those interactions put on the plan.

Although a considerable amount of progress has been made in the area of computer-aided fixture design, some serious limitations still exist. One major problem, as Chang also pointed out in [4], is that most of the research work has neglected the effect of fixture design on process planning. Most research on set-up planning has ignored the fact that set-up planning is dependent upon fixture design, that is, whether a fixture can be economically designed for a candidate set-up in order to satisfy the precision requirements. Very often, precedence relations arising from reference and clamping considerations are ignored in these systems when planning the sequence of machining operations.

Ease of use is another issue that has not been adequately addressed by the existing systems. Many systems do not support interaction between the user and the system. Some systems are not integrated with CAD systems and thus must rely on the user to manually define the geometry of a part.

In this paper, a computer-aided fixture design system will be described. The paper will emphasize two issues: (1) design of an interactive fixture design system, and (2) support for the interaction between fixture design and set-up planning by applying fixturing constraints in the development of a set-up plan.

1.3 Approach

Figure 1.1 illustrates our approach to developing a computer-aided fixture design system. The geometry and tolerance specification for a machined part is imported from a CAD system. Currently, the fixture design system is also capable of importing features from an in-house feature-based design system. In either case, the system provides facilities to the user for defining or modifying machining features.

Candidate set-up orientations are automatically generated by the system based on the tool approach directions of the machining features. The system allows the user to view the part in each of the orientations and to interactively select and position fixture elements such as locators and clamps. Analysis programs can be invoked for the part in each orientation and for the fixture elements specified.

The set-up plan is determined automatically, based on the feedback from the fixture design. The feedback consists of a set of precedence constraints due to reference and clamping requirements. These constraints are currently interactively specified by the user during interactive fixture design. In addition to these constraints, some other precedence constraints are generated automatically. For example, the order among machining operations performed on the same feature is stored as feature-related knowledge and is easily retrieved.

In the following sections, these subjects will be described in detail.

1.4 Modeling of Workpiece and Fixture Elements

The representation of the machined part for fixture design is based on the concepts of features and feature states. All the information necessary to describe the part geometry is represented within the system using machining and tolerance features. The fixtures that will be designed for the machined part are modeled using locating and clamping elements which we consider to be objects within the system. The part and fixture design are composed of machining and

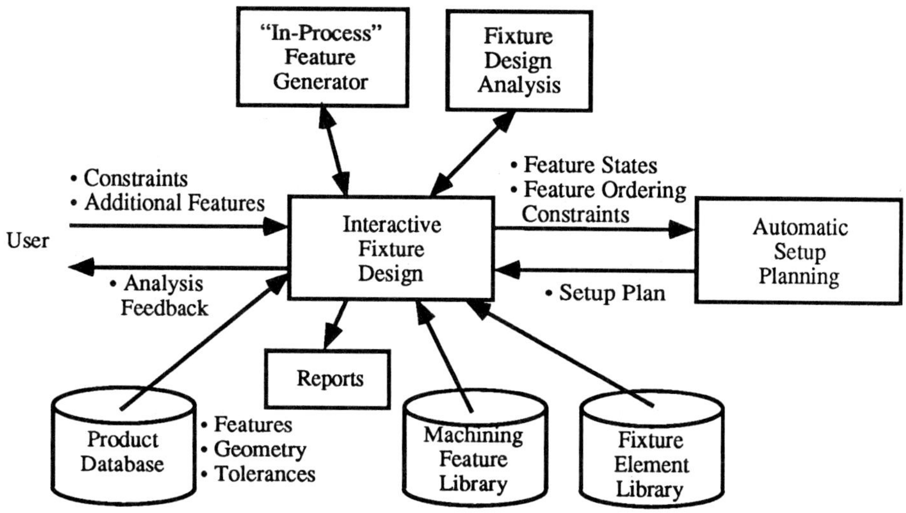

Figure 1.1: An Interactive Fixture Design System

tolerance features and the groups of locating and clamping elements associated with each set-up. This section discusses machining features, tolerances, and fixture elements and their roles in representing the part and fixture design information.

1.4.1 Machining Features and Intermediate Feature States

In order to fixture the part, the part geometry must be expressed in terms of machining features. The machining features, shown in Figure 1.2, are used to represent the part's geometry within the fixture design system. These features are face, 2-D profile, through hole, blind hole, through slot, blind slot, pocket, open step, blind step, and circular groove. In addition to geometry information, each machining feature also holds information and algorithms used to determine the machining operations and sequence of operations necessary to produce a feature of the specified geometry and tolerance.

Often more than one machining operation is necessary to produce a machining feature in a part. For example, a hole may need to be drilled and then bored, or a face may need to be end-milled and then ground. For each operation necessary to produce the feature, there exists an intermediate feature state. For example, suppose a through hole is produced by a drilling and a boring operation. In this example, the through hole has three different feature states: (1) a "null" feature state prior to being drilled, (2) a drilled feature state, and (3) a bored, or final, feature state.

As a part is machined, it passes through a series of intermediate workpiece states. Each of these intermediate workpiece states is composed of a set of intermediate feature states. Figure 1.3 illustrates the intermediate workpiece states that a part passes through as it is machined. The part's first intermediate workpiece state is composed of stock material and a

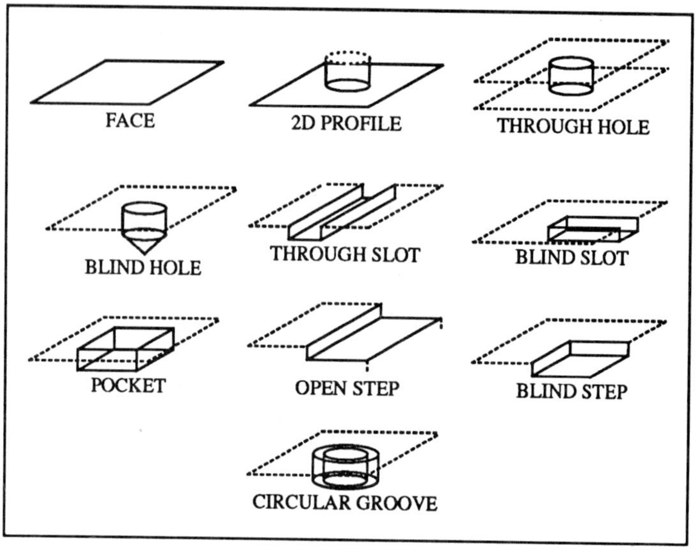

Figure 1.2: Machining Features

set of "null" feature states. Next, the part passes through a series of intermediate workpiece states, each of which is associated with a set of intermediate feature states. Finally, the workpiece reaches the finished state where the geometry of the finished state corresponds to the machining features used to describe the part. The fixture design system uses information from the intermediate workpiece states and intermediate feature states to determine the condition of the workpiece at each of the intermediate workpiece states to enable the selection of the appropriate locating and clamping surfaces for each set-up.

Examples of how the machining feature *through hole* and the intermediate feature states of *through hole* are defined for the fixture design system using the ICAD System's ICAD

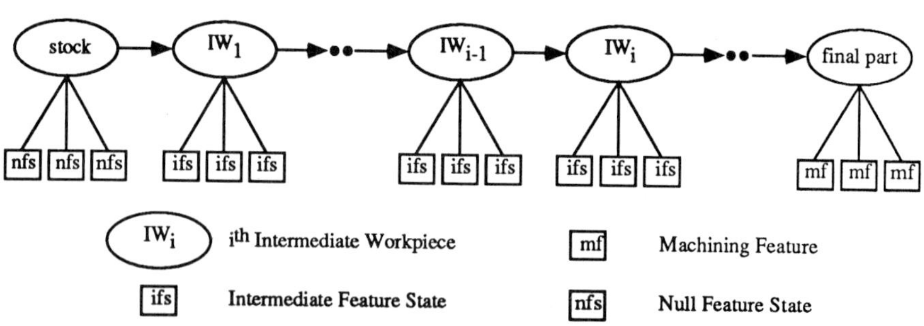

Figure 1.3: Workpiece and Feature States

Design Language[1] (IDL) are discussed in the following paragraphs and source code for the examples appears in Appendix A.

The *through-hole-machining* object, shown below, is defined as a specialization of the *through-hole-intermediate-machining* object with the *:current-state* attribute set to *final* and the *:built-from* attribute set to itself. The through-hole-machining object also has a method, *:tool-approach-directions*, which determines the allowable tool approach directions for the feature and another method, *:feature-machining-ordering-list*, whose value determines the appropriate feature states for the object and the order that the intermediate feature states should be machined.

through-hole-machining

Attributes

:built-from
:current-state
:location
:orientation
:depth
:radius

Methods

:get-tolerance-references
:feature-machining-ordering-list
:tool-approach-directions

The through-hole-intermediate-machining object, shown below, represents all of the intermediate feature states associated with the through hole feature. This is accomplished by having the values of the attributes for the through-hole-intermediate-machining such as *:radius* and *:depth* depend on the value of the attribute *:current-state.*

through-hole-intermediate-machining

Attributes

:current-feature-state

Methods

:depth
:radius
:location
:orientation

[1]The ICAD System and The ICAD Design Language are trademarks of ICAD, Inc.

1.4.2 Tolerances

The machining features do not provide all of the information necessary to determine the order in which the part's features should be machined. Necessary additional information can come from the tolerances associated with the part. Tolerance information can be used to determine the order in which some of the features should be machined. Tolerances are represented as features within the part model. The following tolerance features are included in the fixture design system: size, angle, location, straightness, flatness, circularity, cylindricity, perpendicularity, parallelism, position, concentricity. An example of a tolerance feature, *size-tolerance*, is shown below. The IDL source code for size-tolerance is given in Appendix A.

size-tolerance

Attributes

:name
:feature-1
:feature-2
:lower
:upper

1.4.3 Locating and Clamping Elements

The fixture design system assists the user in creating a set of locating and clamping element designs that can fixture the part for each of the set-ups contained in the machining plan. The locating elements used in the design of the part fixtures are one tool point, two tool points, three tool points, one narrow plate, two narrow plates, one large plate, short pin, long pin, and tapered short pin. The fixture design system uses a simple clamping element that applies a force normal to a designated face. The locating elements are shown in Figure 1.4. An example of the locating element, *long-pin*, is shown below. The IDL source code for long-pin is given in Appendix A. An interesting aspect of the locating element is that all of the locating elements base their *:orientation* and *:position* on the location and position of the machining feature that they are positioned against.

long-pin

Attribute

:locating-feature
:height
:radius

Methods

:constrained-degrees-of-freedom
:orientation
:position

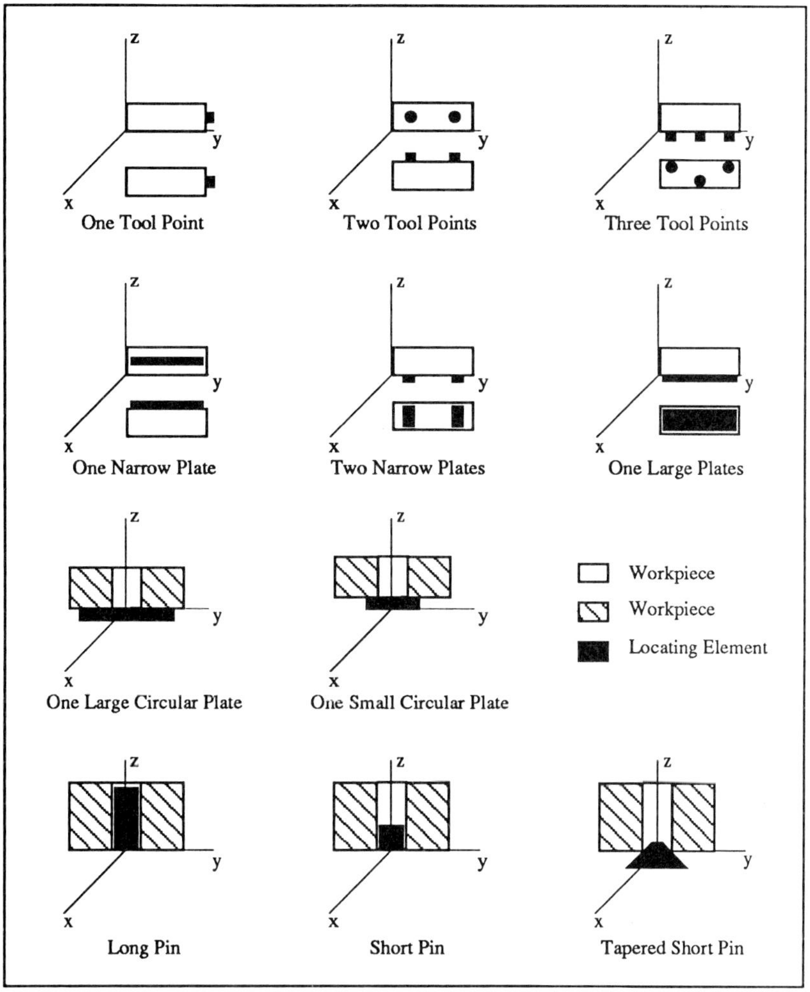

Figure 1.4: Locating Elements Used in Fixture Design

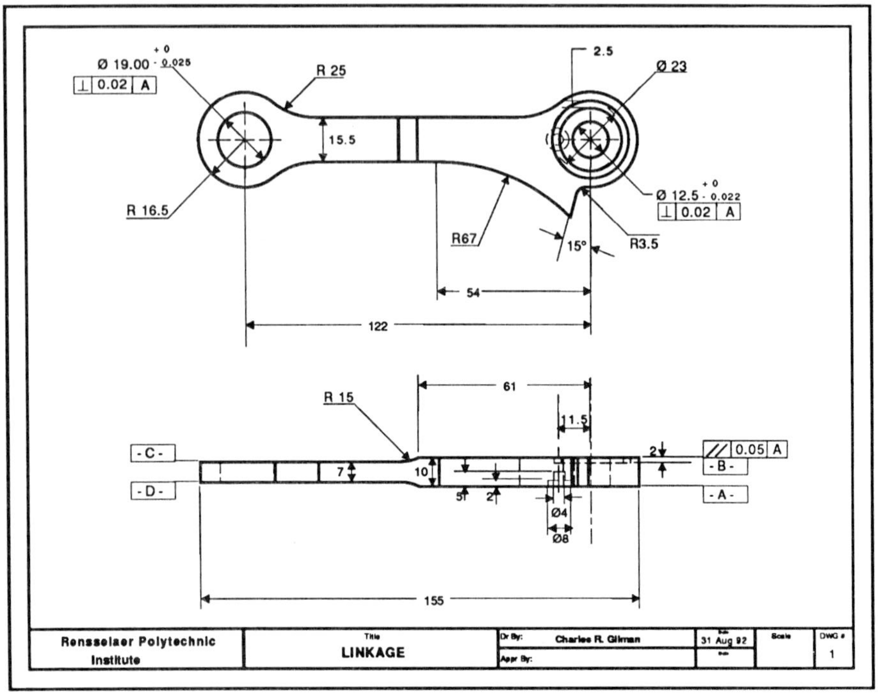

Figure 1.5: Example Part

1.5 Interactive Fixture Design

This section presents an example of how the computer-aided fixture design system supports the fixture designer in the design of a set of fixtures that can be used to machine a part. The sample part selected for this example is the linkage shown in Figure 1.5. This example is taken from Chang [4].

The fixture design process is composed of three steps. The first step is producing the machining feature representation of the part. The second step of the fixture design process is the specification of the locating and clamping elements and constraints for each of the machining set-ups. The third and final step of the fixture design process is the generation of a report that describes the part machining plan and the type and placement of the fixture elements for each of the machining set-ups in the machining plan. The remainder of this section describes these steps in greater detail.

1.5.1 Creating Machining Features

Other than importing machining features from an external product database or CAD system, machining features used by the fixture design system can be created in two ways: interactively defining them within the system, or converting them from design features which can be obtained either from a product database or a CAD system.

Machining features may be created within the computer-aided fixture design system either to produce a machining feature model from scratch or to augment an imported feature model. A machining feature is selected from a menu and the user is presented with a template containing all the attributes for the feature, such as the through-hole feature template in Figure 1.6. The user completes the template and the feature is created within the system. When the geometry of the part is imported from an external CAD system using IGES, a machining feature can be created with respect to the imported part geometry.

When design features on the part have been created in a CAD system and stored in a product database, machining features can also be generated from these design features through feature conversion. Since design features do not always correspond closely to machining features, the feature conversion process can be quite complicated so that some degree of user intervention may become necessary. We are currently conducting research in this area.

The part shown in Figure 1.5 is represented using the machining features and displayed in the computer-aided fixture design system in Figure 1.7. The linkage is represented by ten machining features: the top 2-D profile of the part, two through holes, a circular groove on the top of the linkage, four faces, a blind hole, and a blind hole that acts as a counterbore. The linkage also has eight tolerance features: four datum surfaces, two size tolerances, and two perpendicularity tolerances. Once the user is satisfied that the system has a complete set of machining and tolerance features, the fixture design system determines the intermediate feature states for each machining feature using process planning knowledge contained within the definition of each machining feature.

The intermediate feature states for the sample part are shown in Table 1.1.

1.5.2 Specification of Locating and Clamping Elements for Each Set-up

When the user decides to create all or a part of the fixture design for a particular set-up, he is presented with graphics depicting the in-process part for that set-up as well as a menu of locating and clamping elements. The user then selects the appropriate locating and clamping elements for the in-process part. For the set-up 2 of the sample part, shown in Figure 1.8, the user has selected two short pin locators, two small circular plate locators, and a clamp to create the fixture design.

At any time during the fixture design, the user can check whether the fixture design satisfies locating requirements by selecting the fixture design analysis menu. The analysis dialog box presents the results of an analysis of the fixture design for that set-up. Currently the computer-aided fixture design system supports analysis that detects when a fixture design for a set-up under- or overconstrains the workpiece.

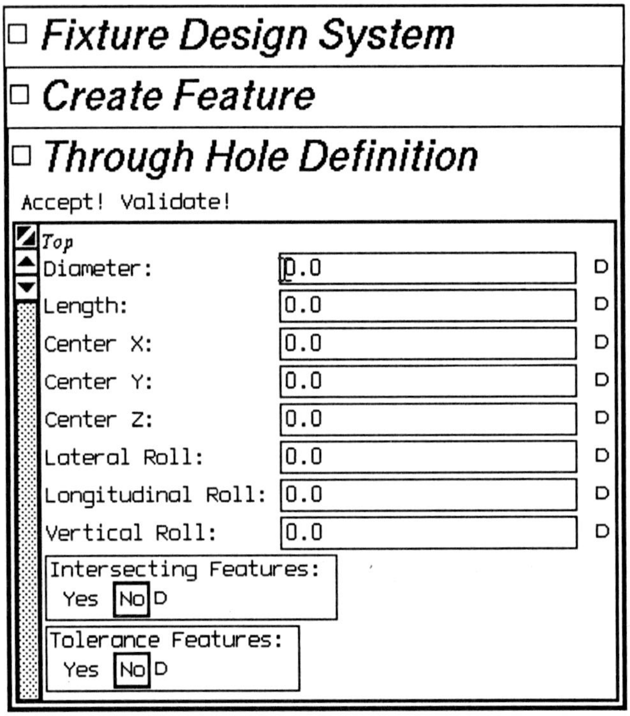

Figure 1.6: Feature Creation Template

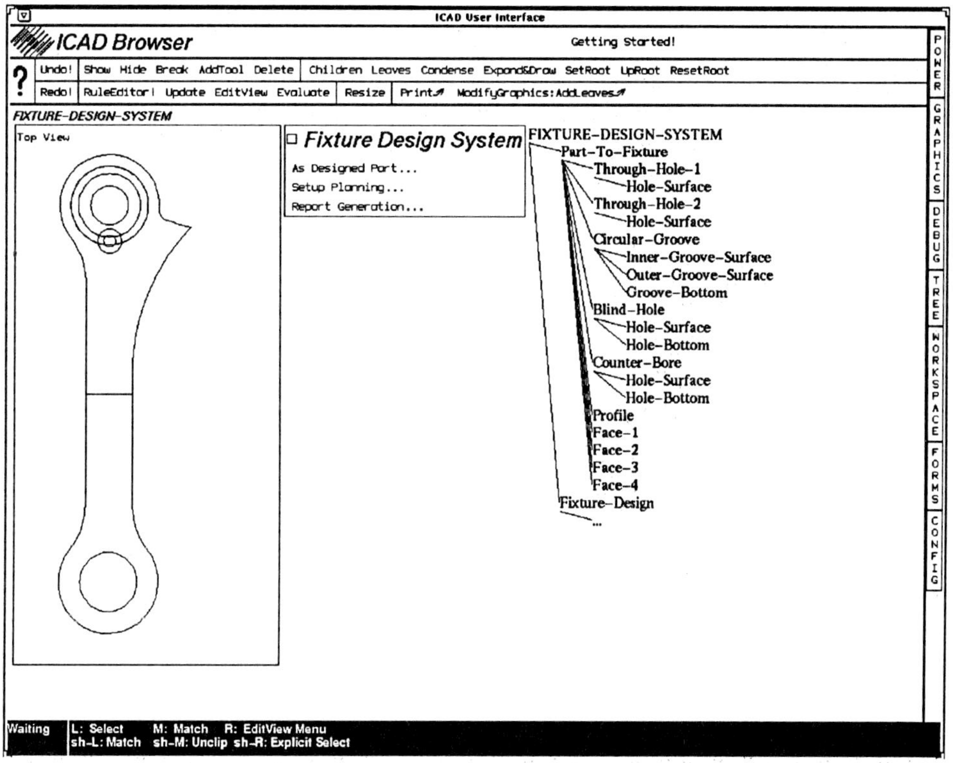

Figure 1.7: Sample Part Represented in Fixture Design System

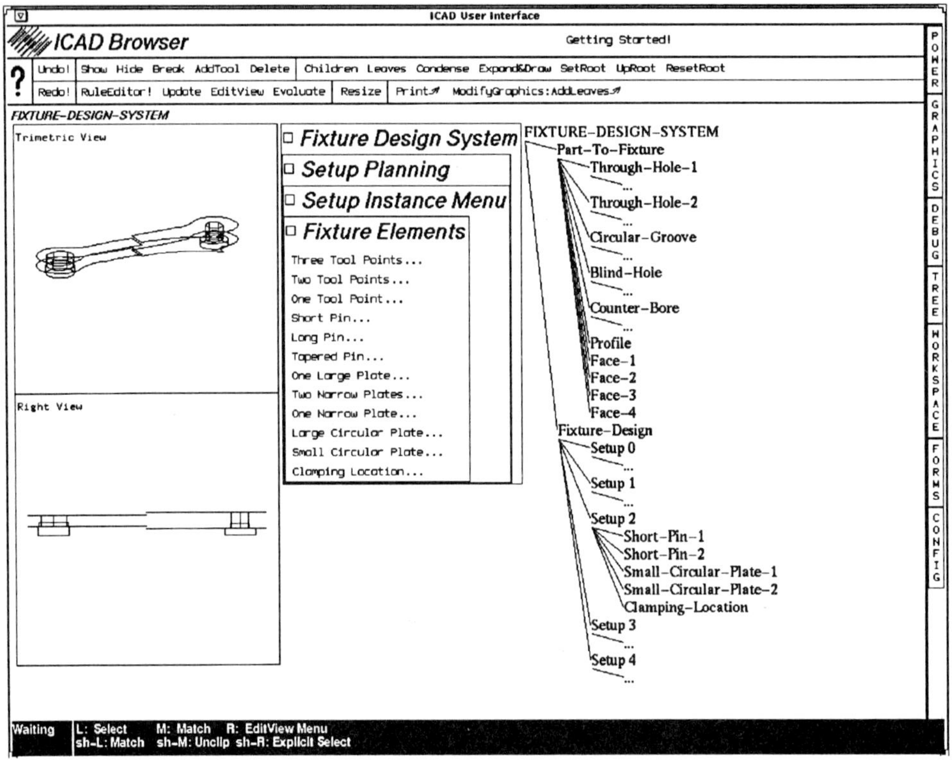

Figure 1.8: Fixture Design

Table 1.1: Intermediate Feature States for the Sample Part

ID	Feature State
FS1	Rough-mill Profile
FS2	Finish-mill Profile
FS3	Drill D12.5 hole
FS4	Bore D12.5 hole
FS5	Drill D19.0 hole
FS6	Bore D19.0 hole
FS7	Rough grooving
FS8	Finish grooving
FS9	Face-mill surface B
FS10	Face-mill surface C
FS11	Face-mill surface A
FS12	Face-mill surface D
FS13	Drill D4 hole
FS14	Counterbore D8 hole

1.5.3 Generation of Fixture Design Report

The final step of the fixture design process involves producing a report that documents the fixture design created within the computer-aided fixture design system. The fixture design report includes the number of set-ups required to produce the part, the intermediate feature states that will be produced during each set-up, the locating and clamping elements for each set-up, their position and the locating faces that they are associated with, and the analysis results for each set-up's fixture design.

1.6 Set-up Planning

In this system, set-up planning consists of two stages: (1) generation of alternative set-up sequences for a machined part and (2) selection of an optimal sequence. During the first stage, sequences that satisfy both static and dynamic constraints are generated. During the second stage, these sequences are evaluated according to some predefined criteria and the optimal one is selected.

The sample part in Figure 1.5 will be used to illustrate the algorithm.

1.6.1 Formulation of Constraints

To correctly describe the precedence constraints among feature states, it is important to distinguish between a feature state and the same feature state with a particular orientation (which we will refer to as *feature-state orientation*). Some precedence constraints exist between two feature states, while others exist between a feature-state orientation and certain feature states. An example of the former is that a feature state corresponding to drilling must precede a feature state corresponding to the boring operation on the same feature. An example of the latter is that a feature state in a certain direction (or orientation) must be preceded by certain feature states which provide the reference surfaces for the machining operation that generates the feature state.

For the sample part, four process geometric constraints are produced automatically by the system from the feature states given in Table 1.1:

C1: FS1 → FS2
C2: FS3 → FS4
C3: FS5 → FS6
C4: FS7 → FS8

where → denotes the precedence relation and FSi → FSj means that feature state FSi must be established before feature state FSj.

Dynamic constraints concerning reference and clamping are specified interactively by the user in the interactive fixture design mode, described in the previous section. For the sample part shown in Figure 1.5, after studying the part specification and examining alternative fixture layouts, the user specifies five additional constraints:

C5: null → (+Z FS1 FS9 FS10)
C6: (FS3 FS5 FS11 FS12) → (+Z FS2)
C7: (FS1 FS9 FS10) → (+Z FS3 FS5 FS11 FS12)
C8: (FS2 FS11 FS12) → (+Z FS4 FS6 FS7 FS8)
C9: (FS2 FS9 FS10) → (-Z FS13 FS14)

Note that while constraints 1 to 4 are specified between feature states, constraints 5 to 9 are specified between feature-states and feature state orientations. Requirements for provisions of reference and clamping surfaces can be easily formulated by the second type of precedence relations. Specifically, provisions for reference surfaces can be formulated as

(<list of feature states>) → (<orientation> <list of feature states>)

and provisions for clamping surfaces can be formulated as

(<orientation> <list of feature states>) → (<list of feature states>).

1.6.2 Generation of Alternative Sequences

To generate set-up sequences, the algorithm makes use of the set of orientation groups used by the user to specify fixturing constraints. Each orientation group consists of an orientation and a set of feature states to be generated in that orientation. For example, in constraint 5, there is one orientation group specified, that is, (+Z FS1 FS9 FS10). Similarly, four more orientation groups can be extracted from constraints 6 to 9: (+Z FS2), (-Z FS3 FS5 FS11 FS12), (+Z FS4 FS6 FS7 FS8), and (-Z FS13 FS14).

The algorithm starts with two arbitrary orientation groups and orders them in a sequence so as to satisfy all the constraints specified on these two groups. If more than one sequence exists, all valid sequences will be stored.

Then the algorithm picks another orientation group from the remaining set and tries to insert it into all the possible positions on every valid sequence. An insertion is successful if all the relevant constraints are satisfied. A constraint is relevant if it relates a feature state in the new group to a feature state in a group already in the sequence. There are three cases of relevant constraints:

- One side of the constraint is the new group and the other side contains a feature state already in the sequence.
- One side of the constraint is a group already in the sequence and the other side contains a feature state in the new group.
- One side of the constraint contains a feature state in the new group and the other side contains a feature state in a group already in the sequence.

All sequences satisfying these constraints will be kept.

The algorithm is summarized as follows:

```
procedure generate_setup_sequence(o_groups,cons,seq)
  begin
    seq = append_seq(o_groups[1],null);
    for i from 2 to num_o_groups
      begin
        r_cons = find_r_cons(cons,seq,o_groups[i]);
        seq = insert_group_all_seq(seq,o_groups[i],r_cons);
      end
  end

function insert_group_all_seq(seq,group,cons)
  begin
    result = null;
    for i from 1 to num_seq
      begin
        new_seq = insert_group_one_seq(group,seq[i],cons);
        result = append_seq(new_seq,result);
      end
    return result;
  end
```

```
function insert_group_one_seq(group,seq,cons)
  begin
    result = null;
    for i from 1 to num_seq_group
      begin
        new_seq = position_group(group,seq,i);
        if satisfy_cons(new_seq,cons,i) append_seq(new_seq,result);
      end
    return result;
  end
```

In the procedure above, the inputs to procedure **generate_setup_sequence** are the orientation groups (**o_groups**) and all the constraints (**cons**). The output is all the set-up sequences (**seq**) that satisfy **cons**.

Several functions used in the procedure deserve some explanation. Function **insert-_group_all_seq** tries to insert a given orientation group into *all* the sequences and returns only the sequences that satisfy those constraints. Function **insert_group_one_seq** is similar to **insert_group_all_seq** except that it inserts a group into only *one* sequence. In addition, function **find_r_cons** returns all the constraints relevant to the orientation group being inserted. Function **append_seq** appends two sequences together and returns the result. Function **position_group** inserts a given group at the *i*th position of the given sequence and returns the new sequence. Function **satisfy_cons** checks if a sequence with a new group inserted at its *i*th position satisfies all the relevant constraints. Note that **num_o_groups** is the number of orientation groups, **num_seq** is the number of valid sequences generated after the insertion of certain number of orientation groups, and **num_seq_group** is the number of orientation groups in a given sequence.

To illustrate the algorithm, we start with (+Z FS1 FS9 FS10), so the initial sequence is ((+Z FS1 FS9 FS10)), shown in Figure 1.9a. Then, we try to insert orientation group (+Z FS2). Since the new group can be inserted either before or after the initial group, two possible sequences are generated, as shown in Figure 1.9b. The relevant constraint for the new group is C1. According to constraint C1, group 1 must precede group 2, so only one sequence is valid, that is, ((+Z FS1 FS9 FS10) (+Z FS2)), shown in Figure 1.9c.

Now we insert (-Z FS3 FS5 FS11 FS12) into the sequence. The relevant constraints are C6 and C7. Although the new group could be inserted at three places in the sequence, only one sequence is valid after applying the constraints. The valid sequence is ((+Z FS1 FS9 FS10) (-Z FS3 FS5 FS11 FS12) (+Z FS2)).

Next, (+Z FS4 FS6 FS7 FS8) is inserted. The relevant constraints are C2, C3, and C8. Again, out of the four possible sequences, only one is valid: ((+Z FS1 FS9 FS10) (-Z FS3 FS5 FS11 FS12) (+Z FS2) (-Z FS4 FS6 FS7 FS8)).

Finally, (-Z FS13 FS14) is inserted. The only relevant constraint is C9. Out of the five possible sequences, two remain valid: ((+Z FS1 FS9 FS10) (-Z FS3 FS5 FS11 FS12) (+Z FS2) (+Z FS4 FS6 FS7 FS8) ((-Z FS13 FS14)) and ((+Z FS1 FS9 FS10) (-Z FS3 FS5 FS11 FS12) (+Z FS2) ((-Z FS13 FS14)) (+Z FS4 FS6 FS7 FS8)), depicted in Figure 1.10.

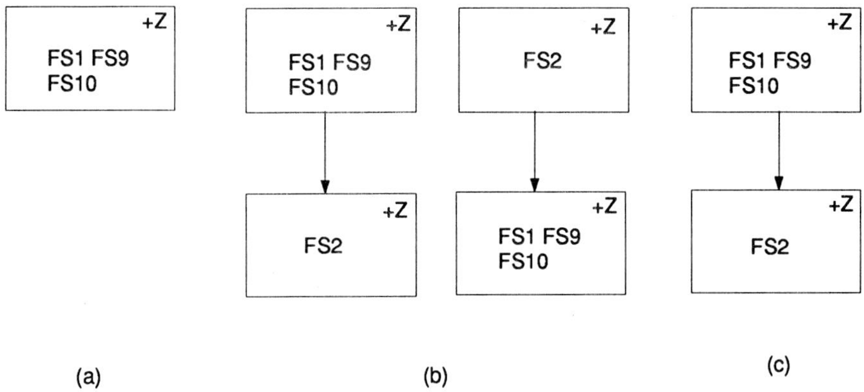

Figure 1.9: Generation of Alternative Set-up Sequences

Theoretically, the maximum number of sequences is $N!$ where N is the number of orientation groups and the minimum number of sequences is 1. In practical situations, the number of valid sequences will be rather small because of the pruning done every time a group is inserted. Also, when loose tolerances are specified on a part, many feature states can be generated on the same set-up, which in turn makes N very small.

1.6.3 Evaluation of Sequences

When several candidate sequences are produced, we need to be able to select the optimal one based on some predefined criteria. Possible criteria include:

1. Minimize the number of set-ups

2. Minimize the cost of fixtures

3. Minimize the number of changes between machines

4. Minimize the number of orientation changes

In our current work, we assume that the number of set-ups and the cost of fixtures are the same for all valid sequences, so we are using criteria 3 and 4 for the selection of the final sequence. The rationale for criterion 3 is apparent as it reduces the nonproductive time and cost associated with transferring a part between machines. Criterion 4 is desirable when similar fixture elements can be used for two consecutive set-ups in the same orientation, as it will reduce the time for assembling the fixture.

Our current strategy is to use criterion 3 as the primary criterion and criterion 4 as the secondary criterion. For the sample part, because all of the machining operations are done on the same machine, criterion 4 will be applied. The sequence ((+Z FS1 FS9 FS10) (-Z FS3 FS5 FS11 FS12) (+Z FS2) (+Z FS4 FS6 FS7 FS8) ((-Z FS13 FS14)) requires three

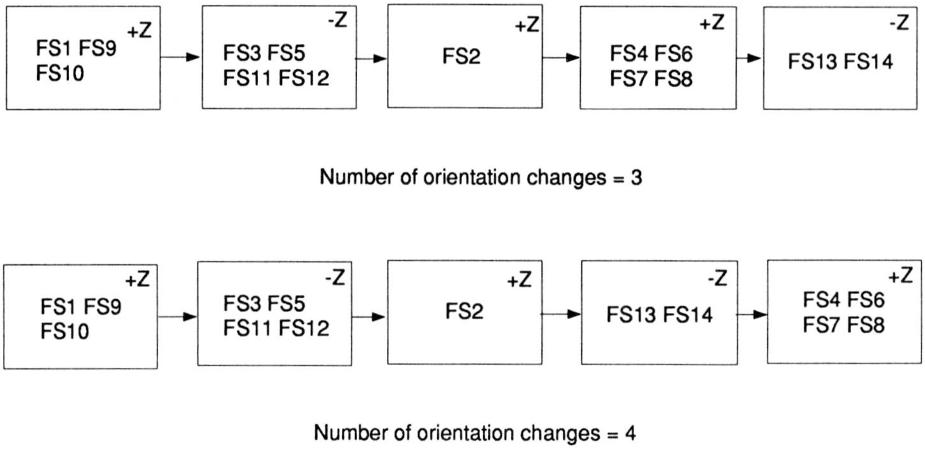

Figure 1.10: Evaluation of Valid Set-up Sequences

orientation changes while the sequence ((+Z FS1 FS9 FS10) (-Z FS3 FS5 FS11 FS12) (+Z FS2) ((-Z FS13 FS14)) (+Z FS4 FS6 FS7 FS8)) requires four (see Figure 1.10). Therefore, the sequence ((+Z FS1 FS9 FS10) (-Z FS3 FS5 FS11 FS12) (+Z FS2) (+Z FS4 FS6 FS7 FS8) ((-Z FS13 FS14)) will be selected for the final set-up plan.

1.7 Software Development

This section discusses implementation details of the computer-aided fixture design system described in the previous sections.

1.7.1 Software Design

The fixture design system is constructed using an object-oriented methodology. Each of the major parts of the fixture design system is represented as a class in the object-oriented model of the fixture design system. The system architecture for the fixture design system is shown in Figure 1.11. This diagram was created using the object-oriented analysis (OOA) notation of Coad and Yourdon [7]. The diagram shows the relationships between the various classes of objects which are represented as double rounded boxes. In the diagram, the shaded arrow indicates that objects in the class at the origin of the arrow request data from objects of the class at the destination of the arrow; the line with the triangle that connects two classes indicates that the class on the base side of the triangle is a part of the class at the apex side of the triangle; and the line with the semicircle indicates that the class on the flat side of the semicircle is a specialization of the class on the rounded side of the semicircle.

In Figure 1.11, the different types of machining features, tolerance features, locating element objects, and clamping element objects are not shown. If locating element objects were included, then they would simply form a single level hierarchy of specialization under the locating element class and the clamping element class. The different tolerance classes would also form a single-level hierarchy of specialization under the tolerance class. The machining features and intermediate feature states would create more complicated relationships if they were included in the figure. All of the classes associated with each feature's geometry would form a single-level hierarchy of specialization below the feature geometry object. Then all of the classes associated with the intermediate feature states would be specializations of both the intermediate feature state class and one of the specializations of the feature geometry class. Likewise, the different machining features would be represented as specializations of the machining feature class and specializations of one of the intermediate feature state specializations. By following the part-of relationships starting with the fixture design system class, the IDL part hierarchy used in the implementation of the computer-aided fixture design system can be identified.

1.7.2 Implementation Details

The fixture design system described in this paper has been implemented using the ICAD System on a Sun Microsystems SPARCstation 1[2]. The ICAD System and Allegro Common LISP form the software development platform for the computer-aided fixture design system. The ICAD Surface Designer[3] provides the capability to create surface models of the machining features and the intermediate feature states. The ICAD System also provides the structure under which to develop the fixture design system and the capability to import surface data from external CAD systems stored in accordance with the standard IGES data format. All of the other algorithms described in this paper, such as the set-up planning algorithm, the determination of intermediate feature states, and the analysis algorithms are all implemented in Allegro Common LISP.

1.8 Summary

A computer-aided system for designing machining fixtures has been described. The system utilizes a feature-based representation to model a machined part. This representation can either be automatically generated from a feature-based design system or interactively defined from the part geometry. Concepts of feature states and feature state orientations have been developed to describe the intermediate shape information about a feature and its relation to a particular set-up. Both static and dynamic constraints are used in determining set-ups and their sequences. While static constraints are automatically generated by the system using process knowledge, dynamic constraints arising from reference and clamping requirements are interactively specified by the user. Although potentially dynamic constraints can also be

[2]SPARCstation is a registered trademark of SPARC International, Inc. and is licensed exclusively to Sun Microsystems, Inc.

[3]ICAD Surface Designer is a registered trademark of ICAD, Inc.

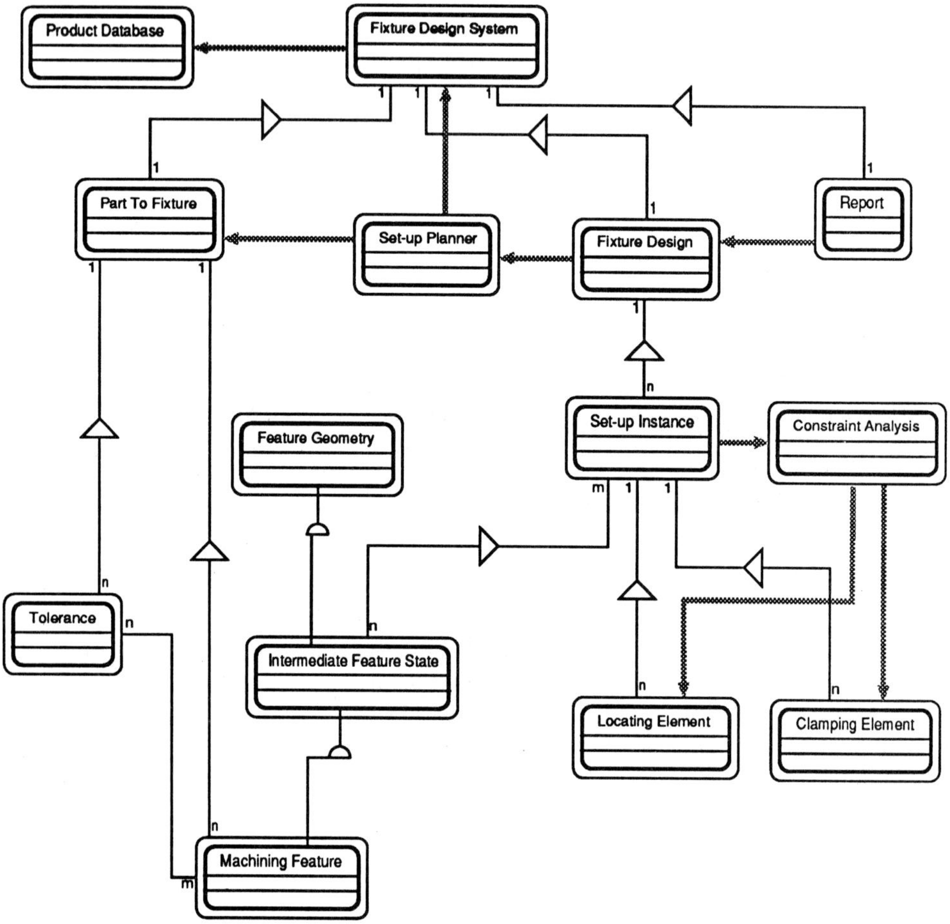

Figure 1.11: Object-oriented Analysis Drawing of Fixture Design System

automatically derived, we feel that a more practical solution at this stage is to have them interactively specified by the user.

Constraints are formulated as precedence relations. Two types of precedence relations have been proposed. The first type relates feature states and the second type relates feature states and feature state orientations. The second type can be effectively used to specify precedence relations due to reference and clamping requirements.

Optimal sequence planning is difficult to achieve because the time complexity of such algorithms is usually exponential. In our current system, we rely on the user to choose the preferred orientations in which to machine a feature, as well as to specify reference and clamping requirements. Once these constraints are specified, the system is able to automatically generate alternative set-up sequences that satisfy these constraints. The ability of the system to generate alternative set-up sequences is important in a flexible manufacturing environment; it makes it possible for a production scheduling system to use an alternative sequence when the optimal sequence presents scheduling problems. The algorithm for generating alternative set-up sequences is simple to implement and effectively makes use of the constraints specified to prune invalid sequences. Because only relevant constraints are used in the pruning task, the search becomes more efficient.

The ICAD system is used as the platform for the software development. It provides tools for constructing object-oriented models and for building customized graphical user interface. Also, commercial software is supported and more acceptable to industry.

For future research, two possible improvements can be made. The first improvement is to automatically formulate the constraints due to fixture requirements. This capability will require computer programs for automatic analysis and synthesis. For example, programs will be needed for selecting locating and clamping surfaces. In addition, a program for positioning error analysis is required to determine whether or not two features can be machined in the same set-up using the same set of locating surfaces.

The second improvement is to evaluate all the orientations in which a feature can be made and choose the optimal one. This is where the computation complexity is likely to be prohibitive. In order to reduce the search space, one possible solution is to prune as many orientations as possible for each feature by using some evaluation function. This approach is currently being investigated.

1.9 Acknowledgments

This project was supported by the National Science Foundation Strategic Manufacturing Grant #DDM-8914172, provided in part by the United States Air Force Materials Laboratory, Manufacturing Research Group, Wright Patterson Air Force Base. Additional support is acknowledged from the NSF Grant ISP79–20240, the Industrial Associates Program of the Rensselaer Design Research Center, and ICAD, Inc. Any opinions, findings, and conclusions or recommendations expressed in this publication are those of the authors and do not necessarily reflect the views of the sponsors.

The authors would like to thank Dr. Warren DeVries for many valuable discussions throughout this project.

References

[1] Asada, H., and By, A.B. (1985). "Kinematic Analysis of Workpart Fixturing for Flexible Assembly with Automatically Reconfigurable Fixtures," *IEEE Journal of Robotics and Automation*, Vol. RA-1, No. 2, pp. 86-94.

[2] Bidanda, B., and Cohen, P.H. (1990). "Development of a Computer Aided Fixture Selection System for Concentric, Rotational Parts," in *Advances in Integrated Design and Manufacturing, DE-Vol. 23-1, Proceedings of 1990 ASME Winter Annual Meeting*, San Francisco, CA, ASME, New York, pp. 151-162.

[3] Boerma, J., and Kals, H. (1988). "FIXES: A System for Automatic Selection of Set-Ups and Design of Fixtures," *CIRP Annals*, vol. 37/1, pp. 443-446.

[4] Chang, C.-H. (1992). "Computer-Assisted Fixture Planning for Machining Processes," *Manufacturing Review*, Vol. 5, No. 1, pp. 15-28.

[5] Chang, T.-C. (1990). *Expert Process Planning for Manufacturing*, Addison-Wesley, Reading, MA.

[6] Chou, Y.C., Chandu, V., and Barash, M.M. (1989). "A Mathematical Approach to Automatic Configuration of Machining Fixtures: Analysis and Synthesis," *ASME Transactions, Journal of Engineering for Industry*, Vol. 111, No. 4, pp. 299-306.

[7] Coad, P., and Yourdon, E. (1991). *Object-Oriented Analysis*, Yourdon Press, Prentice Hall Building, Englewood Cliffs, NJ.

[8] Dong, X., DeVries, W.R., and Wozny, M.J. (1991). "Feature-Based Reasoning in Fixture Design," *CIRP Annals*, Vol. 40/1, pp. 111-114.

[9] Ferreira, P.M., and Liu, C.R. (1988). "Generation of Workpiece Orientations for Machining Using a Rule-based System," *Robotics & Computer-Integrated Manufacturing*, Vol. 4, No. 3/4, pp. 545-555.

[10] Hayes, C., and Wright, P. (1989). "Automating Process Planning: Using Feature Interactions to Guide Search," *Journal of Manufacturing Systems*, Vol. 8, No. 1, pp. 1-15.

[11] Jiang, W., Wang, Z., and Cai, Y. (1988). "Computer-Aided Group Fixture Design," *CIRP Annals*, Vol. 37/1, pp. 145-148.

[12] Mani, M., and Wilson, W.R.D. (1988). "Automated Design of Workholding Fixtures Using Kinematic Constraint Synthesis," in *Proceedings of the 13th NAMRC*, pp. 427-432.

[13] Markus, A., Markusz, Z., Farkas, J., and Filemon, J. (1984). "Fixture Design Using Prolog: An Expert System," *Robotics & Computer-Integrated Manufacturing*, Vol. 1, No. 2, pp. 167-172.

[14] Menassa, R.J., and DeVries, W.R. (1989). "Locating Point Synthesis in Fixture Design," *CIRP Annals*, Vol. 38/1, pp. 165-170.

[15] Menassa, R.J., and DeVries, W.R. (1990). "A Design Synthesis and Optimization Method for Fixtures with Compliant Elements," in: *Advances in Integrated Design and Manufacturing, DE-Vol. 23-1, Proceedings of 1990 ASME Winter Annual Meeting*, San Francisco, CA, ASME, New York, pp. 203-218.

[16] Nee, A.Y.C., Bhattacharyya, N., and Poo, A.N. (1987). "Applying AI in Jigs and Fixture Design," *Robotics & Computer-Integrated Manufacturing*, Vol. 3, No. 2, pp. 195-200.

[17] Nee, A.Y.C. (1991). "A Framework for an Object/Rule-Based Automated Fixture Design System," *CIRP Annals*, Vol. 40/1, pp. 147-151.

[18] Sakurai, H., and Gossard, D.C. (1990). "Geometric Modeling in Setup Planning and Fixture Design," in *Product Modeling for Computer-Aided Design and Manufacturing, Selected and Expanded Papers from the IFIP TC5/WG5.2 Working Conference on Geometric Modeling for Product Engineering*, Rensselaerville, June 17-21, pp. 299-313.

Appendix A Example of ICAD Defparts

A.1 Through-hole-machining Feature Example

```
(defpart through-hole-machining (through-hole-intermediate
                                 through-hole-u-i)
  :attributes
  (:current-state :final
   :built-from-feature (the)
   :design-length (the :u-i-length)
   :design-radius (half (the :u-i-diameter))
   :design-x (the :u-i-x)
   :design-y (the :u-i-y)
   :design-z (the :u-i-z)
   :design-roll-lateral (the :u-i-roll-lateral)
   :design-roll-longitudinal (the :u-i-roll-longitudinal)
   :design-roll-vertical (the :u-i-roll-vertical)
   :tolerance (get-tolerance-references (the) (the :parent))
   :tool-approach-directions (calculate-tool-approach-directions
                               (the :orientation)
```

```
  :feature-machining-ordering-list (built-feature-order-list
                                    (the :tolerances)
                                    (the :design-length)
                                    (the :design-radius)
                                    )
  ))
```

A.2 Intermediate-through-hole-machining Feature Example

```
(defpart through-hole-intermediate-machining (through-hole-geometry
                                               feature-machining)
  :inputs
  (:feature-state-list)
  :attributes
  (:current-state (last (the :feature-state-list))
   :built-from-feature (first (the :feature-state-list))
   :depth (if (eql (the :current-state) :null)
              0 (the :built-from-depth))
   :radius (ecase
             (the :current-state)
             (:final (the :built-from-radius))
             (:drilled
              (get-hole-drill-radius
               (the :built-from-feature :built-from-radius)))
             (:bored
              (get-hole-bore-radius
               (the :built-from-feature :built-from-radius)))
             (:reamed
              (get-hole-ream-radius
               (the :built-from-feature :built-from-radius)))
   :x-center (the :built-from-feature :design-x)
   :y-center (the :built-from-feature :design-y)
   :z-center (the :built-from-feature :design-z)
   :roll-lateral
    (the :built-from-feature :design-roll-lateral)
   :roll-longitudinal
    (the :built-from-feature :design-roll-longitudinal)
   :roll-vertical
    (the :built-from-feature :design-roll-vertical)
   :self-center (make-point
                 (the :x-center) (the :y-center) (the :z-center))
   :self-orientation (roll
```

```
                        :lateral
                        (the :roll-lateral)
                        :longitudinal
                        (the :roll-longitudinal)
                        :vertical
                        (the :roll-vertical))
  ))
```

A.3 Size Tolerance Example

```
(defpart size-tolerance (tolerance-object
                          base-spec-sheet)
  :spec-attributes
  (:name (:domain :string :default "size")
   :feature-1 (:domain :object)
   :lower (:domain :number)
   :upper (:domain :number)
   :feature-2 (:domain :object)
   )
  :spec-controls
  (:title "Size Tolerance"
   :graphics-before t
   ))
```

A.4 Long Pin Locating Element Example

```
(defpart long-pin (box long-pin-specs)
  :inputs
  (:locating-feature)
  :attributes
  (:length (the :locating-feature :radius)
   :width (the :locating-feature :radius)
   :height (the :locating-feature :length)
   :constrained-dof (long-pin-constrained-dof (the :orientation))
   )
  :parts
  ((pin
     :type cylinder
     :position (:bottom 0)
     :orientation (:rotate :right)
     :radius (* (the :width) 0.9)
     :length (* (the :height) 0.75))))
```

Chapter 2

Automated Generation of Sculptured Surface Models

JAMES H. OLIVER and PHILIP THERUVAKATTI
Department of Mechanical Engineering,
Iowa Center for Emerging Manufacturing Technology, Iowa State University

2.1 Introduction

In manufacturing industries, the reduction of process and product development time is recognized as a crucial factor for maintaining or improving overall competitiveness. Thus many companies are striving to reduce the number of physical prototypes and design iterations associated with the traditional design process. For such changes to occur, it is necessary to incorporate design and manufacturing constraints as early as possible in the evolving product model. This trend has motivated the recent flurry of interest in the development of new analytical design tools and the theoretical foundations of the product design and development process itself. It is evident that in order to realize truly significant productivity enhancement, design tools as well as processes must be improved in parallel. In the past, the development of new design tools often precipitated an associated change in the design process. However, more recently, it appears that the necessary changes in the design process are defining the requirements and driving the development of new design technologies. For example, the recent development of feature-based modeling technology can be viewed as a manifestation of the growing emphasis on the philosophy of concurrent product and process design [35] which requires a close association between design geometry and production processes to ultimately reduce both physical prototyping and design changes late in the development process. As this technology matures it will undoubtedly continue to influence design practice and methodology.

The same forces driving solid geometric modeling toward feature-based paradigms are also impacting one of the oldest "computer-aided" design tools namely, sculptured surface modeling. Although this technology is used extensively in the development of a wide variety of consumer and industrial products, sculptured surface models are created today in much the same manner as was done over 20 years ago [3, 14, 17]. Traditionally, surface models are generated by various interpolation techniques, oftentimes based on the coordinate data sampled from a physical model. Consequently, the quality of the surface representation is dependent on the information content of the physical model at the time of data acquisition. Physical models are very expensive and it is very difficult to accurately incorporate changes and refinements of the physical model into the corresponding mathematical representation without completely recreating it. This leads to a rigid design process which inhibits the introduction of simultaneous engineering practices. However, as product development cycles are compressed and the associated number of physical models are reduced accordingly, sculptured surface designers are being asked to capture more design information content per prototype. The technology for modification of existing sculptured surface models is well understood [7, 37] as is the link between such models and automated manufacturing process planning [27, 50]. Furthermore, recent advances in solid geometric modeling are leading to the robust incorporation (i.e., representation) of sculptured surfaces as an integral part of solid geometry [5, 44]. Unfortunately, relatively few methods exist for the creation of surface models subject to constraints derived from spatial, aesthetic, or other design and manufacturing requirements. Recent research at the Iowa Center for Emerging Manufacturing Technology is aimed at addressing this need. This chapter describes the latest research results toward automated generation of sculptured surface models in a spatially constrained environment, subject to specified intrinsic shape properties.

2.1.1 Research Motivation

Automated surface modeling capabilities could be applied in a wide variety of industries. In the case of mass produced items, a design is rarely created entirely from scratch. Frequently, existing components are incorporated from previous designs or from third-party suppliers. Even in the case of a completely new design, exterior shape must accommodate the spatial and functional requirements dictated by interior components.

A typical example of changing design needs is provided by the automotive industry, in which the traditional development process has been serial in nature, driven to a large extent by the body stylist. A design starts with a set of theme sketches which are progressively refined and modified to generate an initial small-scale physical model. Production decisions are often made on the basis of these early sketches and models which may incorporate only very rough dimensional estimates of the proposed design. Development proceeds with the creation of a full-scale clay model of the proposed design. This is typically a working model which is continuously refined based on engineering, manufacturing, and aesthetic criteria. For instance, coordinate measuring machines are used to acquire cross sections of point coordinate data to be used for preliminary packaging verification or various other rough engineering analyses. As the design progresses and more detailed analyses are required, full-scale wooden or plastic models ("bucks") of entire portions of the vehicle are developed based on the (clay) model of the outer body.

Ideally, the final shape of the vehicle converges when the clay model has been validated with respect to the various design criteria (e.g., engineering, manufacturing, marketing) and it satisfies a subjective aesthetic review. At this point no further changes to the clay model are allowed; it is digitized and modeled with sculptured surfaces. This mathematical model then becomes the basis for all further detailed analyses and tooling development. Unfortunately, this process rarely proceeds according to plan. Often, during detailed analysis or testing, design deficiencies are discovered which necessitate changes that affect the outer body. Changes due to manufacturability constraints are also common. Somewhat less frequently, aesthetic, marketing, management, or other factors will impact the design after model release. Outer body changes are first incorporated into the clay model; then the surface model is updated and rereleased. Since the body is typically a smooth shape, changes in one area may affect several adjacent panels. Such model changes are time consuming and often have a ripple effect causing substantial reanalysis and redesign. Furthermore, the later in the development process a change occurs, the greater its expense.

In response to increased competition, the automotive industry is striving to fundamentally reverse the current serial process of developing the outside of the vehicle first, then attempting to accommodate all structural, mechanical, and occupancy requirements. Future vehicle development will be much more parallel in nature and will allow the design of interior components to influence the exterior shape. This change reflects the trend toward incorporating more third-party and "carry-over" components in automotive design. In addition, since aesthetic design is a relatively "fine-grained" activity, it stands to reason that packaging, engineering, and manufacturing constraints take primary precedence through the conceptual design phase. However, this revolutionary change in the development process places new demands on body designers who must not only meet aesthetic design requirements, but also must simultaneously guarantee that a variety of other design criteria are satisfied.

Though the automotive industry is used as an example, such problems and challenges are not limited to automotive design. Similar difficulties are encountered in many other design environments where sculptured surface modeling techniques are employed. For example, the aerospace and shipbuilding industries rely heavily on surface technology and typically incur large expenses in the design process. In the consumer products industries, surface modeling is used extensively in the design of a diverse variety of products, from telephones to tape dispensers. Like the automotive industry, many of these companies are under intense pressure to streamline the product development cycle to remain competitive. Given the wide spectrum of applications and the enormous potential payoff, the development of new strategies for surface synthesis is imperative.

2.1.2 Research Objective

The overall objective of this research is to study and develop methods for the automatic creation of sculptured surface models utilizing and incorporating design and process information which typically exists (or can be derived) during the conceptual stage of product development. In this regard, the surface design problem will be formulated as a combinatorial optimization problem which incorporates both the desired shape properties of the surface as well as a distance metric which relates the surface to polyhedral obstacles. Constraints on surface shape originating from aesthetic, functional, and manufacturability requirements will be accommodated by characterizing their relationship with intrinsic surface properties.

In contrast to traditional methods of generating a surface by interpolating a grid of selected points, the designer would instead work with solid models of known or envisioned components to define obstacles which the surface must avoid by specified spatial tolerances. Additional input may include boundary or character curves and specified bounds on intrinsic surface properties generated from design constraints on functionality and manufacturability. Figure 2.1, a two-dimensional view of an automotive hood design with several constraints, depicts such a scenario for surface synthesis.

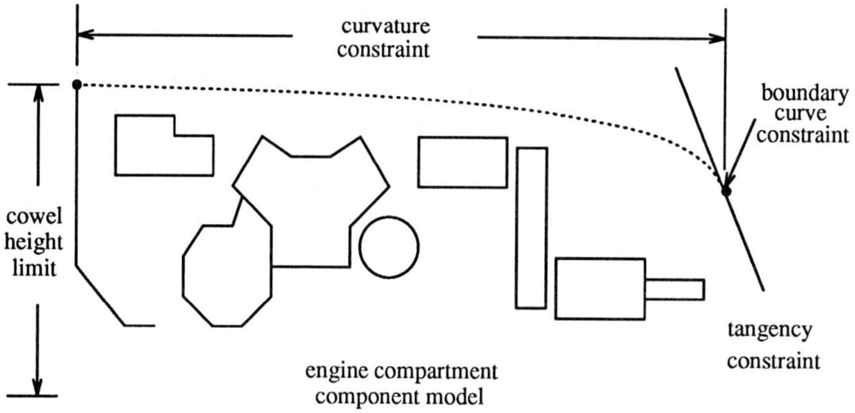

Figure 2.1: Surface Design Synthesis (Surface Shown as Dashed Line)

This new technique will form the basis of a completely automated system for the synthesis of sculptured surface models. Such a tool is not envisioned to provide a highly precise, production-ready surface model, but rather an initial model that globally satisfies all specified constraints. This would provide surface designers with a rapid analytical prototyping facility, which would foster experimentation with novel configurations and enhance creativity.

The following section describes current design synthesis techniques and provides a brief outline of the topic. Then an introduction to current surface design methodology is presented, including traditional surface modeling techniques and recent research in automated surface smoothing. A brief description of energy-based surface generation is also included. Next, the general sculptured surface synthesis problem is formulated as a combinatorial optimization problem, which includes both geometric and intrinsic shape constraints. This section describes the formulation of various constraints, as well as the development of the simulated annealing algorithm for surface synthesis. Several application examples are presented to demonstrate the features of this surface synthesis technique. Finally, some possibilities for future research are described.

2.2 Design Synthesis

In current practice, a designer spends substantial time and effort manually generating candidate shapes or configurations for a design. These designs are generally analyzed or their operation is simulated to either modify or eliminate the various candidates. This apparent inefficiency has spurred research activity in the automated synthesis of component shape. The goal is to relieve the designer of the task of manually creating design candidates by automatically generating configurations which are formulated on the basis of the specified solution space and design constraints. Design synthesis research to date has primarily concentrated on the structural aspects of mechanical elements, for example, trusses, frames, and brackets. From the standpoint of automated surface synthesis, these techniques are not directly extensible, although certain aspects may be applicable.

One approach to the shape synthesis problem is shape optimization, which is essentially a combination of optimization and finite element analysis techniques. Shape optimization methods have been studied extensively [2, 20] and typically employ the iterative application of finite element analysis to optimize characteristic dimensional parameters of an initial configuration with respect to some desired structural behavior (e.g., weight, deflection, cost). A disadvantage of these techniques is that the choice of structural topology, which must be specified at the outset, severely constrains the solution. Imam [23] suggests an adaptive remeshing scheme to address this problem, while Shah [46] presents a combined algorithmic and heuristic technique to generate good candidates for initial form.

The shape optimization problem can be related to the surface synthesis problem. Williams [49] uses a structural analogy between curved shell finite elements and parametric surface patches to formulate a technique for smoothing surfaces by minimizing strain energy. However, this method is computationally expensive since it involves the solution of nonlinear differential equations. Also, like the other shape optimization techniques this approach does not synthesize geometry, but rather modifies it in some local sense. Within this framework, there is no clear way to incorporate global aspects of the surface synthesis problem, such as its proximity to, and avoidance of, obstacles.

The application of knowledge-based system technology to the design synthesis problem is one of the most active areas of design research. In a manner analogous to the shape optimization techniques, early efforts [12] focused on the generation of acceptable design parameters given a general configuration and pertinent domain knowledge. More recently the focus has shifted to automating the generation of structural configuration (topology) as well as parametric dimensions. Welch and Dixon [48] describe a system for the design of sheet metal parts with separate but interacting modules for configuration design and parameter generation. The configuration level allows design storage and backtracking, and both levels employ iterative redesign so that a best-first search of the design space is accomplished. Related techniques that deal with the topological synthesis of 2-D frame structures have also been developed [22, 34].

However, these systems are limited in their domain by the extent of the knowledge base. Generally knowledge-based solutions do not scale up well, that is, techniques that work well in a relatively small, well-defined domain may not be feasible in a much larger domain such as the surface synthesis problem. Such an approach, however, may be useful for solving specific subproblems within the larger framework. A knowledge-based approach may well be suited for finding the initial topological distribution of surface patches — a logical collection of triangular and rectangular patch topologies, for instance.

2.3 Surface Design

With roots in both the automotive and aircraft industries, sculptured surface modeling techniques emerged in the early 1960s in response to the growing capabilities of numerically controlled machining technology and the subsequent need to represent smooth surfaces numerically. The technology of parametric sculptured surfaces evolved as a generalization of the space curve representations and often exploited physical analogies. Both Böhm and colleagues [3] and Farin [14] provide excellent historical and contemporary surveys of curve and surface modeling. In the following section, a general familiarity with various surface representation and interpolation schemes is assumed.

2.3.1 Traditional Surface Modeling

The need to translate a physical model or drawing into a mathematical representation motivated early surface modeling techniques. This process typically involves the discretization of the object into a large set of point coordinates. Various techniques have been developed to interpolate such data to produce smoothly varying bivariate surface functions. In addition, alternate techniques have been devised for "free-form" surface design, that is, the conceptualization of surfaces completely analytically without initial physical models. These two design philosophies form the basis for a general categorization of surface modeling techniques as either transfinite methods (typically associated with S. A. Coons) or tensor product methods (attributed to P. Bezier). Although these techniques were developed contemporaneously and either one can be applied to both design paradigms, transfinite methods are most often applied to generate surfaces from physical models, while tensor product methods are generally used for free-form surface styling [3].

The transfinite surface representation builds on the idea of a ruled surface which, in its simplest form, is a linear interpolation of two space curves which have the same parametrization. Given four boundary curves $\mathbf{p}_0(u)$, $\mathbf{p}_1(u)$, $\mathbf{q}_0(v)$, $\mathbf{q}_1(v)$, Coons observed that a smoothly varying surface between them could be formulated as the "Boolean sum" of the two ruled surfaces, $\mathbf{P}(u,v)$ and $\mathbf{Q}(u,v)$, formed from opposite boundary curves. Thus, a Coons surface (patch) is defined as

$$\mathbf{S}(u,v) = \mathbf{P} \oplus \mathbf{Q} = \mathbf{P}(u,v) + \mathbf{Q}(u,v) - \mathbf{PQ}(u,v) \tag{2.1}$$

where $\mathbf{PQ}(u,v)$ is a doubly ruled surface which simply blends the four curve end-points. The piecewise implementation of this technique often led to problems in maintaining continuity across adjacent patches. Subsequently, the Gordon surface was developed as a generalization of the Coons patch over a network of intersecting curves. A typical modeling procedure (for a topologically rectangular surface) involves, as a first step, the interpolation of curves in both parametric directions through a regular grid of digitized data points. Due to variations in the data acquisition and interpolation techniques the parametrically orthogonal curves rarely intersect in Euclidean space. Thus the most tedious part of the modeling task is to manipulate the curves until they mutually intersect within machine precision. When this is accomplished, the model is complete, since a Gordon surface is completely defined by its bounding curves.

The tensor product surface can be viewed as a generalization of curve interpolation techniques. Although they can be described in terms of any interpolative basis, the Bernstein basis of Bezier (and more generally B-spline) curves is used here due to its intuitive nature. Consider a cubic Bezier curve in the plane of the paper defined by four control points $\mathbf{p}_i$ and the function

$$\mathbf{p}(u) = \sum_{i=0}^{3} \mathbf{p}_i \mathbf{B}_i(u) \qquad u \in [0,1] \tag{2.2}$$

where $\mathbf{B}_i(u)$ are the cubic Bezier blending functions. If this curve is swept normal to the paper and each control point is allowed to trace out a Bezier curve $\mathbf{p}_i(v)$ such that

$$\mathbf{p}_i(v) = \sum_{j=0}^{3} \mathbf{c}_{ij} \mathbf{B}_j(v) \qquad v \in [0,1] \tag{2.3}$$

then the resulting tensor product (Bezier) surface is defined by

$$\mathbf{T}(u,v) = \sum_{i=0}^{3} \sum_{j=0}^{3} \mathbf{c}_{ij} \mathbf{B}_i(u) \mathbf{B}_j(v) \qquad u,v \in [0,1] \tag{2.4}$$

This technique of surface design is typically implemented by allowing the user to interactively create a series of approximately parallel curves. These could be created by simply allowing the user to select and modify control points, or the system may initially interpolate curves (i.e., calculate the governing control points) which pass through a series of data points. Assuming that the series of roughly parallel curves has the same parametrization, the tensor product surface is obtained by interpolating the corresponding control points from each curve in the opposite parametric direction. In the piecewise construction of tensor product surfaces, the user must take special measures to ensure derivative continuity. Thus with a relationship analogous to that between Gordon and Coons surfaces, the B-spline surface evolved as a generalization of the Bezier surface patch.

2.3.2 Automated Surface Generation

Much of the research toward automated surface generation has followed a procedure analogous to the shape optimization approach, that is, given an initial model which meets most design constraints, generate a slight perturbation to optimize it with respect to some other constraint, typically curvature. For example, Andersson and colleagues [1] minimize control point change with a linearized distance metric, subject to linearized normal curvature constraints, to ensure convexity. An iterative solution is proposed in which the surface is sampled to find a set of points which violate the constraint, then the linear programming problem is solved with Karmarkar's algorithm [25] and the process is repeated until a stable solution is reached. Lott and Pullin [30] reverse this formulation by using a measure of surface curvature as objective function, subject to a constraint based on the distance between the original and modified surfaces. Both objective and constraint functions are linearized in the surface parameter space and the simplex method [42] is used to minimize the curvature metric. Both these approaches reduce the degrees of freedom of the problem by restricting the control points to deviate only along normals of the original surface. In contrast, Ferguson [18] develops convexity conditions on the governing control point net to eliminate spurious inflections from B-spline curves. This approach is extended to surfaces [19] by projecting the control point net onto a user-specified plane. General perturbations are allowed, and a full nonlinearly constrained optimization problem is formulated and solved with successive quadratic programming. It is interesting to note that these three approaches address a similar problem from three different applications: Andersson deals with automotive design, Lott and Pullin with ship-building, and Ferguson with aircraft surfaces.

A similar approach is surface smoothing based on reflection lines [26, 40]. These techniques simulate the reflection of lights on a reflective surface to assess its quality. Such methods are not as "finely tuned" as the curvature-based methods described above, since reflection lines are based on a first-order interrogation, while curvature involves second-order interrogation. However, they may be more useful in detecting global imperfections [14].

2.3.3 Energy-based Methods

Terzopoulos [47] and Platt and Barr [39] present techniques for physical system simulation based on classical mechanics. Celniker and Gossard [6] use a similar approach in the development of an energy-based system for free-form surface design. This system allows the designer to control surface shape by imposing boundary conditions and external loads. The surface generated is not a sculptured surface in the traditional sense but rather a wireframe which connects points that behave according to the physical model. Applications for this system include representation of noncritical surfaces in design or "shrink-wrapping" solid models [21].

Current surface synthesis methods impose, on the designer, the task of ensuring that the surface not only accommodates the global environment in which it must function, but also all smoothness criteria derived from aesthetic, performance, manufacturability, or other design requirements. Obviously, most of the effort toward automation has focused on the latter of these tasks. The energy-based technique, by providing control via external loading, comes the closest in spirit to global surface synthesis. But the resulting surface model is inadequate for the precise representation and processing required by many products.

2.4 Problem Formulation

This study focuses on the synthesis of nonuniform rational B-spline (NURBS) surfaces. This representation has numerous attractive attributes with respect to design functionality, and is emerging as the de facto standard in geometric modeling applications. For example, NURBS in general provide high-order derivative continuity for parametric curve and surface design. In addition NURBS are well suited for interactive design due to the local nature of control point influence, that is, modification of a single control point generally affects only a local vicinity of the surface, while the remainder is unaffected. In computer-aided design applications substantial numerical manipulation is common; thus the superior numerical stability of the Bernstein basis relative to the monomial (or power series) basis is significant [16]. Another useful property of NURBS is the convex hull property, which states that the surface is always contained within the convex hull of its control polygon [14]. Finally, NURBS can be used to precisely represent common quadric surfaces such as spheres, cones, ellipsoids, and so on [36]. General descriptions of B-spline curves and their various forms can be found in most geometric modeling textbooks (e.g., Farin [14] and Mortenson [33]) or any of several survey articles on the topic (e.g., Böhm et al. [3] and Piegl [38]). A brief description follows to facilitate the problem formulation.

B-spline curves and surfaces are based on Bernstein basis functions which are typically generated via the Cox-de Boor algorithm [9, 10]. Given knot vector $\mathbf{T} = \{t_0, ..., t_i, t_{i+1}, ..., t_m\}$ a monotonically increasing sequence of real numbers, the ith B-spline basis function of degree p (order $p+1$), denoted $N_{i,p}(t)$, is defined by the recursive relationship

$$\mathbf{N}_{i,0}(t) = \begin{cases} 1 & \text{if } t_j \leq t < t_{j+1} \text{ and } t_j < t_{j+1} \\ 0 & \text{otherwise} \end{cases} \tag{2.5}$$

$$\mathbf{N}_{i,p}(t) = \frac{t - t_i}{t_{i+p} - t_i}\mathbf{N}_{i,p-1}(t) + \frac{t_{i+p+1} - t}{t_{i+p+1} - t_{i+1}}\mathbf{N}_{i+1,p-1}(t) \tag{2.6}$$

where it is understood that $0/0 = 0$. A B-spline surface of degree $(p,\, q)$ is specified by an $(m+1, n+1)$ grid of control points $\mathbf{P}_{ij}$ arranged in a topologically rectangular array and knot vectors $\mathbf{U}$ and $\mathbf{V}$ of length $(m+p+2)$ and $(n+q+2)$, respectively. The surface, denoted as $\mathbf{S}(u,v)$, is thus defined as the tensor product of the control point array and the B-spline basis functions are defined over each knot vector

$$\mathbf{S}(u,v) = \sum_{i=0}^{m}\sum_{j=0}^{n} \mathbf{N}_{i,p}(u)\mathbf{N}_{j,q}(v)\mathbf{P}_{ij} \tag{2.7}$$

The knot vector governs the relationship between parametric and spatial variation, and its entries represent the parameter values at the segment joints (knots). The knot vector for a uniform B-spline surface contains integer entries ranging monotonically from zero to the number of segments in each parametric direction. A nonperiodic B-spline is characterized by a knot vector in which the first and last knot values are repeated p (degree) times. This results in a surface that interpolates the control points on the edge of the rectangular array. A rational B-spline allows assignment of a weight factor to each control point. Homogeneous coordinates are used to represent a rational surface in r-dimensional space as a polynomial in $(r+1)$-dimensional space. A projective mapping is used to recover the resulting r-dimensional surface. Finally, the parametric derivatives of a NURBS surface (denoted by $\mathbf{S}^u(u,v)$, $\mathbf{S}^v(u,v)$, $\mathbf{S}^{uu}(u,v)$, $\mathbf{S}^{vv}(u,v)$, and so on) can be compared precisely [4, 29].

Using this terminology the general surface synthesis problem is described as follows. Consider the simple design environment depicted in Figure 2.2. Suppose that a sculptured surface model is desired in the vicinity of a single polyhedral obstacle. Suppose also that the

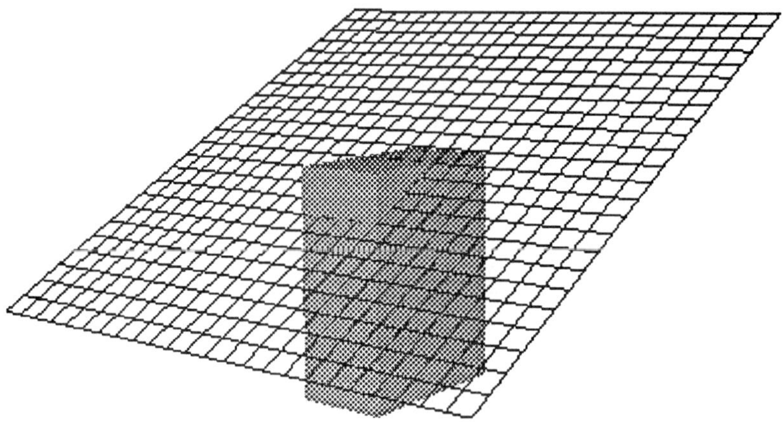

Figure 2.2: Surface Synthesis, Initial Configuration: Obstacle and Surface

general spatial location of the surface relative to the obstacle is known and is approximated initially by a planar distribution of control points as shown in the figure. To characterize the desired functional behavior of the surface, a set of constraints is formulated to induce the surface to avoid the obstacle by at least some tolerance δ. These constraints characterize the global behavior of the surface. In addition, suppose that a number of other design constraints, based on surface functionality and manufacturability, for instance, can be formulated in terms of surface intrinsic properties. These constraints generally have a subtle local influence on the overall shape of the surface. To maintain the rectangular topology of a surface patch, some of the control points (generally the corners) are designated as fixed. The coordinates of the remaining control points are the design variables of the surface synthesis problem, that is, a surface is synthesized by finding suitable positions for the free control points such that all the constraints are satisfied.

2.4.1 Simulated Annealing Algorithm

To simultaneously address the independent goals of global obstacle avoidance and local control of intrinsic shape properties, the surface synthesis problem is formulated as a combinatorial optimization problem and solved via simulated annealing (SA). Simulated annealing is a probabilistic combinatorial optimization technique based on an analogy to the statistical mechanics of disordered systems [28]. In the physical process of annealing, a material is heated and allowed to cool slowly at incrementally decreasing temperatures, so that it reaches thermal equilibrium at each temperature. Consequently, its atoms will reach a state of minimum energy, the ground state, despite any local minima.

In SA, the objective (cost) function to be minimized is analogous to the total energy of the system. From an algorithmic point of view, SA is essentially an iterative improvement strategy augmented by a criterion for occasionally accepting higher cost configurations [43]. Given an objective (cost) function $C(\mathbf{x})$ and an initial state vector $\mathbf{x}_0$, iterative improvement seeks to improve the current state by randomly perturbing $\mathbf{x}_0$. If the new state $\mathbf{x}_i$ yields a lower cost, then it replaces the current state and the perturbation process continues from $\mathbf{x}_i$. If the perturbed state produces a higher cost than the original state then it is rejected, and the perturbation continues from the original state. This procedure is continued until no further improvement in the cost can be obtained. The drawback of iterative improvement is the possibility of converging to a local minimum. One could restart the process using a number of different initial configurations and take the best result, yet there is no guarantee that a good solution will be found. In addition, such an approach can be extremely inefficient.

If a higher cost state is generated in the SA algorithm, the state is accepted with a probability based on the current temperature. Since the probability of accepting a higher cost state decreases with temperature, the SA algorithm mimics the physical process of annealing. It has been described as a "hill-jumping" technique because it can tolerate temporary degradation of the objective function to "jump" out of the vicinity of a local minimum. The behavior of the SA algorithm has been characterized as first following the gross behavior of the objective function to find an area in its domain where the global minimum should be present, irrespective of local minima found on the way. It progressively develops finer details, ultimately finding a good, near-optimal local minimum if not the global minimum itself [8]. This approach has been successfully applied to problems in VLSI circuit design [11], structural truss design [13], mechanism synthesis [24], and robotic path planning [45]. Given its global nature and suitability for problems with a large number of variables, simulated annealing is ideal for application to problems in geometric model synthesis. In fact, Malhotra and colleagues [31] demonstrate the feasibility of this approach to geometric curve synthesis problems.

In an algorithm to simulate the distribution of atoms in equilibrium at a given temperature T, Metropolis and colleagues [32] modeled this behavior with a Boltzman distribution as follows:

1. Randomly perturb $\mathbf{x}_i$ to some new state $\mathbf{x}_{i+1}$, and calculate the corresponding change in cost $\Delta C = C(\mathbf{x}_{i+1}) - C(\mathbf{x}_i)$.

2. If $\Delta C < 0$, accept the new state.

3. If $\Delta C > 0$, accept the new state with probability $P = \exp\left(\frac{-\Delta C}{T}\right)$

These three steps are referred to as the Metropolis algorithm and represent the core or the "inner loop" of the SA algorithm. The acceptance criterion is typically implemented by generating a random number $R \in [0, 1]$ and comparing it to P; if $R < P$ then the state is accepted otherwise it is rejected. The number of inner loop iterations is generally formulated as a function of the current temperature (control parameter) and the number of degrees of freedom in the system. The inner loop is repeated at incrementally decreasing values of temperature. The "outer loop" of the algorithm is referred to as the cooling schedule and specifies the method (i.e., linear, exponential) by which the temperature is decreased. Typically, the initial temperature is chosen such that 80 percent of all states generated are accepted. If the initial temperature does not yield the desired percentage of acceptance then it is increased and the inner loop is repeated. The algorithm is terminated when the cost function remains approximately unchanged (to within some tolerance) for a number of consecutive outer loop iterations.

Implementation of the simulated annealing algorithm requires the specification of a problem configuration (i.e., domain over which the solution will be sought), as well as the definition of a neighborhood to govern the nature and magnitude of the allowable perturbation, a cost function which characterizes the function to be minimized, and a cooling schedule to control the rate of temperature decrement and the number of inner loop iterations. The work presented here builds on the results of Malhotra and colleagues [31]. In particular, although the problem configuration has been generalized from a two- to a three-dimensional Euclidean space, the same neighborhood and cooling schedules have been applied. The primary distinction is in the formulation of a cost function for surface synthesis.

2.4.2 Initial Cost Function Formulation

The following discussion provides a chronological description of the research, experimentation, and analysis which has guided the formulation and development of the general surface synthesis technique. The initial study focuses on the solution of the synthesis problem outlined above. First, a cost function which characterizes only the global proximity problem is formulated, and its implementation in a simple design configuration is demonstrated. Attempts to extend this simple cost function to more general synthesis problems revealed its limitations. These results guided the development of several subsequent reformulations which are described in detail. The viability of the final cost function formulation is described with an example which demonstrates the synthesis of a NURBS surface which is periodic (and closed) in one parametric direction.

Since the general cost function for this problem must accommodate components derived from several diverse contributing factors, a reasonable implementation is to sample the surface at regular parametric intervals and sum the contribution of each point to each cost component. For a given configuration the total cost contributed by each point on the surface is defined initially as the sum of the following components:

$$C_T = C_o + C_a \tag{2.8}$$

where C_o is the cost due to obstacle proximity or interference and C_a is the cost due to excessive surface area. In this initial cost formulation, the individual components of the total cost are defined as functions of surface point coordinates $\mathbf{S}(u,v)$. The cost function components also depend on the number of samples M and N in the u and v parametric directions, respectively. Thus, M and N are used as input parameters; they have the expected trade-off effect between accuracy and convergence on the one hand and computational efficiency on the other.

The individual components of the cost function are defined as summations of surface characteristics sampled at regular parametric intervals, that is,

$$\begin{aligned} u_0 &= 0 \\ u_i &= u_{i-1} + \tfrac{1}{M-1}, \qquad i = 1,2,...,M-1 \end{aligned} \tag{2.9}$$

$$\begin{aligned} v_0 &= 0 \\ v_i &= v_{i-1} + \tfrac{1}{N-1}, \qquad i = 1,2,...,N-1 \end{aligned} \tag{2.10}$$

The obstacle cost component C_o requires a function to determine the distance from an arbitrary point to an obstacle model. The sophistication of such a distance function is dictated by the complexity of the obstacle model. Since surface synthesis is intended as a tool for conceptual design, this study considers only convex polyhedral obstacle models. Given an obstacle model $\mathbf{Q}$ comprised of V vertices, a distance function $d(\mathbf{S}(u,v),\mathbf{Q})$ is formulated by first finding the vertex closest to the surface point in question, $\mathbf{P}_c$. Then the signed distance from $\mathbf{S}(u,v)$ to each of the faces which reference $\mathbf{P}_c$ is calculated, and the minimum (absolute magnitude) of these is returned as d. This $O(V)$ distance algorithm is a three-dimensional extension of the one developed in Malhotra and colleagues [31]. It is formulated such that $d(\mathbf{S}(u,v),\mathbf{Q})$ yields a negative value if $\mathbf{S}(u,v)$ is inside $\mathbf{Q}$ and a positive value otherwise. The obstacle proximity cost component is thus defined as

$$C_o = \sum_{i=0}^{M-1}\sum_{j=0}^{N-1} C_{ij}, \tag{2.11}$$

with

$$C_{ij} = \begin{cases} K_1 & \text{if } \quad d(\mathbf{S}(u_i,v_j),\mathbf{Q}) \leq 0 \\ K_2(\delta - d(\mathbf{S}(u_i,v_j),\mathbf{Q})) & \text{if } \; 0 < d(\mathbf{S}(u_i,v_j),\mathbf{Q}) \leq \delta \\ 0 & \text{if } \quad \delta < d(\mathbf{S}(u_i,v_j),\mathbf{Q}) \end{cases} \tag{2.12}$$

where K_1 and K_2 are positive constants used to control the relative emphasis of C_o with respect to the total configuration cost. Note that when the distance is greater than the assigned tolerance then the associated cost is zero. When the surface point is within the tolerance region then the cost is computed such that it is greatest at the obstacle boundary and zero on the outer boundary of the tolerance zone.

To counteract the expansion effect of C_o, an area penalty is formulated as follows. The surface area of a parametric surface is defined [33] by

$$Area(\mathbf{S}(u,v)) = A = \int\int |\mathbf{S}^u(u,v) \times \mathbf{S}^v(u,v)| du dv \tag{2.13}$$

and can be calculated approximately via Gaussian quadrature. Thus a simple area penalty cost component is formulated as

$$C_a = K_3 A \tag{2.14}$$

where K_3 is a positive constant.

2.4.3 Computational Experiments

To test this cost function formulation a simple surface synthesis problem was considered. The problem configuration consisted of a single surface in the vicinity of an obstacle, in an initial relative orientation as shown in Figure 2.2. The surface model considered was quite simple: a nonrational bicubic B-spline surface defined by a 4×4 grid of control points (i.e., a Bezier surface). The initial positions of the control points were specified as equally spaced in the planar configuration shown in Figure 2.2. The control points lying on the edges of the surface were constrained (fixed), while the positions of the four interior control points were considered as the design variables. This case was run with the cost function parameters set to $K_1 = 150.0$, $K_2 = 300.0$, and $K_3 = 5.0$. The resulting surface is shown in Figure 2.3. The calculation of this solution required 8,620 cost function evaluations and executed in 830 seconds on a Silicon Graphics Indigo/Elan workstation.

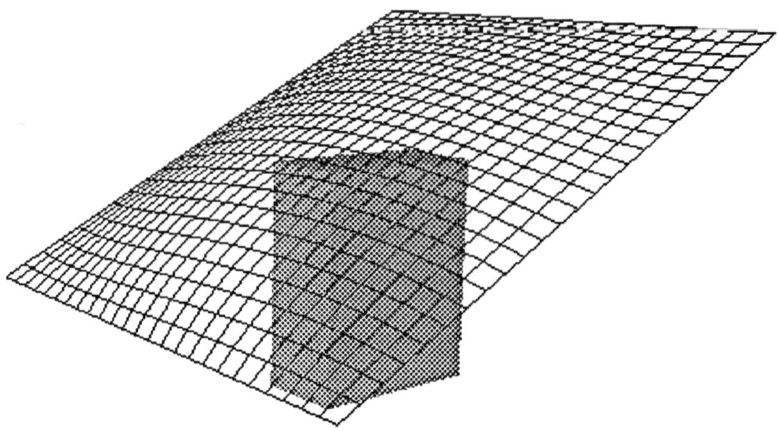

Figure 2.3: Surface Synthesis, Final Configuration

An examination of this formulation revealed that a large portion of the computation time was expended in calculating the area cost, C_a. To reduce this computational burden, an alternative component was developed which generates a similar effect with far fewer computations. In a parametric surface model, if one parameter is held constant while the other parameter is varied through its range, a curve is generated which lies in the surface. A cost function component C_l which penalizes the length of constant parameter curves is thus defined as

$$C_l = K_4 \left\{ \sum_{j=0}^{N-1} |(L_{u_j} - L_j)| + \sum_{i=0}^{M-1} |(L_{v_i} - L_i)| \right\} = K_4 C_{arc} \tag{2.15}$$

where L_j is the initial arc-length of constant v curves and L_i is the initial arc-length of constant u curves, respectively, and

$$\begin{aligned} L_{u_j} &= \sum_{i=0}^{M-2} |\mathbf{S}(u_{i+1}, v_j) - \mathbf{S}(u_i, v_j)| \\ L_{v_i} &= \sum_{j=0}^{N-2} |\mathbf{S}(u_i, v_{j+1}) - \mathbf{S}(u_i, v_j)| \end{aligned}$$

Note that at each sample point, C_a requires evaluation of both surface parametric derivatives and calculation of their cross products, while C_l is a function only of surface point evaluations. In addition, since surface point evaluations are necessary for C_o, they need not be recalulated for C_l. The simulated annealing (SA) algorithm typically requires many cost function evaluations. Thus any improvement in the efficiency of the cost function should have a dramatic effect on the overall computation time. This was demonstrated when the simple experiment described above was repeated with C_a replaced with C_l. With cost parameters set to $K_1 = 150.0$, $K_2 = 300.0$, and $K_4 = 1.0$, a solution essentially the same as the one shown in Figure 2.3 was reached in 90 seconds and 3,862 cost function evaluations.

2.4.4 Generalized Cost Function Formulation

The cost function formulation described above proved to be adequate for simple surface synthesis problems. However, when more general problems were attempted with this cost function, the SA failed to converge to a reasonable solution. For example, Figure 2.4 depicts a design environment with three obstacles and a single uniform, nonrational B-spline surface defined by a 7×7 grid of control points in an initially planar configuration. In this experiment the edge control points were specified as fixed, and the 25 interior points were allowed to vary.

Using the initial cost function formulation, many experiments were conducted with a variety of combinations of different values for cost component constants, neighborhood magnitude (i.e., maximum perturbation size), and SA algorithm parameters. In spite of these attempts, all these experiments resulted in divergent solutions similar to the one shown in Figure 2.5.

It is apparent from Figure 2.5 (and the results of many similar experiments) that there is sufficient freedom in the system such that the control points are allowed to cross spatially which results in self-intersections and the folded appearance of the surface. Thus, an

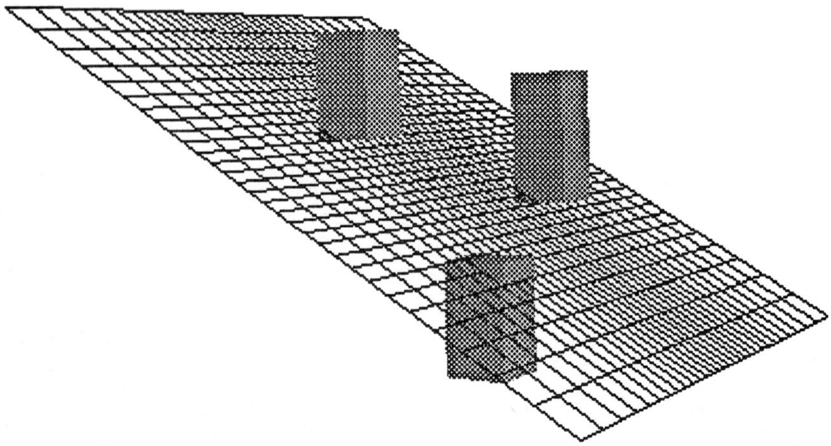

Figure 2.4: NURBS Surface Synthesis, Initial Configuration

additional cost function component was developed to inhibit this behavior. Since it is generally desirable to maintain the topological structure of the surface, a penalty on topological change was desired. A first attempt was to enforce the orthogonality of surface tangent

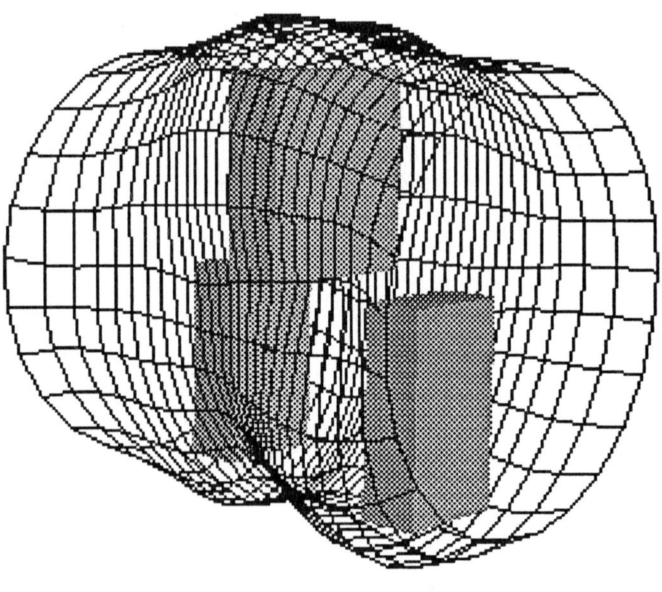

Figure 2.5: Typical Divergent Solution

vectors. Thus, a tangent vector orthogonality cost was defined as

$$C_{ortho} = \sum_{i=0}^{M-1} \sum_{j=0}^{N-1} C_{or_{ij}}, \tag{2.16}$$

where

$$C_{or_{ij}} = \begin{cases} 0 & \text{if} \quad 0 \leq |\mathbf{S}^u(u_i, v_j) \cdot \mathbf{S}^v(u_i, v_j)| \leq \delta_0 \\ K_5 & \text{if} \quad |\mathbf{S}^u(u_i, v_j) \cdot \mathbf{S}^v(u_i, v_j)| > \delta_0 \end{cases} \tag{2.17}$$

This cost component penalizes u and v direction tangent vectors which approach parallel or antiparallel relative orientations. The parameter δ_o is a threshold parameter which allows for the specification of a degree of nonorthogonality to be tolerated without penalty. It was hoped that this penalty would enforce the rectangular topology of the surface and inhibit the folding behavior exhibited by the initial cost function.

Experiments with this enhanced form of the cost function indicated that the orthogonality cost component was insufficient to inhibit the undesirable behavior. In fact, it tended to have a detrimental effect on the overall solution. During the early stages of the SA the neighborhood magnitude is at its largest, the system has high energy, and it is likely to accept higher cost moves. In this condition, it was found that the system would often accept a state with relatively high orthogonality cost and, on the next perturbation, reach a lower cost state with the tangent vectors in reversed orientation (i.e., the surface normal would change directions). As temperature decreases, the tendency to accept higher cost moves decreases, so the creases and folds so induced tended to "freeze" into the surface. It was observed that the orthogonality cost had the desired effect of enforcing a general rectangular topology, but generally not until the global solution had been reached, that is, its effect was relatively subtle, and noticeable only in the later stages of the SA.

Since the intended global effect was to inhibit an orientation change in surface normal vectors, a second reformulation focused directly on this effect. A surface normal can be calculated from the cross product of the first parametric derivatives,

$$\mathbf{n} = \mathbf{S}^u(u, v) \times \mathbf{S}^v(u, v) \tag{2.18}$$

In order to detect an orientation change in surface normal direction, some reference orientation must be applied. Two separate cost function components were developed, each based on a different definition for the reference direction. The first one employed a global static reference normal vector. In this formulation, the reference normal $\mathbf{n}_r$ was defined as normal to the plane in which the control points were initially defined (oriented parametrically via Eq. 2.18). The cost component to inhibit surface twist C_{twist} was then implemented by calculating the scalar product $\mathbf{n}_r$ with each $\mathbf{n}_{ij}$ in the $M \times N$ rectangular grid of sampled surface points, that is,

$$C_{twist} = \sum_{i=0}^{M-1} \sum_{j=0}^{N-1} C_{t_{ij}}, \tag{2.19}$$

$$C_{t_{ij}} = \begin{cases} 0 & \text{if} \quad \mathbf{n}_r \cdot \mathbf{n}_{ij} > 0 \\ K_6 & \text{if} \quad \mathbf{n}_r \cdot \mathbf{n}_{ij} < 0 \end{cases} \tag{2.20}$$

Normal vector orientation change, as indicated by sign change in the scalar product, was thus penalized accordingly. Experimentation with this cost component incorporated with C_o, C_l, and C_{ortho} yielded inconclusive results. In some cases the twist penalty had a slightly beneficial effect in that the divergence was somewhat delayed, but none of the experiments yielded acceptable results.

While C_{twist} seemed to have a only a mild effect on the design configuration of Figure 2.4, it proved to be much too restrictive in more general synthesis problems. Consider, for example, a case in which the general configuration of the surface is U-shaped, or even closed in one parametric direction. In such cases, the global initial reference normal is much too restrictive and generally inhibits the desired shape.

To address this problem the second formulation of C_{twist} employed a local incremental reference normal vector. In particular, at each step of the SA, each $\mathbf{n}_{ij}$ was dotted with its corresponding normal from the previous step. It was hoped that this would represent a more accurate penalty on normal orientation change. However, experimentation proved otherwise, as this formulation was also ineffective. The apparent limitation was the situation in which, at the higher temperatures (and thus larger neighborhood magnitudes) between successive SA steps, a legitimate spatial change in normal orientation often occurred simultaneously with an illegitimate reorientation of tangent vectors, thus the reorientation was not detected.

In addition to these discouraging results, it was apparent that both formulations for C_{twist} added substantial computational burden to the overall algorithm, due to the repeated calculation of surface derivatives. The final alternative was to drop the twist cost component altogether and reformulate the length cost. Obviously, the undesirable twisting effect also affected C_l, but repeated attempts at increasing the influence of C_l through manipulation of K_4 proved to be ineffective. So, to realize a quadratic effect similar to C_a, C_l was reformulated as

$$C_l = K_4(C_{arc})^2 \tag{2.21}$$

Combining this component with C_o and C_{ortho} and the constant assignments of $K_1 = 400.0$, $K_2 = 600.0$, $K_4 = 0.7$, and $K_5 = 4.0$, the design configuration of Figure 2.4 was synthesized into the surface model shown in Figure 2.6 in 4,002 seconds and 12,457 cost function evaluations. This combination of cost components has proven to be robust in solving a variety of general synthesis problems and will form the basis for future research in surface synthesis.

2.4.5 Further Examples

To demonstrate the generality of the surface synthesis technique, the following example is presented. A rational B-spline is useful for the exact representation of natural quadrics. In fact, one of the most common uses of the rational form is the representation of circular curves and cylindrical surfaces [36]. Figure 2.7 shows a cylindrical NURBS surface in the vicinity of three obstacles. This surface is defined as quadratic in the circumferential direction and cubic in the axial direction. It was specified by a grid of nine control points in the circumferential direction (distributed about a square of side length equal to the cylinder diameter) by eight control points equally spaced in the axial direction. The knot vector and weight assignments

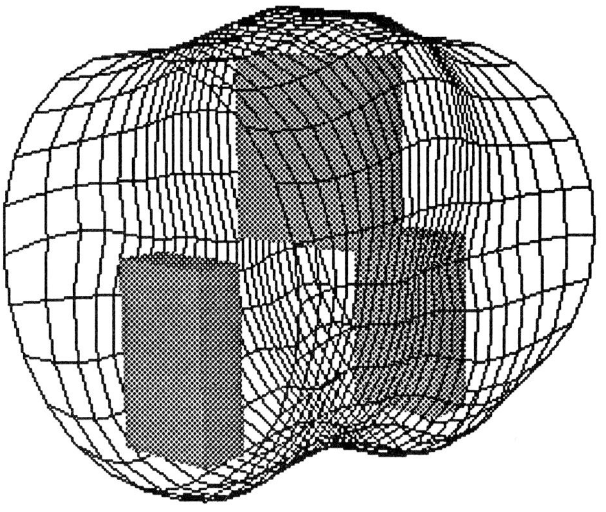

Figure 2.6: NURBS Surface, Final Configuration.

in the circumferential direction were specified to reflect the desired cylindrical shape (see, for example, Piegl and Tiller [36]) while in the axial direction a uniform knot vector was specified and the control point weights were set equal to one.

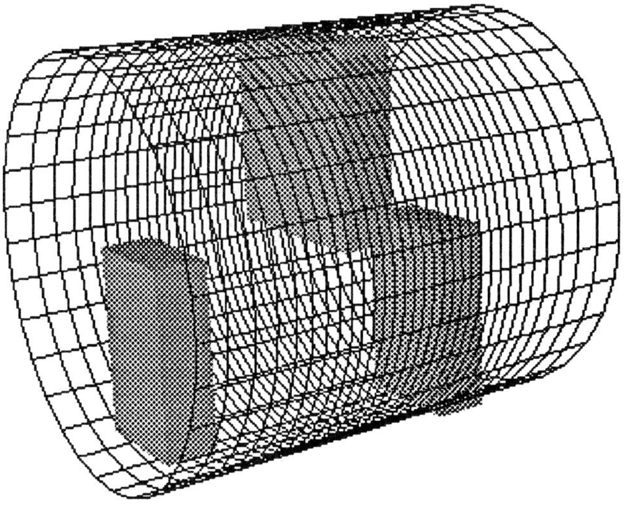

Figure 2.7: Tubular Surface Synthesis, Initial Cylindrical Configuration

The first solution of this example design configuration assumed that the circular cross-section was to be maintained. Thus the control points were restricted to move only in radial planes. (Since this constraint rendered the cost component C_{ortho} ineffective, it was not included in the cost function.) In addition, in each radial plane the control points were perturbed such that the square shape of the control point net was maintained, thus maintaining the circular cross section of the surface (i.e., each square was either randomly enlarged or reduced in size). Finally, the control points on both ends of the cylinder were specified as fixed. Using cost component parameters of $K_1 = 10.0$, $K_2 = 60.0$, and $K_4 = 1.0$, the result of this synthesis example is shown in Figure 2.8. This solution was reached in 1,155 seconds and 4,600 cost function evaluations.

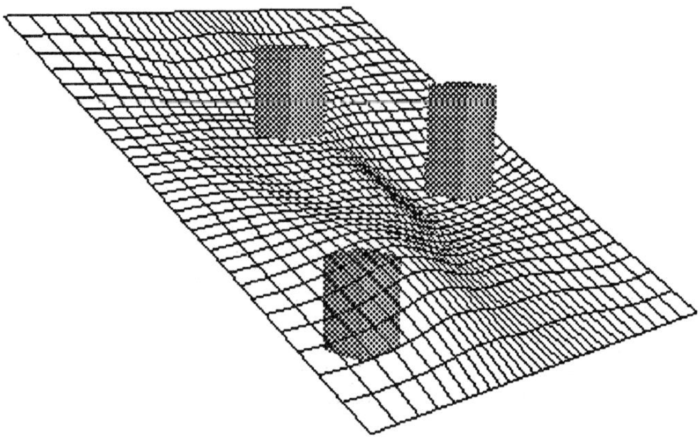

Figure 2.8: Tubular Surface Synthesis, Final Axisymmetric Configuration

The axisymmetric surface in Figure 2.8 is a direct result of the restrictions placed on the mobility of the free control points. In a second experiment with this configuration, the axisymmetric constraint was relaxed, that is, the control points were still constrained to radial planes, but they were free to move anywhere within the plane. The results of this example are shown in Figure 2.9. As expected, this surface exhibits a more localized response to the imposed constraints. This result was reached with cost component parameters specified as $K_1 = 50.0$, $K_2 = 100.0$, and $K_4 = 0.2$, in 3,738 seconds and 12,856 cost function evaluations.

2.5 Future Research

The successful implementation of the general surface synthesis method has opened many potential avenues for further research. One of the most important issues to be addressed is

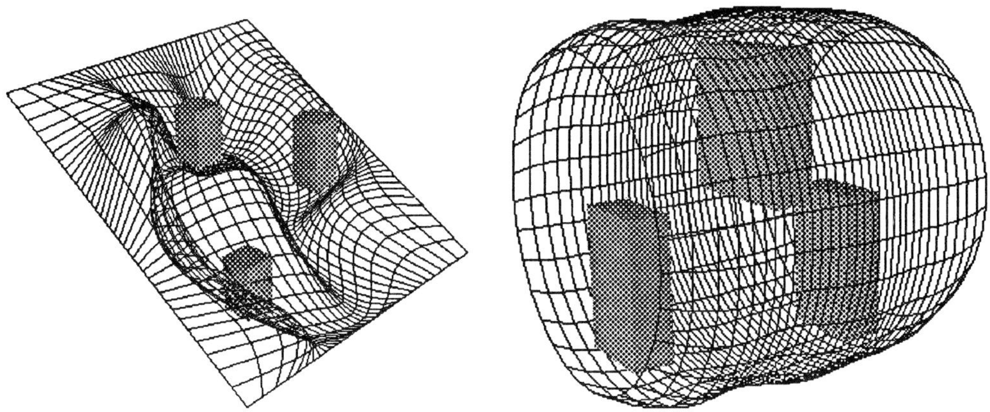

Figure 2.9: Tubular Surface Synthesis, Non-axisymmetric Configuration (two views)

an automated method for selecting constants. In addition to the cost function parameters described above, a number of other parameters associated with the simulated annealing algorithm must be specified for each application. These include parameters to specify the initial temperature, the neighborhood magnitude (and its rate of decrement) as well as the number of inner loop iterations (and its variation with temperature). In the experiments conducted to date these parameter values have been set based on an intuitive understanding of the specific synthesis problem and on experiential knowledge regarding the behavior of the algorithm. Future research will focus on the automatic specification of cost and algorithmic parameters which will ensure high likelihood of an acceptable solution. This will involve the development of heuristic rules based on the geometry of design environment and the constraints specified by the user.

Another major research thrust is the characterization of general design and manufacturability constraints in terms of analytical surface properties. For example, earlier work in curve synthesis demonstrated the incorporation of constraints based on curvature and parametric distribution and described their relationship to machinability [31]. These constraints will be generalized for application to surfacing. In addition, research is underway to characterize a general sheet metal formability constraint which will be incorporated in the general synthesis formulation. Finally, in addition to obstacle proximity, other surface functionality constraints, based perhaps on hydrodynamic, aerodynamic, or even kinematic performance indices, may be developed.

To make this new technique more accessible as a general design tool, further research is necessary on methods to define and interact with the design environment. In particular, methods are required to facilitate the interactive specification of the design environment, including the spatial location of obstacles and surfaces, the number and distribution of various surface topologies (i.e., collections of rectangular and n-sided topologies), and the specification and maintenance of various geometric and parametric constraints. Techniques will also be studied to explore the feasibility of automatically generating a problem configuration given minimal user input and direction.

Finally, research is necessary in the general area of computational efficiency. Although the computation times reported for the examples presented are not generally excessive, efficiency improvements would certainly enhance the viability of this surface synthesis methodology. Research toward algorithmic efficiency is progressing in both the SA implementation and the low-level surface evaluation functions. In addition, a parallel implementation of the surface synthesis technique is under consideration. Since the cost function evaluation involves sampling the surface at many different locations, a processor could be assigned to each location, thus speeding the evaluation to a large degree.

2.6 Acknowledgements

The research described in this chapter was supported by the National Science Foundation (grant numbers DDM-9111122 and DDM-9258114) and the Iowa Center for Emerging Manufacturing Technology. The authors are grateful for this support.

References

[1] Andersson, E., Andersson, R., Boman, M., Elmroth, T., Dalhberg, B., and Johansson, B. (1988). "Automatic Construction of Surfaces with Prescribed Shape," *Computer Aided Design*, Vol. 20, No. 6, pp. 317-324.

[2] Bennett, J.A., and Botkin, M.E., eds. (1986). *The Optimum Shape*, Plenum Press.

[3] Böhm, W., Farin, G., and Kahmann, J. (1984). "A Survey of Curve and Surface Methods in CAGD," *Computer Aided Geometric Design*, Vol. 1, pp. 1-60.

[4] Böhm, W. (1984). "Efficient Evaluation of Splines," *Computing*, Vol. 33, pp. 171-177.

[5] Casale, M.S., and Bobrow, J.E. (1989). "A Set Operation Algorithm for Sculptured Solids Modeled with Trimmed Patches," *Computer Aided Geometric Design*, Vol. 6, pp. 235-247.

[6] Celniker, G., and Gossard, D. (1989). "Energy-Based Models for Free-Form Surface Shape Design," *ASME Advances in Design Automation, B. Ravani, ed.*, DE-Vol. 19-1, pp. 107-112.

[7] Cohen, E., Lyche, T., and Riesenfeld, R.F. (1980). "Discrete B-splines and Subdivision Techniques in Computer Aided Geometric Design," *Computer Graphics and Image Processing*, Vol. 14, No. 2, pp. 87-111.

[8] Corana, A., Marchesi, M., Martini, C., and Ridella, S. (1987). "Minimizing Multimodal Functions of Continuous Variables with the Simulated Annealing Algorithm," *ACM Transactions on Mathematical Software*, Vol. 13, No. 3, pp. 262-280.

[9] Cox, M. (1972). "The numerical Evaluation of B-splines," *Journal of the Institute of Mathematics Applications*, Vol. 10, pp. 134-149.

[10] de Boor, C. (1972). "On Calculating with B-splines," *Journal of Approximation Theory*, Vol. 6, pp. 50-62.

[11] Devadas, S., and Newton, A.R. (1987). "Topological Optimization of Multiple-Level Array Logic," *IEEE Transactions on Computer-Aided Design*, Vol. CAD-6, No. 6, pp. 915-941

[12] Dixon, J.R., Simmons, M.K., and Cohen, P.R. (1984). "An Architecture for Applying Artificial Intelligence to Design," *Proceedings of 21st ACM/IEEE Design Automation Conference*, Albuquerque, NM.

[13] Elperin, T. (1988). "Monte Carlo Structural Optimization in Discrete Variables with Annealing Algorithm," *International Journal for Numerical Methods in Engineering*, Vol. 26, pp. 815-821.

[14] Farin, G. (1988). *Curves and Surfaces for Computer Aided Geometric Design*, Academic Press, San Diego, CA.

[15] Farouki, R.T. (1986). "The Approximation of Non-Degenerate Offset Surfaces," *Computer Aided Geometric Design*, Vol. 3, pp. 15-43.

[16] Farouki, R.T., and Rajan, V. (1987). "On the Numerical Condition of Polynomials in Bernstein Form," *Computer Aided Geometric Design*, Vol. 4, pp. 191-216.

[17] Faux, I.D., and Pratt, M.J. (1979). *Computational Geometry for Design and Manufacture*, Ellis Horwood Ltd., Chichester, UK.

[18] Ferguson, D.R. (1986). "Construction of Curves and Surfaces Using Numerical Optimization Techniques," *Computer Aided Design*, Vol. 18, No. 1, pp. 15-19

[19] Ferguson, D.R., Frank, P.D., and Jones, A.K. (1988). "Surface Shape Control Using Constrained Optimization on the B-spline Representation," *Computer Aided Geometric Design*, Vol. 5, pp. 87-103.

[20] Gallagher, R.H., and Zienkiewicz, O.C. eds. (1973). *Shape Optimization and Sequential Linear Programming*, Wiley, London.

[21] Gossard, D. (1989). Personal communication.

[22] Graham, P.V., and Ulrich, K.T. (1989). "Structural Synthesis of Sheet Metal Parts: An Analogy to Path Planning using Manufacturability as a Guide," *ASME Advances in Design Automation*, B. Ravani, ed., DE-Vol. 19-1, pp. 289-294.

[23] Imam, H.M. (1982). "Three-Dimensional Shape Optimization," *International Journal for Numerical Methods in Engineering*, Vol. 18, pp. 661-673.

[24] Jain, P., and Agogino, A.M. (1988). "Optimal Design of Mechanisms Using Simulated Annealing: Theory and Applications," *ASME Advances in Design Automation*, S.S. Rao, ed., DE-Vol. 14, pp. 233-240.

[25] Karmarkar, N. (1984). "A New Polynomial Time Algorithm for Linear Programming and Applications," *Proceedings of 16th ACM Symposium on Theory of Computing*, New York.

[26] Kaufmann, E., and Klass, R. (1988). "Smoothing Surfaces Using Reflection Lines for Families of Splines," *Computer Aided Design*, Vol. 20, No. 6, pp. 312-316.

[27] Kim, K., and Biegel, J.E. (1988). "A Path Generation Method for Sculptured Surface Manufacture," *Computers in Industrial Engineering*, Vol. 14, No. 2, pp. 95-101.

[28] Kirkpatrick, S., Gelatt, C.D., and Vecchi, M.P. (1983). "Optimization by Simulated Annealing," *Science*, Vol. 220, No. 4598, pp. 671-680.

[29] Lee, E.T.Y. (1983). "A Simplified B-spline Computation Routine, *Computing*, Vol. 29, pp. 365-371.

[30] Lott, N.J., and Pullin, D.I. (1988). "Method for Fairing B-spline Surfaces," *Computer Aided Design*, Vol. 20, No. 10, pp. 597-604.

[31] Malhotra, A., Oliver, J. H., and Tu, W. (1991). "Synthesis of Spatially and Intrinsically Constrained Curves," *ASME Advances in Design Automation*, G. Gabriele, ed., DE-Vol. 32-1, pp. 145-155, (also to appear in *ASME Transactions, Journal of Mechanical Design*).

[32] Metropolis, N., Rosenbluth, A., Rosenbluth, M., and Teller, A. (1953). "Equations of State Calculations by Fast Computing Machines," *Journal of Chemical Physics*, Vol. 21, pp. 1087-1091.

[33] Mortenson, M. E. (1985). *Geometric Modeling*, Wiley, New York.

[34] Nevill, G.E., and Paul, G.H. (1987). "Knowledge-Based Spatial Reasoning for Designing Structural Configurations," *Proceedings of ASME Computers in Mechanical Engineering Conference*, New York.

[35] Nevins, J.L., and Whitney, D.E. (1989). *Concurrent Design of Products and Processes: A Strategy for the Next Generation in Manufacturing*, McGraw-Hill, New York.

[36] Piegl, L., and Tiller, W. (1987). "Curve and Surface Constructions Using Rational B-splines," *Computer Aided Design*, Vol. 19, No. 9, pp. 485-498.

[37] Piegl, L. (1989). "Modifying the Shape of Rational B-Splines. Part 2: Surfaces," *Computer Aided Design*, Vol. 21, No. 9, pp. 538-546.

[38] Piegl, L. (1991). "On NURBS: A Survey," *IEEE Computer Graphics and Applications*, Vol. 11, No. 1, pp. 55-71.

[39] Platt, J., and Barr, A. (1988). "Constraint Methods for Flexible Models," *Computer Graphics (Proceedings of SIGGRAPH'88)*, Vol. 22, No. 4, pp. 279-288.

[40] Poeschl, T. (1984). "Detecting Surface Irregularities Using Isophotes," *Computer Aided Geometric Design*, Vol. 1, pp. 163-168.

[41] Preparata, F.P., and Shamos, M.I. (1985). *Computational Geometry, An Introduction.* Springer-Verlag, New York.

[42] Reklaitis, G.V., Ravindran, A., and Ragsdell, K.M. (1983). *Engineering Optimization, Methods and Applications*, John Wiley, New York.

[43] Rutenbar, R.A. (1989). "Simulated Annealing Algorithms: An Overview," *IEEE Circuits and Devices*, January, pp. 19-26.

[44] Saia, A., Bloor, M.S., and dePennington, A. (1989). "On the Integration of Parametric Polynomial Surface Representations into CSG Based Solid Modeling," *ASME Advances in Design Automation*, B. Ravani, ed., DE-Vol. 19-1, pp. 121-127.

[45] Sandgren, E., and Venkataraman, S. (1989). "Robot Path Planning via Simulated Annealing: A Near Real Time Approach," *ASME Advances in Design Automation*, B. Ravani, ed., DE-Vol. 19-1, pp. 345-351.

[46] Shah, J.J. (1988). "Synthesis of Initial Form for Structural Shape Optimization," *Transactions of the ASME, Journal of Vibration, Acoustics, Stress, and Reliability in Design*, Vol. 110, pp. 564-570.

[47] Terzopoulos, D., Platt, J., Barr, A., and Fleischer, K. (1987). "Elastically Deformable Models," *Computer Graphics (Proceedings of SIGGRAPH'87)*, Vol. 21, No. 4, pp. 205–214.

[48] Welch, R.V., and Dixon, J.R. (1989). "Extending the Iterative Redesign Model to Configuration Design: Sheet Metal Brackets as an Example," *ASME Design Theory and Methodology - DTM'89*, DE-Vol. 17, pp. 81-88.

[49] Williams, C.J.K. (1987). "Use of Structural Analogy in Generation of Smooth Surfaces for Engineering Purposes," *Computer Aided Design*, Vol. 19, No. 6, pp. 310-322.

[50] Wysocki, D.A., Oliver, J.H., and Goodman, E.D. (1989). "Gouge Detection Algorithms for Sculptured Surface NC Generation," *ASME Computer-Aided Design and Manufacture of Cutting and Forming Tools*, PED-Vol. 40, pp. 39-44, (also, to appear in *Transactions of the ASME, Journal of Engineering for Industry).*

Chapter 3

An Intelligent Approach to Conceptual Design of Mechanical Systems

QUN WANG and MING RAO
Department of Chemical Engineering, University of Alberta

JI ZHOU
Department of Mechanical Engineering, Huazhong University of Science and Technology

3.1 Introduction

The conceptual design of mechanical systems is the creative part of design process. Many important decisions are made in this phase on the overall design of a mechanical system. These decisions, if properly made, can lead to superior system behaviors, including reduced energy consumption, higher raw materials utilization, more profits, reduced environmental impacts, as well as increased flexibility, operability, controllability, and safety of manufacturing processes. Therefore, the quality of a conceptual design dominates the quality of a mechanical system and the profits of a plant. In this chapter, an integrated distributed intelligent system for conceptual design of transmissions (IDIST) is presented. The difficulties and characteristics of conceptual design for mechanical systems are first addressed. A methodology for solving the problems in conceptual design is proposed. The control structures suitable for the conceptual design of mechanical systems are described. A case study is used to illustrate the method.

3.1.1 Conceptual Design

In mechanical design, integrated computer-aided design (ICAD) technology has evolved a new generation of design techniques. This approach can pave the way for implementing computer integrated manufacturing (CIM) [4, 7, 17]. However, one of the primary tasks in accomplishing ICAD, the automation of conceptual design of a mechanical system or product, requires to be studied further [11].

Conceptual design is a very important and very difficult task in mechanical design. The quality and reliability of a mechanical system, as well as the production costs and productivity for making the system, depend not only on the quality of its components, but also on the conceptual design of the system and the coordination of its components.

In the past few decades computer technology has been extensively used in engineering optimization, finite element analysis, reliability design, computer graphics, and simulation. But most of these activities are aimed at detailed design tasks. Very few achievements were made in improving conceptual design. Since the traditional computer schemes cannot handle nonnumerical or nonalgorithmic information efficiently, the use of computers has been limited almost exclusively to finding algorithmic solutions. This leads to many difficulties in applying computer technology to automate mechanical design, especially the conceptual design (or design synthesis) part. The development of the computer-aided design (CAD) system has been hindered. It is crucial to develop new computer techniques to improve the quality and efficiency of conceptual design. The integrated and distributed intelligent system environment presented in this chapter serves this need.

3.1.2 Automation of Conceptual Design

The conceptual design of mechanical systems consists of two main subtasks. (a) Scheme design determines the general specification, performance, functions, and primary parameters of a design and selects the structural form, materials, and configuration of the designed product. The results generated in this design phase provide design information in both numerical and symbolic forms for the following detailed design. (b) Layout (structure) design determines the connections and relative positions of all components for a designed mechanical system. These two subtasks involve nonnumerical or nonalgorithmic information and thus not amenable to algorithmic computation [9]. Solving this type of ill-structured problems requires knowledge reasoning and decision making based upon the domain knowledge and experience of an expert. The expert system technique can meet these special needs of conceptual design.

To automate mechanical design and to develop a powerful integrated distributed intelligent environment for conceptual design, it is imperative to combine the expert system approach with techniques presently used in CAD, including computer solid modeling, database management, system analysis, and optimization [5, 12].

3.2 Problems in Developing an Intelligent Design System

The earlier research in expert system applications indicated that the power of expert systems is most apparent when the problem at hand is sufficiently complex [13]. In mechanical design, most problems are quite complex, and many numerical computation tools for finding solutions to these problems, for example, the optimization package, are already available. While (artificial intelligence) AI emphasizes symbolic processing and nonalgorithmic inference, the utilization of numerical computation with expert systems will make these tools more powerful in dealing with mechanical design problems [10].

Most existing expert systems were developed for specific applications. Many were implemented with Lisp-based (or Prolog-based) tools, and production rules were used to represent domain expertise. Such expert systems can process only symbolic information and conduct heuristic inference. A lack of numerical computation and coordinated multiple applications limited their capability of solving real design problems.

Conceptual design is a hybrid engineering problem of performing numerical computation and symbolic inference alternately. The problem solving depends not only on inferences but also on algorithms. Thousands of practical expert systems in manufacturing engineering have been developed in the United States, Germany, and worldwide. However, most of these systems are used in fault diagnosis and production planning, and only a few are used in mechanical design. In addition, hundreds of the tools (or shells) for building expert systems are commercially available; many of these shells are dedicated for applications in diagnosis and planning and are not suitable for use in design. The primary reason leading to this situation is that the development of AI tools and expert systems for design are very demanding tasks. The development of these expert systems is confronted mainly with the following problems:

- Difficulty in identifying optimal design from multiple design solutions. Generally speaking, problem solving in diagnosis follows a "multiple input/single output" problem solving pattern. One conclusion (or output) can be inferred from some evidence with an inference engine (sometimes more than one). In contrast, design problem solving is "single input/multiple output" in nature, that is, several results that all meet the given requirements may be produced. Two major obstacles make the single input/multiple output design task very difficult: (a) large decision space and (b) comprehensive evaluation of the quality of design. The problem is how to find all feasible designs that satisfy design requirements and how to choose an effective decision making method for selecting the best design among all feasible candidates.

- Handling multiple task levels and multiple objectives. A design needs to be performed at various subtask levels. For instance, the design for machine tools involves many aspects, such as transmission, control system, hydraulic components and system, power utility, ease of operation and so on. These subtasks may be implemented at different levels and controlled by metaknowledge [6]. These, in turn, lead to new problems: How to automatically plan and resolve a design task; how to represent the relationship among the subtasks; how to handle conflicts; and how to choose the appropriate problem-solving strategy to handle a subtask.

- Requiring an intelligent design environment for performing both computation and inference. Conventional expert systems and AI tools emphasize symbolic processing and nonalgorithmic inferences while solving a design problem requires both symbolic reasoning and numerical calculation. An intelligent distributed intelligent environment should be developed to integrate a variety of existing analysis and simulation packages and to exchange information with a database management system.

- Complexity of knowledge representation and problem-solving strategy. The product design requires various problem-solving methods, including reasoning, calculation, table look-up, and graphics, as well as different types of knowledge representations. During the development of an intelligent system serving design tasks, the knowledge base, database, and control strategy must be segregated to allow users to organize different design models and domain expertise efficiently. Each of these problem-solving components can be designed and modified separately.

- Knowledge representation and inference of geometric information. The outcome of a product design, including conceptual design and detailed design, should be represented on mechanical drawings that contain 80 percent of design information. In addition to the documentation of various geometric information, the drawings also specify the structural and layout designs. These designs touch upon knowledge representation and inference of geometric information. Compared with the conventional symbolic inference, knowledge inference based upon geometric information is more difficult. The method for describing and coping with the geometric, functional, and topological information of design components must be developed.

- Complexity of redesign and combinatorial explosion of design problems. Redesign is an inevitable part of design. When the design results are unsatisfactory to users, the intelligent system for conducting mechanical design has to carry out redesign. With the increase of the system size and problem complexity at the top level, redesign is very difficult. The problems are how to store and apply the "failure" information to direct redesign; how to select an optimum problem-solving strategy when multiple tasks conflict; and how to implement expertise knowledge.

3.3 Solving the Problems of Intelligent Design

3.3.1 Design-Analysis-Evaluation-Redesign Model

An effective and practical model for conceptual design, called the Design-Analysis-Evaluation-Redesign (DAER) model, is introduced here. The model summarizes present practices in mechanical design, rather than introducing a new methodology [1], as illustrated in Figure 3.1. The model authentically reflects the thought of a design expert in solving design problems. The approach can be used to conduct empirical design. The IDIST system presented in this chapter was developed based upon the DAER model.

Design has been recognized as an art and a creative process [2]. In conceptual design, a designer often approaches the design problem as a painter develops a painting. The initial design is preceded by the development of a pencil sketch, where the designer want to suppress

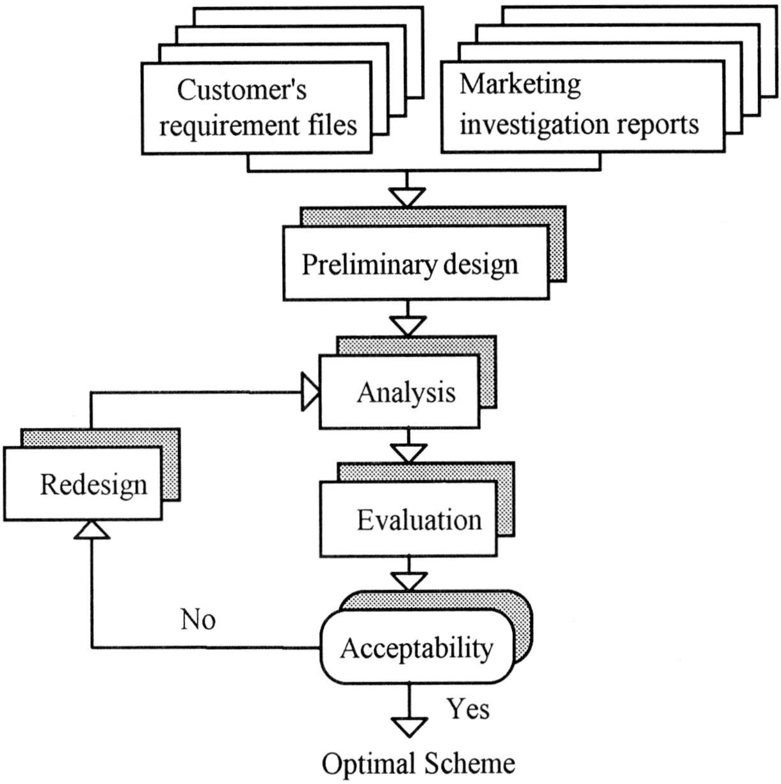

Figure 3.1: DAER Model for Conceptual Design of Mechanical Systems

all but the most significant parts of the design. In other words, the designer wants to discover the most important parts of the design which determine the final performance and cost of the product (or system) under design. When the pencil sketch is finished, an artist usually evaluates the preliminary drawing and makes modifications, using gross outlines of the subjects. Similarly, a designer evaluates the initial design and generates a number of design alternatives. In this way, the designer hopes to generate a "fine looking" rough design before many design details are added.

In short, a design process can be divided into two stages: (1) generating a preliminary design scheme and (2) repeating the "analysis-evaluation-redesign" process until all design requirements and constraints are met and a feasible design is generated. The DAER model follows this design process. All parts of DAER model are independent modules with different functions. These functions include:

Preliminary design: Preliminary design is usually used in generating the sketch of a new product or system. Problem-solving strategies used in the preliminary design include: (i) generating a scheme similar to an existing design scheme; (ii) generating a new scheme that meets all requirements and constraints; (iii) generating a scheme that satisfies only some of the constraints, and (iv) randomly generating a scheme. These strategies can be used alone or in combination. The process of preliminary design depends upon the specific domain knowledge.

Analysis: The objective of analysis is to provide detailed data and information for the evaluation of a design. Generally, two types of information are produced in a design process: symbolic information generated from knowledge reasoning and numerical information generated from computation. In design analysis, various modern analysis tools are used. These include finite element analysis, numerical optimization, reliability study, and various mathematical models of the mechanical component (or system).

Evaluation: Evaluation involves comprehensive consideration of all the design targets to identify an optimal or near optimal design scheme. The process requires both analysis models and experiences from a domain expert.

Acceptability test: The process is used to find whether all design requirements and constraints are satisfied.

Redesign: Redesign uses feedback from the analysis and evaluation modules to generate a modified design scheme.

3.3.2 Problem-Solving Strategy

The problem-solving strategy developed for the conceptual design of mechanical systems and represented by the DAER model consists of the following stages. These stages are also shown in Figure 3.2.

In the first stage, the function of a design is defined based upon the application environment (from application environment to functions). The applied function knowledge

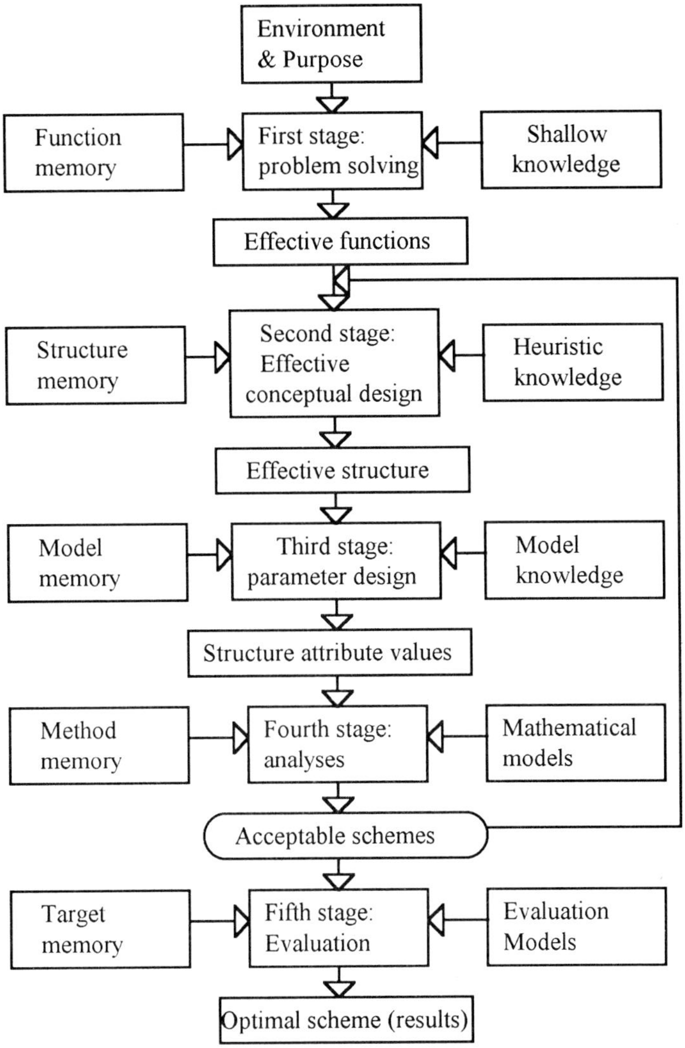

Figure 3.2: Problem-solving Strategy

is chosen from the expertise function memory according to the application environment. For example, the location where the designed mechanical system is used influences the conceptual design since the utilities, such as the temperature of natural cooling water available on site, depend on the geographical location of the plant. The knowledge for function definition is shallow knowledge [3].

The second stage is an effective conceptual design stage (from function to structure). The structures to execute identified functions are selected from structure memory (can be viewed as another part of the knowledge base). The communication between functions and structures is not a "one-to-one" mapping. Such a "multiple-to-multiple" mapping configuration indicates that a function can be accomplished by many different structures, and a structure may have many functions. For example, the function for cooling can be implemented by several structures: cooling water, air, oil, or others. This "multiple-to-multiple" mapping makes the pattern design alternatives more completed and diversified. Each design alternative has a design scheme. If there is more than one design alternative (pattern), a near optimal solution is to be identified. In the case that no design alternatives are available, new design techniques will be used (since no existing structure can be used for the needed functions). If existing design alternatives fail to satisfy application requirements, the design has to be improved. The knowledge used for forming effective structure concepts is heuristic knowledge which can be represented by heuristic rules. The IDIST system developed in our work can generate a set of design patterns (alternatives). This pattern set is decomposed into individual patterns such that each individual pattern forms a design scheme. The process is called scheme resolving.

The third stage is a parameter design stage (from structure to parameter). At this stage, the detailed description of structures is completed using the design models stored in model memory according to the characteristics of effective structure concepts. The knowledge to determine structural attributes is deep knowledge, namely model knowledge, which is represented by object-oriented frames.

The fourth stage is the analysis stage (from parameter to analysis). Because functions and structures share a "multiple-to-multiple" mapping configuration, numerous design schemes are usually produced. After parameters are given, all design schemes will be analyzed by selecting numerical computation methods (e.g., engineering analysis, optimization) from method memory. Conventional computer-aided analysis techniques can be utilized.

The last stage is used for comprehensive evaluation (from analysis to evaluation). According to the results of the analysis, a proper evaluation system from target knowledge memory and a comprehensive mathematical model from evaluation models will be chosen to evaluate the selected scheme. Techniques of fuzzy mathematics and system engineering are used in the evaluation.

3.4 System Configuration

A good problem-solving strategy must be implemented using a good program structure to ensure the quality and efficiency of the software. The control structure of an IDIST system is shown in Figure 3.3. The structure simulates the behavior of an engineer conducting conceptual design. The prototype system can be used as a general framework for developing an integrated distributed intelligent system for the conceptual design of general mechanical systems/products. The system was developed with multiple function modules, consisting of a menu, 15 subsystems (or modules), and several different knowledge bases. The following briefly describes these subsystems.

Menu Management System —It can guide users to select various function modules and to monitor the performance of these modules. The menu system employs a tree structure. Each subsystem has its own submenu. An object-oriented programming technique for man-machine interface is used. A user can select and execute a function module from the screen menu.

Task Definition Module —The module is a window for providing input information to the system. A user can specify design tasks, working environment of applications, and design specifications using this window.

Function Concept Design Module —The module functions as the first step of the problem solving. It expands the facts and information in the knowledge base based upon the existing specifications and determines global variables (parameters) that can be shared by many subsystems. In general, to identify these global variables the entire system must be considered. This is a difficult task for a new designer. In addition, a user never knows whether he/she has received correct and complete information. When important data are missing from the inputs, the subsystem should supply these data using domain knowledge.

Structure Conceptual Design Module —This module determines the types of machine and the potential structural configuration of the machine to satisfy the specified functions and key constraints. This module works as the second step in problem solving and supports parameter design with a variety of features. Its knowledge base contains experiences and heuristic knowledge of experts, which are represented in the form of heuristic rules. The constraint-reasoning engine is used in the inference [11].

Parameter Design Module —Its function is similar to the third step of design. With the structures obtained from concept design and the facts provided by the user, the attribute values (parameters) of a system/product are determined. Parameter design usually deals with local variables by calculations focused on specific interests. The local variables are usually well understood by engineers, since these variables are simply associated with one subsystem, rather than the overall system or several subsystems. The knowledge of parameter design is expressed using the data structure of the object-oriented frame. The frame is operated by a problem solver [11].

Layout Design Module —Based upon given geometric parameters, the layout design module determines the position and orientation of each mechanical component, and

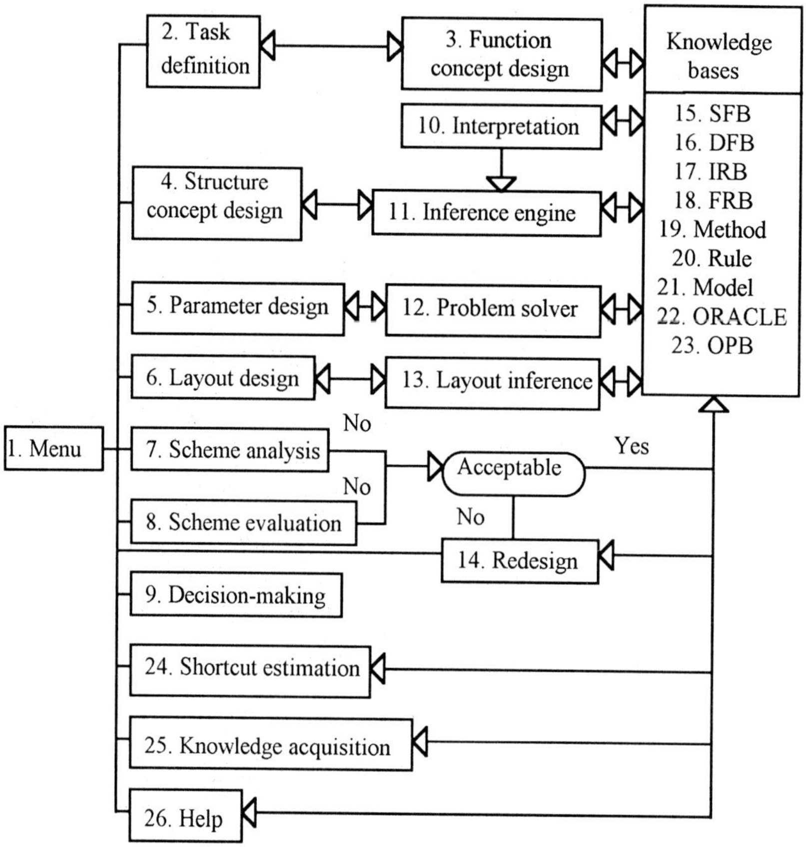

Figure 3.3: Control Structure of an IDIST

produces a drawing. The other function of this module is to perform simulation for identifying geometric interference among different components.

Scheme Analysis Module —The module is used in the fourth step of design problem solving. Based upon the results from parameter design and layout design, this module analyzes each preliminary design scheme and provides data for evaluation. The module compares alternative designs generated and identifies the superior scheme, using analysis and simulation software packages. If no one scheme satisfies the design requirements, redesign will be activated.

Scheme Evaluation Module —This module is equivalent to the fifth step of problem solving (comprehensive evaluation). The function of this module is to perform a comprehensive evaluation on the generated design schemes. The evaluation is conducted using various engineering analysis tools, and the performances of different design schemes are judged with the help of fuzzy mathematics [8, 14, 15, 16]. Various design indices for conducting the evaluation (such as costs of material and manufacturing) and the weights put on these indices are determined by domain experts. The evaluation module leads to two paths. If the results satisfy the design specifications provided by the user, the design information is sent to the following decision-making module. Otherwise, a global redesign will be activated.

Decision-making Module —In most cases, multiple design schemes are generated during the design problem solving. Since the index system solicits various opinions from different domain experts, conflicting solutions may be generated. The decision-making module will select the best scheme among these candidates and avoid conflicts.

Interpretation Module —The module interfaces with the inference engine and interprets the concepts and reasoning paths of the intelligent system to its user to allow a user to know the reasoning processes and to control the system.

Inference Module —An inference engine is used to execute the knowledge reasoning based upon fired data. To solve large and complex mechanical design problems, a simple inference engine that provides one simple reasoning technique becomes ineffective. The integration of different inference mechanisms becomes necessary.

Problem Solver —The problem solver operates a method knowledge base with a variety of problem-solving strategies, including reasoning, calculation, table look-up, interpolation, and analogy. In the conceptual design, these design methods are often used together. For example, reasoning can be used to determine the empirical coefficients, table look-up and interpolation can be used to provide coded information, and analogy can be applied in choosing structure types.

Layout Inference Engine —Layout inference is used to conduct knowledge reasoning based upon the shape, position, and orientation of mechanical components, as well as the space constraints and interferences among different components. The development of a layout inference engine is much more difficult than a regular symbolic inference engine due to these demanding tasks.

Redesign Module —The redesign module uses the information provided by the design analysis and evaluation modules as guidelines to improve the weak parts of the current design and to generate a new design solution.

Static Facts Base (SFB) —The base stores task definitions and specifications provided by a user. The information in SFB contains the essential conditions (constraints) for various design functions and the structure of a concept design. The facts in SFB are expressed using lists.

Dynamic Facts Base (DFB) —In knowledge reasoning, a user needs to continuously provide detailed facts and data regarding the design. These data will be stored in DFB. In an IDIST system, the DFB interfaces only with the inference engine and supports the interpretation module.

Intermediate Result Base (IRB) —It stores intermediate results in symbolic reasoning and numerical computation. The IRB has two functions: (a) storing intermediate results in the continuous reasoning process and (b) supporting back track (redesign) using the stored information when a design fails.

Final Results Base (FRB) —The base stores all acceptable design schemes. The decision-making module selects and refines the best design scheme among all feasible candidates in FRB.

Method Base —The method base records the structural descriptions of all problem-solving methods. Separated from the problem solver, the method base partially describes the problem-solving procedure. When handling new parameters, a user may add additional descriptions to the method base without changing the problem solver. All methods are described using object-oriented frames.

Rule Base —The rule base consists of many files of production rules. These files of production rules were created by the knowledge acquisition module for specific subtasks.

Model Base —It stores the parametric models of mechanical components for forming the assembly. The part parameters include position (x, y, x) and orientation (vx, vy, vz). A component model can be placed at any assembly location when their parameters and a position transformation matrix are provided.

ORACLE —ORACLE is a commercial database under the UNIX operation system and functions as a center for exchanging information. The database formed using ORACLE works as a bridge between design and manufacturing in computer integrated manufacturing.

Optimization Base —This base contains seven optimization packages, including linear programming, nonlinear constrained optimization, discrete optimization, and so on. An optimal design model, formulated in the method base, can automatically be linked to the optimization programs to generate an executable file. The results of optimization are stored in the Intermediate Result Base.

Cost Estimation Module —Scheme quotation is an important part of product design. It also forms the foundation of scheme evaluation and final design scheme selection. Analysis of production costs, prediction of demand, calculation of raw material costs, and estimation of profits are important to the development of mechanical products.

Knowledge Acquisition Module —This module creates files of production rules for the rule base containing design knowledge. The module also maintains the rule base by supporting rule searching, adding, deleting, and modification.

Help —The help module supports a help window that explains each of the system commands. The help window can be selected at any time from any module.

3.5 Case Study

3.5.1 Functions of the IDIST System

The prototype integrated distributed intelligent system for conceptual design of transmissions (IDIST) combines expert systems with conventional computer-aided design and analysis techniques. The IDIST can be used to form the general design scheme of transmissions based upon customer's specifications. The system can be used to conduct conceptual design, layout design, and cost estimation for various types of transmissions.

The IDIST system supports various design tasks such as design of a power utility and generation of a layout sketch. It automates routine design tasks such as determination of connecting methods and selection of bearings and sealing materials. The system also determines various design parameters, including parameters of material property, design function, part geometry, and component position/orientation.

In addition, the system can carry out strength analyses of important parts, evaluate design schemes generated, and estimate manufacturing costs. The design information in both numerical and symbolic forms, generated by the design functions at earlier stages, provides a platform for the following detailed design task. These functions also create an ideal environment for an integrated CAD/CAM system. Another function of the system is to conduct layout design. The rough assembly drawings can be automatically generated. During the formation of design, a user is allowed to modify the drawings and to influence the system's decision making.

To maintain the knowledge base and to add new knowledge, the system provides a special knowledge acquisition module for modifying and expanding the knowledge base.

3.5.2 Design Principles of the IDIST System

The two primary problems in developing the IDIST system are (a) how to combine expert systems with conventional computer-aided design and analysis techniques used in transmission design and (b) how to meet the design specifications in the environment of computer integrated manufacturing (CIM). The following basic principles are followed:

- The development of the system must satisfy the requirements of CIM.
- Two or three domain experts are designated as the sources for exacting design and manufacturing knowledge.
- The system should have totally separate knowledge bases and inference engines to allow different knowledge models from various expert domain to be built efficiently.
- A knowledge acquisition subsystem should be included to maintain and upgrade the knowledge base efficiently.
- The implementation of the system should support a multiple programming language (Common Lisp, FORTRAN, Pascal, and C) environment for processing symbolic and numerical information.
- The system should be able to exchange information with the ORACLE database management system.
- A user-friendly menu system should be provided.
- A module for interpreting the final and intermediate design solutions to users should be available.
- The system should be able to serve as a platform for developing general integrated distributed intelligent systems for conceptual design of mechanical systems or products.
- The system should contains various problem-solving strategies.
- The prototype system needs to be evaluated using real industrial design problems.

3.5.3 Features of the IDIST System

The prototype IDIST system was developed based upon the integrated distributed intelligent system methodology and the metasystem technique [6]. An integrated distributed intelligent system is a large knowledge integration environment which consists of several symbolic reasoning systems, numerical computation packages, neural networks, a database management subsystem, computer graphics programs, an intelligent multimedia interface, and a metasystem. The integrated software environment supports execution of programs written using different languages and data sharing among these programs using a common database. The separate intelligent systems, numerical analysis packages, and modeling programs are under the control of a supervising intelligent system, namely, the metasystem. The metasystem manages the selection, coordination, operation and communication of these subprograms. The IDIST system provides the following features:

- Coordinates all symbolic reasoning systems, numerical computation routines, and computer graphics programs in an integrated environment.
- Distributes knowledge among separate expert systems and numerical routines to allow regular users (rather than its original developers) to easily update the knowledge base of the system.
- Acquires new knowledge efficiently.
- Identifies a near optimal outcome for the designs with conflict solutions at different task levels.
- Supports parallel processing in the integrated distributed intelligent system.
- Interfaces with commercial packages, AI tools and programs written using different programming languages.

3.5.4 Illustration

The IDIST system was implemented on a Sun SPARC 1+ workstation and formed a heterogeneous knowledge-integration environment. The symbolic reasoning system was developed using Common Lisp, while the geometric conceptual design and the simulation system were programmed in C. The evaluation module, optimization package, and analysis module were developed using FORTRAN. The CGI graphical environment provided by Sun was used in the implementation.

The process of design formation using the IDIST system is illustrated using the following tables and figures and brief discussion.

1. A user starts a design by completing several pages of design requirements of the transmission as shown in Table 3.1.
2. At the conceptual design stage, various structure forms of the transmission, materials, and so on are determined. A typical structure form provided by the system and its specific matrix are shown in Table 3.2. The automatically generated matrix can also be modified by the user.
3. After the conceptual design, the system provides two design schemes for the transmission. The rough layout forms of these designs are shown in Figure 3.4. The speed diagram illustrating speed changes is shown in Figure 3.5. A more detailed drawing that includes data and information for analysis and evaluation is given in Figure 3.6. The resulting side view of the layout design, which determines the position and orientation of shafts, is shown in Figure 3.7.

Table 3.1: Specifications of Design

Tab–Surface–W(mm):	1000
Tab–Surface–W(mm):	3000
Space–Width(mm):	500
Spacd–Height(mm):	500
Space–Length(mm):	500
Table–Load(N):	8000
Max.–Feed–V(mm/min.):	1000
Min.–Feed–V(mm/min.):	10
Rapid–Traverse(mm/mini.):	0
Spindle–Power–(kw):	15
Max.–Spindle-V(r/min.):	630
Min.-Spindle-V(r/min.):	50
Feed-Travel(mm):	0
Move–Part–Weight(kg):	0
Millsprindle–Max–T(N.m):	0
Cutter–Diameter(mm):	0
Mill–Head–Number:	1
FB-Modle:	FB001
Order–No:	MUN001
Desibner:	Engineer
Plane-Type:	Small
Feed–Box–Type:	Mill–H–F
Driven–Type:	T–Lead–SC
I/O–Shaft–D:	Orthogonal
Mill–Head–D:	Mill–Head–H
Mill–Head–P:	Left

Table 3.2: The Matrix for Representing Scheme 1

Structure Form: 1[3]-2[4]-3[3]-4[1]						
shaft1	d_gear 3	dn_gear2	d_gear1	0	0	0
shaft2	d_gear41	d_gear1	dn_gear1	0	d_gear41	0
shaft3	d_gear 4	0	0	0	dn_gear 41	d_gear 5
shaft4	0	0	0	0	0	dn_gear5

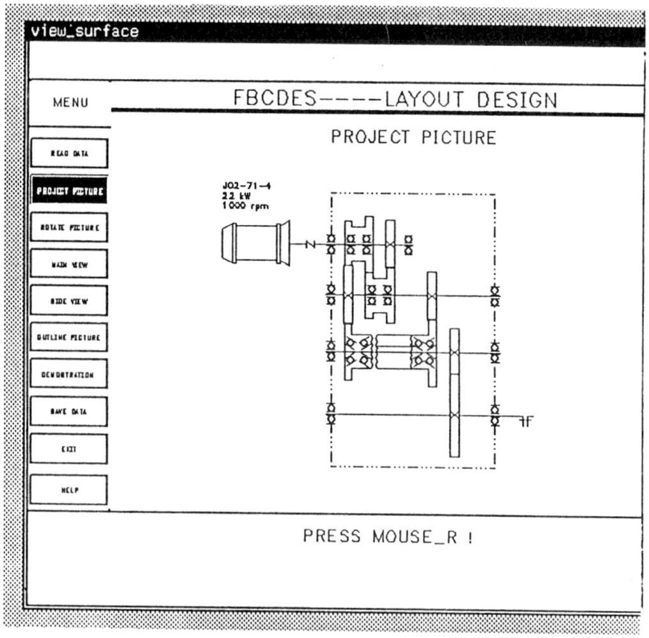

a) Scheme 1

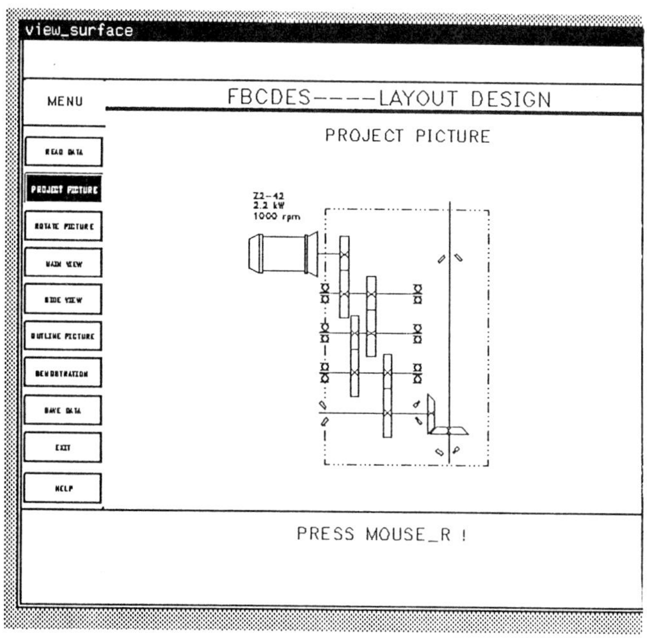

b) Scheme 2

Figure 3.4: Sketch of two transmissions

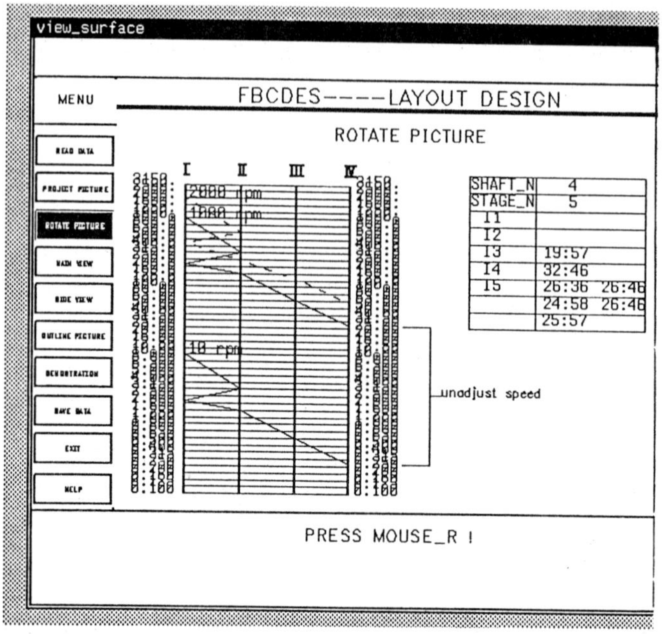

a) Scheme 1

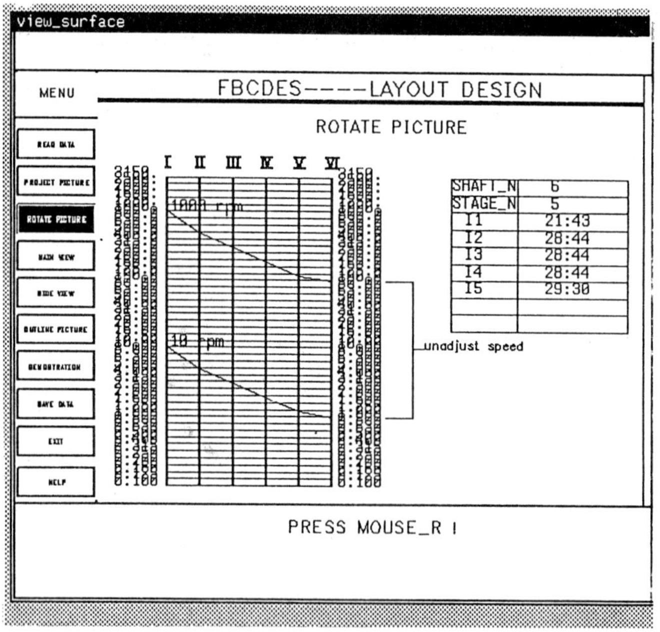

b) Scheme 2

Figure 3.5: Speed diagrams

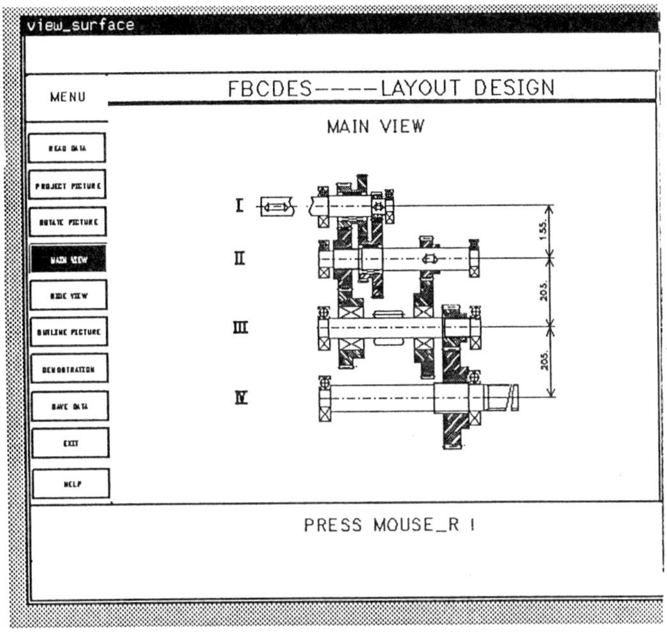

a) Scheme 1

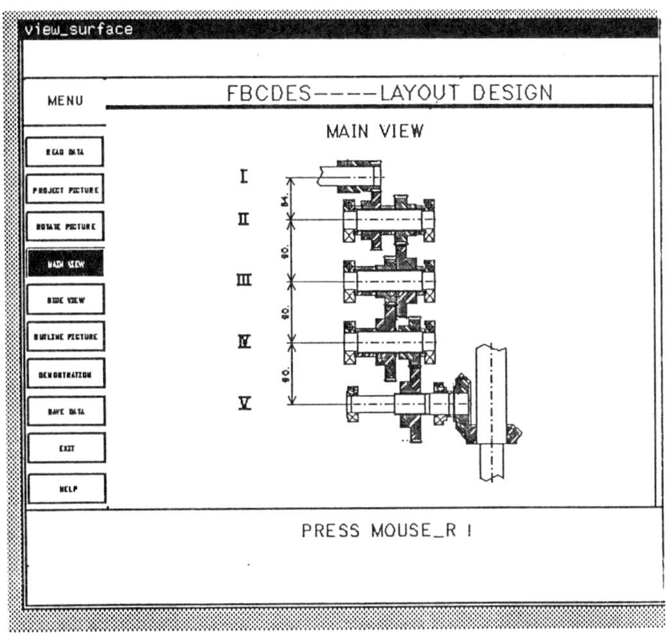

b) Scheme 2

Figure 3.6: Structure drawings

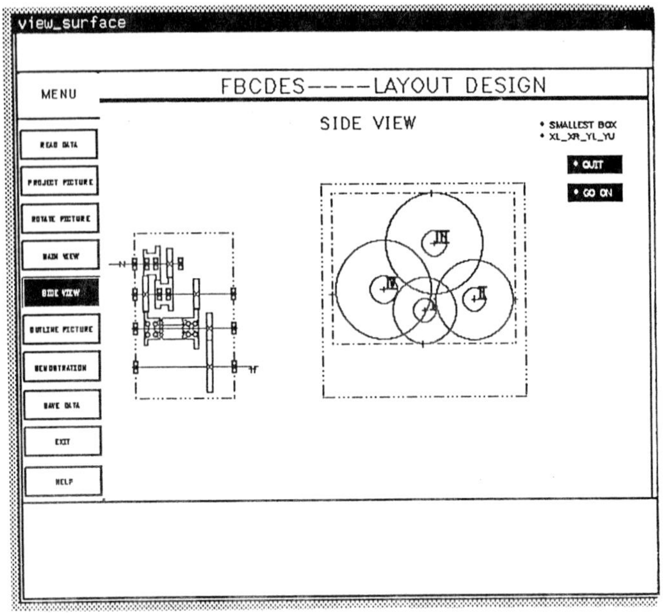

a) Scheme 1

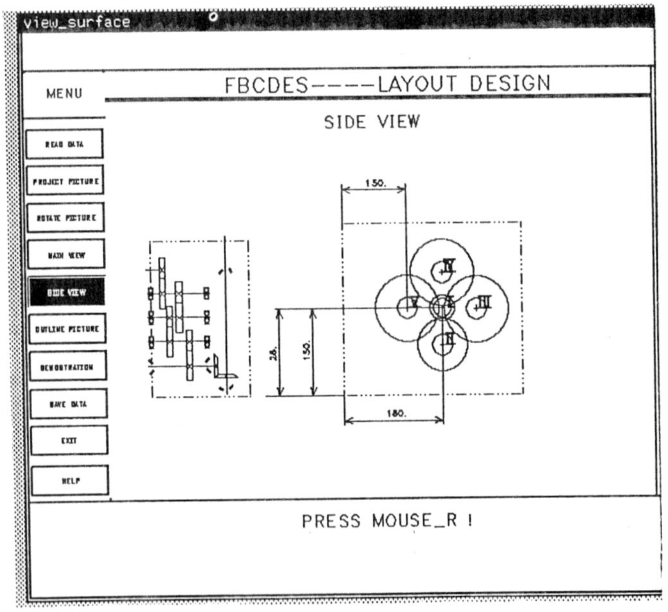

b) Scheme 2

Figure 3.7: Side view of the drawings

Table 3.3: Numbers of Stages, Shafts, and Gears

Types	Stages	Shafts	Gears
Values	5	4	11

Table 3.4: Parameters of Motor

Power Source Type	Power (kW)	Max speed (rpm)	Min speed (rpm)
D.C motor	10	1000	10

4. Tables 3.3, 3.4, 3.5, 3.6, and Table 3.7 represent the results of parameter design for scheme 1, including the numbers of transmission stages, shafts, and gears; parameters of the selected motor; a variety of drive ratios; materials; geometric parameters; machining requirements of shafts, gears, box and so on. These results provide important information for scheme analysis and evaluation.

5. If the design results following scheme 1 satisfy the conditions of the working environment, design analyses are conducted. These consist of the strength and stress analyses on the designed gears, shafts, bearings, and box. The results are shown in Table 3.8.

6. Table 3.9 shows the results of the comprehensive evaluation. Four items have been chosen as the final targets: working efficiency, volume, manufacturing time, and cost. Similarly, this procedure can also be applied to the design following scheme 2. When all are completed, the best conceptual design can be identified based upon the results of the comprehensive evaluation.

3.6 Acknowledgments

The research is financially supported by the Natural Sciences and Engineering Research Council (NSERC) of Canada and the National Natural Science Foundation (NNSF) of China.

Table 3.5: Drive Ratio

Level	1	2	3	4	5
Drive Ratio	3.0	3.0	18/13	23/18	57/25

Table 3.6: Shaft Design

Shaft No.	Type	Materials	Diameter (mm)	Speed (rpm)	Torque (Nm)
Shaft1	C Shaft	45 Steel	45	2000	48.4
Shaft2	C Shaft	45 Steel	70	335	285.1
Shaft1	C Shaft	45 Steel	80	262	364.3
Shaft1	C Shaft	45 Steel	90	115	830.1

Table 3.7: Gear Design

Gear No.	Shape	Model	No. of Teeth	Material	Precision	Treatment
Gear1	SG	4	19	45CR	8_DC	Teeth_G48
Gear2	SG	4	57	45CR	8_DC	Teeth_G48
Gear3	SG	4	32	45CR	8_DC	Teeth_G48
Gear4	SG	4	46	45CR	8_DC	Teeth_G48
Gear5	SG	5	26	45CR	8_DC	Teeth_G48
Gear6	SG	5	36	45CR	8_DC	Teeth_G48
Gear7	SG	5	46	45CR	8_DC	Teeth_G48
Gear8	SG	5	25	45CR	8_DC	Teeth_G48
Gear9	SG	5	25	45CR	8_DC	Teeth_G48
Gear10	SG	5	57	45CR	8_DC	Teeth_G48
Gear11	SG	5	23	45CR	8_DC	Teeth_G48

Table 3.8: Results of Gear Analysis

Gear	Actual Stress (MPa)	Allowed Stress (MPa)	Contact Stress (MPa)	Allowed Con. Str. (Mpa)	Bending Stress (MPa)	Allowed Ben. Str. (MPa)	Is the design OK ?
Gear 1	246.3	1370.0	22.48	354.0	17.75	354.0	Yes
...	...	...	...	...	...	...	...

Table 3.9: Results of Evalution

	Efficiency	Volume (M^3)	Machine Time (h)	Costs (Yuan)
Calculated	58.56%	0.0237	111.91	1892.71
Allowable	50.00%	0.0250	150.00	2500.00

References

[1] Dixon, J.R., and Simmons M.K. (1983). "Computer in Design: Expert System for Mechanical Engineers," *Computers in Mechanical Engineering*, Vol. 4, No. 10, pp. 120-134.

[2] Douglas J.M. (1988). *Conceptual Design of Chemical Design*, McGraw-Hill, New York.

[3] Kapp, D., et al. (1989). "An Analysis of Distinction between Deep and Sallow Expert Systems," *International Journal of Expert System*, Vol. 2, No. 1, pp. 1-34.

[4] Kidd, P. T. (1990). "Organization, People and Technology: Advanced Manufacturing the 1990s," *Computer-Aided Engineering*, Vol. 7, No. 9, pp. 149-152.

[5] Manoocheiti, S., and Seireg A. (1987). "Computer-Aided Generation of an Optimum Machine Topology for Specified Tasks," *CIME*, Vol. 6, No. 3, pp. 10-24.

[6] Rao, M. (1991). *Integrated System for Intelligent Process Control*, Lecture Note Series in Control and Information Science, Springer Verlag, Berlin.

[7] Rao, M., Cha, J., and Zhou, J. (1990). "New Software Platform for Intelligent Manufacturing," *Proceedings of AAAI 90 Workshop Intelligent Manufactures*, ASME, New York, pp. 62-65.

[8] Rao, M., Jiang, J.S., and Tsai, J.P. (1988). "IDSCA: An Intelligent Direction Selector for Controller's Action in Multi-loop Control System," *Applied Artificial Intelligence*, Vol. 2, pp. 285-305.

[9] Rao, M., Jiang T.S., and Tsai J.P. (1989). "Combining Symbolic and Numerical Processing for Real-Time Intelligent Control," *Engineering Application of Artificial Intelligence*, Vol. 2, pp. 19-27.

[10] Rao, M., Tsai, J.P., and Jiang, J.S. (1988). "An Intelligent Decision Maker for Optimal Control," *Applied Artificial Intelligence*, Vol. 2, pp. 19-27.

[11] Wang, Q., Yang, H., Zhou, J., and Yu, J. (1990). "QUINT: A Problem-Solving Strategy for Mechanical System Concept Design," Part I and Part II, *Advance in Design Automation*, ASME, New York, pp. 331-343.

[12] Yu, J., Zhou, J., and Wang, Q. (1987). "Mechanical Product General Scheme Design CAD Based on Expert System Technology," *Advance in Design Automation*, ASME, New York, pp. 310-314.

[13] Wang, Q., Zhou, J., and Yu, J. (1989). "A Method of Product Conceptual Design Using AI Technology," *Proceedings of the 5th International Conference CAPE*, Edinburgh, U.K., pp. 200-210.

[14] Wang, Q., Zhou J., and Yu J. (1989). "Decision Making System of Mechanical Product General Scheme Design," *Proceedings of ICED'89*, pp. 705-714.

[15] Zhang, Z., and Rice, S.L. (1989). "Conceptual Design: Perceiving the Pattern," *Computer in Mechanical Engineering*, Vol. 11, No. 7, pp. 58-60.

[16] Zhou, J. Duan J. and Wang, Q. (1989). "A Study of Evaluation Subsystem or Scheme Design Expert System," *Journal of HUST*, Vol. 17, No. 2, pp. 10-24.

[17] Zhou, J., Wang, Q., and Yu, J. (1989). "Mechanical Product General Scheme Optimization Design and Intelligent CAD," *Journal of Mechanical Engineering*, Vol. 3, pp. 56-61.

Chapter 4

Developing an Intelligent CAD System

TETSUO TOMIYAMA, DEYI XUE, and HIROYUKI YOSHIKAWA
Department of Precision Machinery Engineering, The University of Tokyo

4.1 Introduction

Present computer-aided design (CAD) systems are used mainly for representing geometric information. The functions of generating two-dimensional (2-D) and three-dimensional (3-D) drawings for various mechanical, electrical, and architectural applications are widely available in commercial CAD systems. However, representing geometric information is only one part of the complex design task. To provide a general design tool covering all aspects of design activity, especially the critical conceptual design part, a CAD system needs to be built with more design-oriented functions and with "intelligence" for conducting design. To this end, the study on *intelligent CAD* was initiated.

Most of the present intelligent CAD systems have been developed using the expert system approach [11]. In these systems, problem-solving techniques such as reasoning and searching are used for conducting designs. Since the essence of design is not well understood, these intelligent CAD systems were developed in an ad hoc manner [26].

The intelligent integrated interactive CAD (IIICAD) system discussed in this chapter is based on the result of theoretical study on design processes, design objects, and design knowledge [16, 25, 33]. The IIICAD project was initiated at the Centre for Mathematics and Computer Science (CWI) in Amsterdam with the participation of the author [26]. At present, this project is being carried out at both CWI and the University of Tokyo. Recent progress from this project has been reported [2, 27, 30, 33]. The content of this chapter is based mainly on the research recently conducted at the University of Tokyo.

The research on developing the IIICAD system was started with theoretical study on design. A *cognitive design process model* describing how design is performed was acquired through experimental study on design process [25]. A *computable design process model* is then formed from this cognitive design process model using logical techniques. A *metamodel mechanism* was developed to maintain the consistency of different design object models through the theoretical study on design objects [16]. A design knowledge representation language called the Integrated Data Description Language (IDDL), a crucial tool for describing both design process knowledge and design object knowledge, was introduced [27, 31, 33].

Based on the results of these theoretical studies, the architecture of IIICAD system was proposed [33]. In this IIICAD system, the design process is presented by the computable design process model; design objects are organized by the metamodel mechanism; and design knowledge is represented by the IDDL language.

In this chapter, Section 4.2 gives a brief survey of current development on CAD systems and introduces the concept of IIICAD. Section 4.3 illustrates the result of the theoretical study on design processes, design objects, and design knowledge. Section 4.4 presents the architecture of the IIICAD system and the design knowledge representation language IDDL. Section 4.5 describes the implementation of the IIICAD system and gives an example to show how to conduct a design using the system. Section 4.6 summarizes the chapter.

4.2 Research on CAD System Development

4.2.1 Aspects of Design

In order to clarify the method adopted in the IIICAD system, the fundamental concepts for developing CAD systems are described in this section. The three aspects of design—design activities, design object models, and design processes, as shown in Figure 4.1—are briefly introduced first [26].

Design activities consist of both technical tasks and management tasks such as calculation, drawing generation, document preparation, planning, meeting. Design objects are represented by different models, such as the mathematical model of the mechanism, geometric model of the shape, finite element method model of stress. The design process is generally divided into conceptual design, basic design, detailed design, and production design.

4.2.2 Historical Overview

The concept of CAD first appeared in Sutherland's SKETCHPAD [23] which was aimed at building a computer-aided drawing system to replace the conventional drawing boards. The development of CAD systems has experienced three major steps:

- 1960s: The first generation experimental systems were developed.
- 1970s: Two-dimensional drawing systems were available to industries. Research on three-dimensional solid modeling was started.

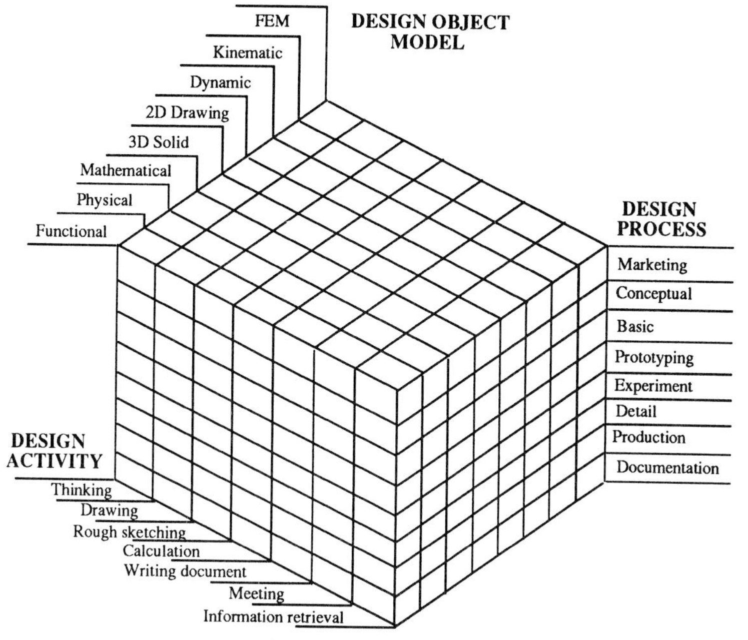

Figure 4.1: Design Activities, Models, and Processes

- 1980s and 1990s: Three-dimensional solid modeling systems became widely available. Techniques either derived from or influenced by knowledge engineering, such as constraint management, product modeling, and feature-based representation, were used for assisting design work.

Among these developments, the advance in three-dimensional solid modeling techniques plays a crucial role. Progress in computer hardware/software, new database management methods, and other information processing techniques has also greatly impacted on the development of CAD systems.

With the automated CAD and computer-aided manufacturing (CAM) functions, research on CAD/CAM integration was started. This research was aimed at transforming design information into directly usable manufacturing information and eliminating the gaps among these automated functions [6]. To further improve the efficiency of product development, concurrent engineering, which provides manufacturing and market analysis information to the design phase, has been initiated [21].

Despite these developments, the presently available CAD systems provide only the geometric modeling and drawing generation functions. Much work needs to be done to provide practical integrated and intelligent CAD systems.

4.2.3 Intelligent CAD

Although intelligent CAD is a fairly new concept, the research on developing intelligent CAD systems has also experienced three major stages.

The first-generation intelligent CAD systems are developed using expert system techniques [11]. In most of these systems, design knowledge is represented in the form of production rules. A design function specification is generally divided into several subspecifications, and each of these subspecifications can be accomplished using a specific type of function element. By combining all the generated function elements, a final solution to the entire design is provided. Based on the consideration of design as ***a problem-solving process*** [11], ***a decision-making process*** [19], and ***an optimization process*** [10], expert systems that support designers in problem-solving processes, decision-making processes, and optimization processes are developed. Since the building blocks of a design task model must be defined beforehand, these rule-based design systems are applicable only to "routine design." This approach has a major drawback — lack of flexibility. The systems rely on the previous knowledge of how to combine various functional components. The system also cannot support a designer to perform creative designs to generate new functional elements.

The second generation intelligent CAD systems, although sharing a similar platform with the earlier systems, have the capability of constraint management [15, 24]. In these systems, the solution to a design task is generated by combining subelements under multiple constraints on design attributes, product properties, and so on. To represent such attributes and properties, the concept of product modeling was introduced. Since a certain set of these attributes and properties often corresponds to a useful concept in design (particularly for describing the designer's intention), which is called a "feature" [5], feature-based representation techniques were also introduced. Special techniques for constraint management were developed for implementing these systems.

Both the first- and second-generation intelligent CAD systems aim at solving problems related to design objects, particularly the object geometry by using artificial intelligence techniques. However, to develop a tool for supporting design in every aspect, both representation of design objects and representation/application of design knowledge must be conducted [31].

4.2.4 Problems of Conventional CAD Systems

To identify the direction for developing the third-generation intelligent CAD system, problems of the conventional CAD systems are summarized as follows:

1. Not complete and with limited intelligence

   (a) *Design processes are only partially supported.* Representing geometric information is only part of the design process. The new CAD systems should have the capability of supporting designers at every stage of design, particularly in conceptual design.

(b) *Problem-solving ability is poor.* Intelligent design systems based on expert system technology can perform only routine designs with a predefined problem solving-sequence. The new CAD systems should provide the facilities for conducting creative designs where the problem solving-sequence is unknown.

2. Lack of integration

   (a) *Design object models are not well integrated.* In mechanical design, various models are used for representing the same design object. Tremendous data translating work is needed to maintain the consistency among design object models. In current CAD systems, these interfaces are conducted by the user. The future CAD systems should provide a mechanism which can generate these relations automatically.

   (b) *Representation of design processes and design objects is not unified.* The research on expert system-based knowledge representation and problem solving is performed separately from the research on geometry-based design object representation. The new CAD systems should have the capability for organizing both design process information and design object information in a unified manner.

3. Poor user interface

   (a) *The data input to a CAD system is lengthy.* The amount of (geometric) input data in CAD applications is usually enormous, which results in mistakes and misunderstandings during man-machine communication. The new CAD systems should provide the facilities to understand the functional purposes of the designer's input and translate the input data into the adapted format for computers.

   (b) *Error check is not available.* The mistakes in data input and knowledge selection can result in contradictions in later design. New facilities for resolving the contradictions at both the knowledge-base level and database level should be provided.

Based on these considerations, the concept of the third-generation intelligent CAD system, namely intelligent integrated interactive CAD (IIICAD), was proposed [26]. The development of this system focuses on how to manipulate design processes, design objects, and design knowledge. The theoretical study on design and the architecture of the IIICAD system are discussed in the following sections.

4.3 Theoretical Study on Design

4.3.1 Theoretical Study on Design Processes

Compared to design objects, design process is not well understood and properly represented. A lot of research has been done to understand the nature of design, including the design process [12, 14, 22, 25, 28, 34]. The work can be classified into two major categories. Research in the first category focuses on design methodologies [12, 14, 22]. The nature and experience of specific designs such as mechanical design and VLSI design are analyzed in the work. Research in the second category focuses on how to model a generic design process suitable for various specific designs [25, 28, 34]. The approach proposed in this subsection belongs to the second category [25].

In this subsection, an evolutionary design process model is first introduced. Second, a cognitive design process model for describing the evolutionary design process is presented. This cognitive design process model can be represented by a computable design process model using logical techniques.

Evolutionary Design Process Model

The theoretical study on design process was initiated with *general design theory* [28, 34], which aims at introducing a mathematical description of design process using axiomatic set theory. In general design theory, a design is considered as a mapping from the *functional space* where design specifications are described in terms of functions onto the *attribute space* where design solutions are described in terms of attributes. In other words, a design process begins with descriptions of design objects in terms of functionality and ends up with descriptions of manufacturable objects.

However, in most cases, this mapping cannot be realized directly. Design is a stepwise, evolutionary process. The evolutionary design process property is shown in Figure 4.2. In order to support the evolutionary design process, all information about the design object at each design stage must be stored and organized in a consistent manner. The concept of *metamodel* is used for this purpose. The evolution of a design is described by a directed sequence of metamodels. These metamodels reflect the designer's mental model of the design object. When the design specifications are given, the designer generally thinks of what the design solution should be like, how the solution could be reached, and so on. Regardless of one's preference, the design is always started with a design model. The design model is then evolved to the final solution through a step-by-step process. Various aspect models for evaluating the design from different perspectives, such as the kinematic model and geometric model, are generated from the metamodel.

The evolution of a metamodel is accomplished through a sequence of "unit" design processes. A designer goes through these processes by observing the current status of the metamodel and deciding what operations to perform next. These operations include identifying design problems at this stage, generating candidates for improving the design, acquiring information and evaluating this for the candidates, and finding the optimal solution from these candidates. A designer can improve the current metamodel to a new metamodel. The design process can also be reversed to a previous metamodel for developing the design object in a different direction.

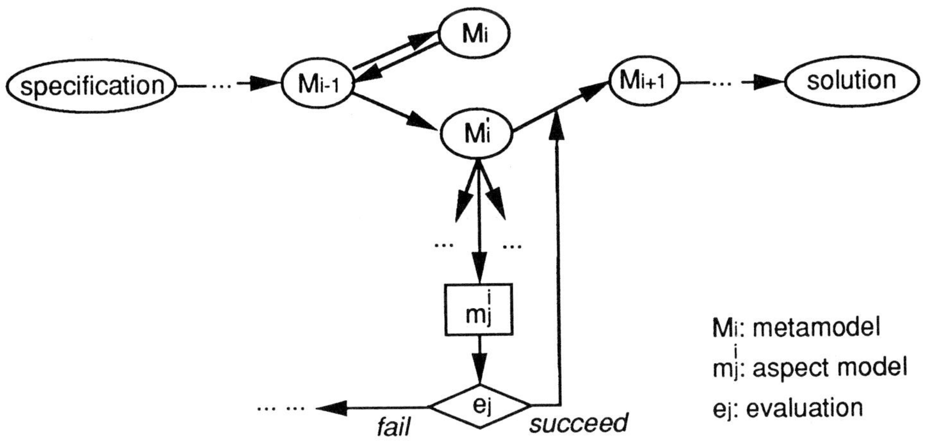

Figure 4.2: Evolutionary Design Process Model

Cognitive Design Process Model

In order to understand how the metamodel of a design evolves, a method called *design experiment* was conducted [25]. During a design experiment, designers were asked to design a certain mechanism with given specifications. The designers related what they were thinking during the design and drew sketches while performing the design tasks. The whole design process was recorded on video tape and a special CAD tool. By analyzing the protocols derived from the design experiments with the protocol analysis method [7], a *cognitive design process model* was acquired [25]. In this model, a design process is composed of many unit *design cycles.* Each design cycle has five subprocesses as follows:

1. *Problem awareness*—to identify a design problem by comparing the design object under consideration and the design specifications

2. *Design suggestion*—to suggest key concepts needed for solving the identified design problem

3. *Concept development*—to generate candidates for the design problem from the key concepts using various design knowledge

4. *Candidate evaluation*—to evaluate the design candidates in different aspects such as structural analysis, performance simulation, cost evaluation.

5. *Conclusion*—to find the best candidate for modifying the design

The design protocols can be classified on two levels: *action level* and *object level.* The action level protocols are those common for all design tasks, including the planning of the design process, and so on, while the object level protocols are related to special design tasks such as gear design or shaft design.

For the protocols representing design knowledge, the distinction between the action level and object level is essential for considering the architecture of the IIICAD system.

Computable Design Process Model

To incorporate the cognitive design process model into the IIICAD system, the model was transformed into a computable design process model using logic techniques as shown in Figure 4.3 [25]. With the computable design process model, the design knowledge and knowledge inference are organized and conducted at both the action level and the object level.

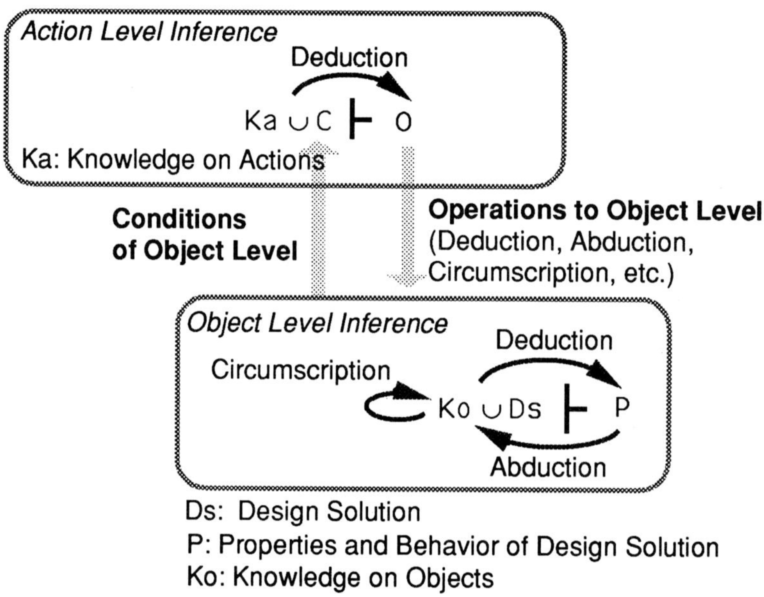

Figure 4.3: Computable Design Process Model

The object level design processes are logically formalized as

$$K_o \cup D_s \vdash P$$

where D_s is a set of logical formulae describing the design solutions that the designer wants to obtain; K_o is the knowledge on object properties; and P is a set of logical formulae describing properties of the design solutions, respectively. Required design specifications are described in P. The cognitive design process model can be represented by this computable model in the following way. First, the problems to be solved are described in P. Design suggestion is an abduction process [8] to obtain D_s from P and K_o. Concept development and candidate evaluation are deduction processes to obtain P from D_s and K_o for getting more information about the design solution. In the concept development and candidate evaluation subprocesses, because of the incompleteness of the design knowledge, contradictions sometimes occur. *Circumscription* [18] can be used for resolving these contradictions. It is a method to solve the contradictions by adding exceptions to the original knowledge; in other words, circumscription is a minimal patching technique for contradictory knowledge bases. For example, we have the following three rules with a contradiction:

$$bird(X) \land \neg ab(X) \rightarrow fly(X)$$
$$ostrich(X) \rightarrow \neg fly(X)$$
$$ostrich(X) \rightarrow bird(X)$$

The first rule is added with an *abnormal predicate* described as $ab(X)$ which represents the exception. By applying circumscription, this abnormal predicate is calculated as

$$ab(X) \equiv ostrich(X)$$

By replacing the abnormal predicate with this solution, we get the modified rule represented as

$$bird(X) \land \neg ostrich(X) \rightarrow fly(X)$$

Introducing this rule is by no means the only way to solve the contradiction, but it gives a minimal patch to do so. If the required properties cannot be concluded because of the modifications of the knowledge base, these properties should be defined as the new problems to be solved. This is a jump to the problem awareness subprocess.

Action level design processes are logically formalized as

$$K_a \cup C \vdash O$$

where, K_a is the action level design process knowledge, that is, the relations between the conditions of the design object and the possible operations; C is a set of the conditions of the design object; and O is a set of the possible operations to the design object, respectively. In the computable design process model, the action level inference is performed to observe the condition of the design object and to propose to users the possible operations using action level design process knowledge. The user selects the most appropriate operation from these proposed operations, that is, the most appropriate object level design process knowledge is selected and object level operations, such as abduction, deduction and circumscription, are performed.

4.3.2 Theoretical Study on Design Objects

Although the representation of design objects is well developed in some areas such as geometric modeling, further studies on a broader representation scheme are needed [16]. First, general knowledge of the physical world is important for evaluating the behaviors of design objects. This kind of knowledge should be included in the representation for conducting creative design. Second, the representation should support automated integration of various design object models and serve as a framework for intelligent design.

In this subsection, qualitative physics [1, 9, 32] for describing the knowledge about the physical world is introduced. The metamodel mechanism [16] for maintaining the consistency of various design object models is described.

Qualitative Process Theory

Qualitative physics [1, 9, 32] is a method for describing knowledge about the physical world. It has two functions: symbolic modeling of a physical system and reasoning about the dynamic behaviors of the system. In the theoretical study on design objects, the basic idea of *qualitative process theory* [9] was utilized. In this theory, an *individual view* represents an entity existing in the physical world and a *process view* represents a physical phenomenon that influences the system. For example, a gear can be represented by an individual view; the physical phenomenon introduced by a gear pair can be represented by a process view. Each individual view or process view consists of two parts: precondition part and result relation part. During qualitative reasoning, individual views and process views are created or removed dynamically. The view parameters may also change following the predefined relations. The behavior of the physical system is represented by a sequence of varying view parameters.

Kiriyama and colleagues developed a qualitative reasoning system [16] based on the qualitative process theory using Smalltalk-80 [13]. The system uses the Assumption-based Truth Maintenance System (ATMS) [4] to obtain and manage reasoning results. The system also serves as the basis for building the metamodel mechanism.

A physical feature modeling system for modeling a design object was also developed by Kiriyama and colleagues [16]. The concept of *feature*, originally used for representing geometric information, is extended for representing teleological knowledge on the relationship between a functional concept and its attributive information [5]. A physical feature is used for describing the knowledge of a physical phenomenon and its attributes. The entities of a physical feature are described in individual views and the physical phenomena are described in process views. For example, a gear pair is a physical feature which consists of two gear entities and a rotation function phenomenon. The construction of a design object model, shown in Figure 4.4, illustrates the combination of physical features.

Currently, a project for building a physical feature knowledge base with the order of about 10,000 knowledge chunks is being conducted at the University of Tokyo. The present knowledge base consists of around 2,000 chunks of knowledge. This effort is similar to the *Cyc* project for building a large-scale, common-sense knowledge base, which is being conducted by Microelectronics and Computer Technology Corp. [17], and the *How Things Work* project for building a knowledge base of physical devices used in engineering design and diagnosis, which is being conducted at Stanford University [3].

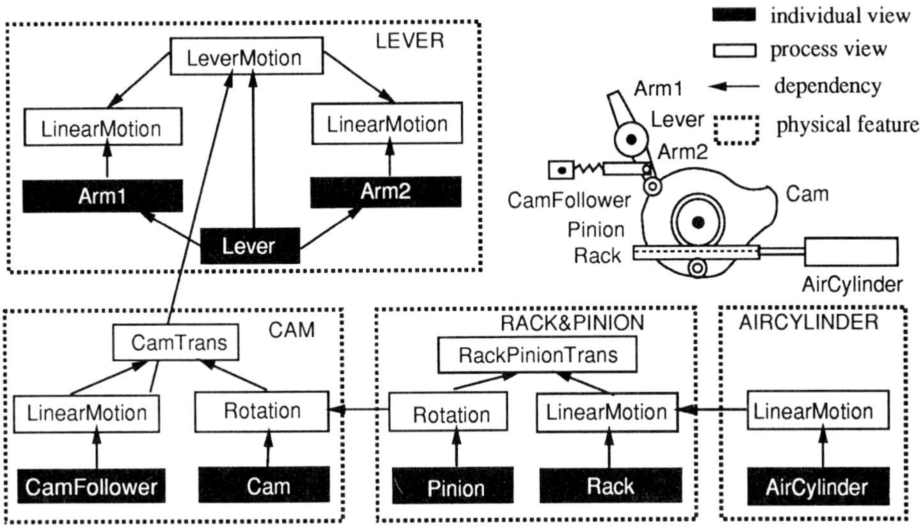

Figure 4.4: Model Construction by the Combination of Physical Features

Metamodel Mechanism

As discussed earlier, a design object can be represented by various *aspect models.* In design process, these aspect models can be generated using the specific knowledge from the metamodel where a general description about the design object is preserved. These aspect models are not independent from each other. Changes in one aspect model may cause changes in other aspect models. Therefore, the consistency of these aspect models must be maintained. In our earlier study [16], it was found that a special theory exists behind each aspect model. These theories are called *background theories.* For example, the background theory for a geometric model is algebraic geometry. By describing these background theories using the unified description of qualitative process theory, the consistency among aspect models can be maintained. This mechanism for maintaining the consistency of various aspect models is called a *metamodel mechanism* [16]. The structure of the metamodel mechanism is shown using an example illustrated in Figure 4.5.

4.3.3 Theoretical Study on Design Knowledge

By analyzing the protocols derived from the design experiments, it was found that design knowledge is used mainly for the following two purposes:

- generating design object models
- evaluating generated design object models

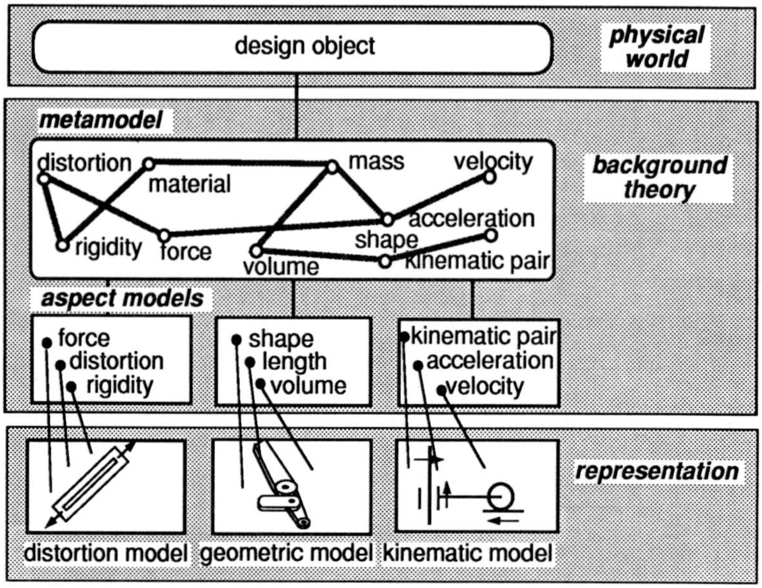

Figure 4.5: Example of the Metamodel Mechanism

In the first part, the design object is evolved from a set of abstract descriptions to a set of more concrete and complicated descriptions using the knowledge about the relations of these descriptions. This knowledge usually exists in the form of design experiences. The knowledge is called *knowledge about concept operations*, or *design process knowledge*. In the evaluation aspect, the design object is explained by more detailed descriptions using the knowledge about the design object. This knowledge is usually independent of the designer's and is called *knowledge about concept definitions*, or *design object knowledge*. For example, a gear pair can be selected for accomplishing the given design function of rotation transform using the knowledge about concept operations. Other design properties of the gear pair can be generated using knowledge about the concept definitions.

Design process knowledge is generally shallow knowledge on design experiences. Production rules are appropriate for representing this kind of knowledge. On the other hand, design object knowledge is user-independent, deep knowledge on the principles of the physical world. Qualitative process theory is appropriate for representing the design object knowledge.

4.4 Intelligent Integrated Interactive CAD System

Based on the theoretical study of design processes, design objects, and design knowledge, the architecture of the intelligent integrated interactive CAD (IIICAD) system was proposed

[33]. In this system, a design process is organized by the computable design process model [25], and a design object is organized by the metamodel mechanism [16]. Design knowledge is represented by a language called Integrated Data Description Language (IDDL) [27, 31, 33]. Since IDDL serves as the kernel language in the IIICAD system, this language is introduced first.

4.4.1 IDDL — A Design Knowledge Representation Language

The IDDL language was designed for implementing the IIICAD system. The *prototypes*, *objects*, *facts*, *scenarios*, and *worlds* are used for describing the design process knowledge, design object knowledge, and evolution of a design.

1. *Prototype* A prototype is used for describing a standard entity or a physical phenomenon, which is called an individual view or a process view in the Qualitative Process Theory [9]. The prototypes, organized in a hierarchical structure, are used for instantiating design object entities (called objects), constructing physical features by their combinations, and serving as the background theories for aspect models.

2. *Objects and facts* Objects and facts are used to describe the evolutionary design object. Objects represent design entities and physical phenomena. Facts represent the relationships among objects. In IDDL, facts are written by predicates, and objects are written as the terms of these predicates. For instance, *gearPair(gear1,gear2)* defines a fact. The two terms used for defining this fact, *gear1* and *gear2*, are two objects. Objects are instantiated from the prototypes. Objects also have their own structure described in both *attributes* and *functions* over these attributes. The object *gear1* may have attributes including the number of teeth, z, gear module, m, and gear diameter, d. The built-in predicate *=(d[gear1],m[gear1]*z[gear1])* is the function over these attributes.

3. *Scenarios* Scenarios are used to describe the design process knowledge. A scenario is a set of production rules. Each rule has the $IF - THEN$ structure. Both the condition part and the result part are described by a set of facts connected with the logical-and (&). Rules can be used in *forward reasoning* for deduction and in *backward reasoning* for abduction. These rules can also be modified by circumscription.[1]

4. *Worlds* Worlds are the workspace for design evolution. The design object is evolved from the world with design specifications to the world with design solutions. A new world can be created using one of the following three approaches: directly initiating an empty world, copying all information from an existing world, or merging several existing worlds.

[1] In IDDL, a circumscription algorithm proposed by Nakagawa and colleagues [20] is used.

4.4.2 IIICAD System Architecture

The architecture of the IIICAD system is illustrated in Figure 4.6. The system has three major components: a *design process handling system*, a *metamodel mechanism*, and a *user interface manager*.

In the design process handling system, the design process knowledge is described in *scenarios*. Scenarios are divided into two levels: action level and object level. The action level scenarios are programmed in an *action level knowledge-base browser*. These scenarios describe the knowledge common to various kinds of designs, such as the general relationship between database status and the appropriate database operations. The object level scenarios are programmed in an *object level knowledge-base browser*. These scenarios describe the knowledge about special designs, such as the relationship between design objects and the properties/behaviors of those design objects.

In the IIICAD system, a design process is managed by a supervisor module as shown in Figure 4.6. This supervisor module has two design process managers: an *action level manager* and an *object level manager*. The action level manager uses action level scenarios to observe the status of the design object database and to suggest appropriate design evolution operations. These operations are used for selecting the most suitable object level scenario from the object level scenario base and the most appropriate reasoning method for the object level manager. In other words, the action level manager accomplishes the design cycle (problem awareness, design suggestion, concept development, candidate evaluation, and conclusion subprocesses). In addition the object level manager solves object level problems using the selected object level scenarios. Three types of inference systems, namely an *abduction system*, a *deduction system*, and a *circumscription system*, are used by the object level manager to perform various operations. This manner of design management forms a design environment called a *design browser*.

Design object knowledge consists of *prototypes* and *physical features* (combination of prototypes). The design object knowledge can be programmed in the *prototype browser* and *physical feature browser*. The evolutionary design object is described in the IIICAD system using a set of *worlds*. These worlds are organized in an environment called *world manager*. Only one world is active and operations are performed in the *active world*. Each world consists of a *primary model* and a *metamodel*. The primary model is built up by the design process handling system where various reasonings are conducted using the IDDL scenarios. The basic elements and configuration of a design object are represented by the primary model using IDDL objects and facts. The metamodel is generated from the primary model using a qualitative reasoning system. The consistency of design object data is provided by the Assumption-based Truth Maintenance System (ATMS).[2] Various aspect models are generated from the metamodel using specific background theories written in IDDL prototypes. These aspect models are used either for evaluating the design object from various perspectives, or for associating a design object with its *external models* (e.g.,

[2] In the IIICAD system, two ATMS systems are used for maintaining the consistency in the primary model and the metamodel. The two ATMS systems are not independent. The primary model and metamodel share the same assumptions.

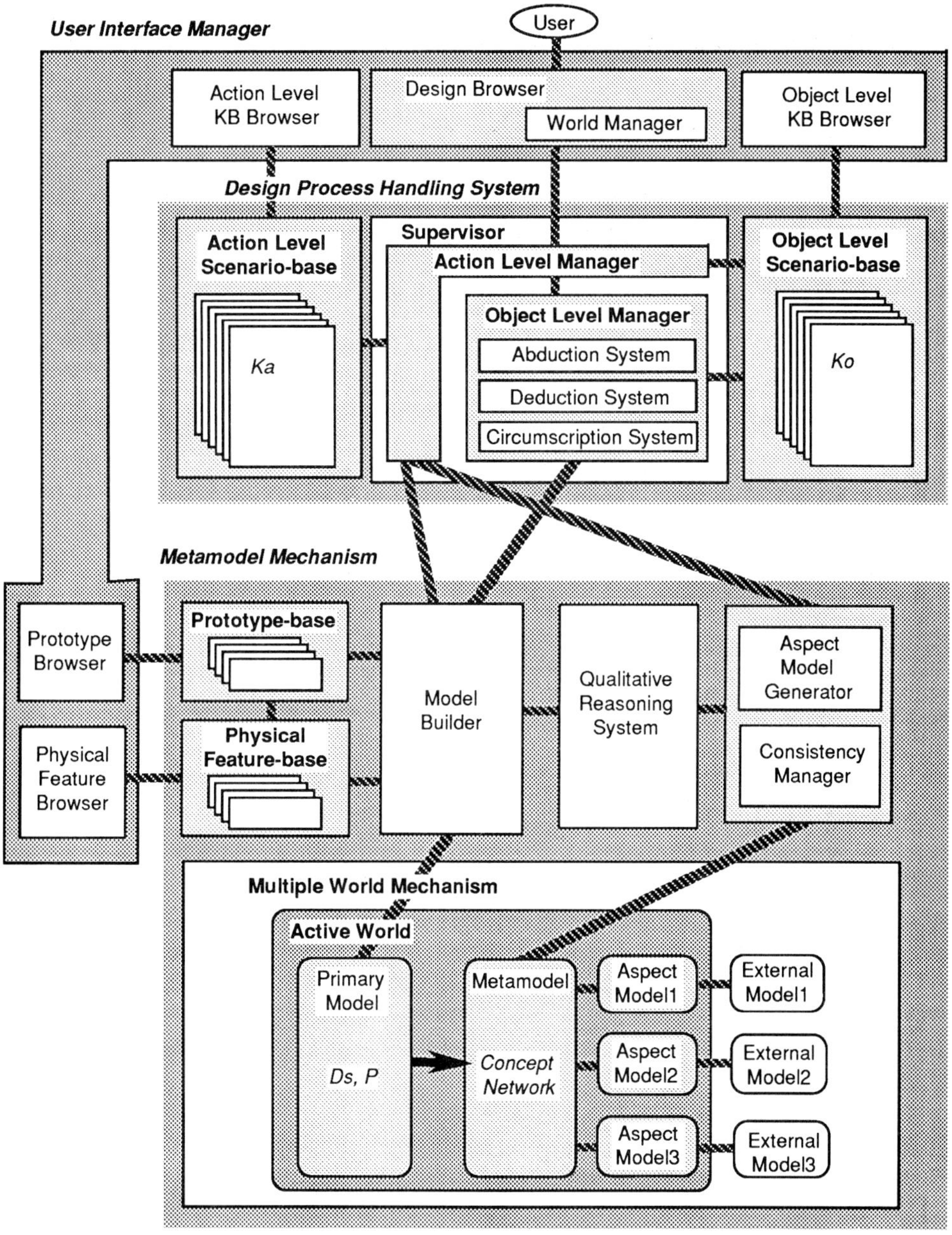

Figure 4.6: Architecture of the IIICAD System

geometric model). Operations on design objects are performed using the three components of the metamodel mechanism: the *model builder* for constructing the primary model and metamodel; the *aspect model generator* for generating various aspect models; and the *consistency manager* for maintaining the consistency among different aspect models.

4.5 Implementation of the IIICAD System

The prototype IIICAD system was implemented using the object-oriented language — Smalltalk-80 [13]. The Smalltalk-80 system was used because of its excellent user interface and rapid prototyping capability. A snapshot of the developed system is illustrated in Figure 4.7.

In the prototype IIICAD system, various interface environments are used for

- editing design process knowledge and design object knowledge described in the form of action level scenarios, object level scenarios, prototypes, and physical features;
- organizing various design activities such as selection of the most appropriate scenarios and operations;
- representing the evolutionary design object written in the form of worlds, primary models, metamodels, and so on.

4.5.1 Representation of Design Processes

The representation of design processes in the IIICAD system is based on the computable design process model [25]. Design process knowledge is described by scenarios. Both the organization of this knowledge and inference using this knowledge are conducted at two separate levels: action level and object level. In the design process, the action level knowledge manager acquires the status of the design object database and suggests possible operations using action level scenarios. The most appropriate object level scenario is selected when a proper action level scenario is executed. In addition, the most appropriate object level operations including abduction, deduction, and circumscription are conducted upon the selected object level scenarios.

In the IIICAD system, progress of a design process using the action level manager is accomplished by

1. *Problem awareness* The design object is evolved from the world with design specifications to the world with design solutions. In each world, both the primary model and the metamodel are described by objects, facts, and attributes of objects. All these data represent the design object at a certain stage of design. A user can work on a manageable design problem defined from the partial design data at a specific design stage.

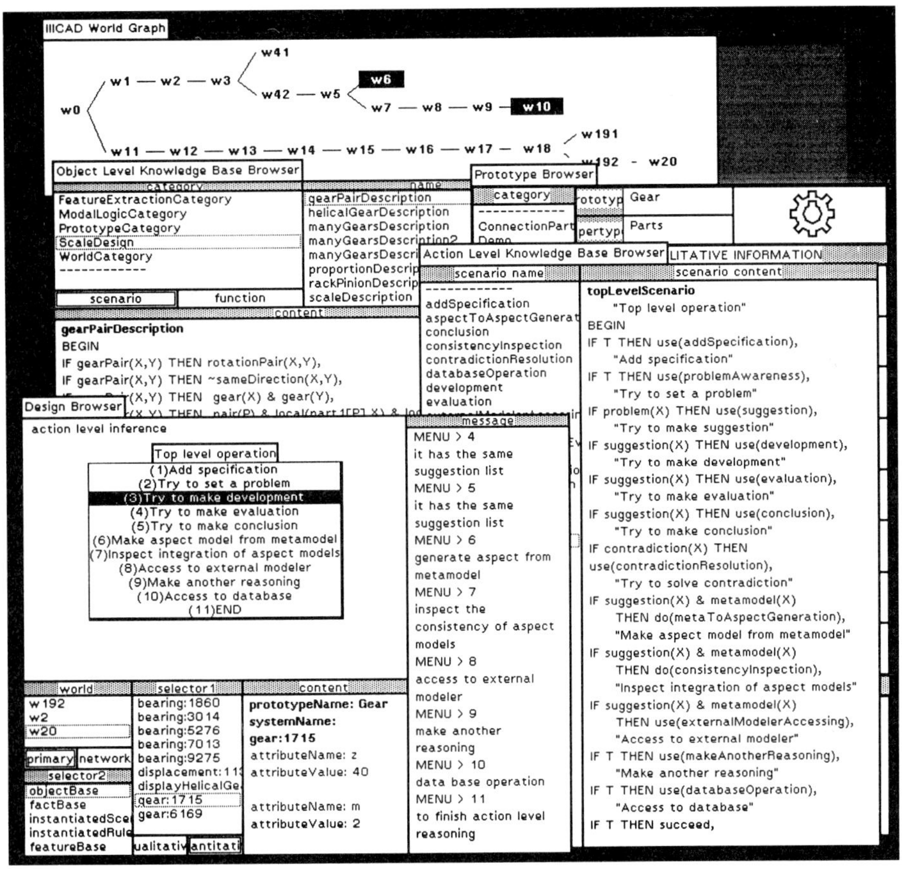

Figure 4.7: Screen Hardcopy of the IIICAD System

2. *Design suggestion* The design suggestion subprocess is used to generate key concepts for solving the focused design problem. The design suggestion is generated using appropriate scenarios. An action level scenario used for design suggestion is shown in Figure 4.8. The scenario for design suggestion has the following form:

IF X is a problem, Y is an object level scenario, and Y can be used for generating a design suggestion to X by abduction,

THEN Scenario Y can be selected and the abduction operation can be performed. The database should be updated to record that problem X has been processed.

or

IF X is a problem, Y is an object level scenario, and Y can be used for generating a design suggestion to X by functional description method [29],

THEN Scenario Y can be selected and the functional description method operation performed. The database should be updated to record that problem X has been processed.

```
suggestion

    "Making suggestion to problem"
BEGIN
IF problem(X) & scenario(Y) & canSuggestByAbduction(Y,X)
    THEN select(Y) & do(suggestionByAbduction) & ~problem(X),
    "Find suggestion by abduction"
IF problem(X) & scenario(Y) & canSuggestByFunctionDescription(Y,X)
    THEN select(Y) & do(suggestionByFunctionDescription) & ~problem(X),
    "Find suggestion by function description"
IF T THEN succeed,
    "END"
END
```

Figure 4.8: Action Level Scenario for Suggestion

In these scenarios, the built-in predicate *select* is used for selecting object level scenarios, and the built-in predicate *do* is used for performing object level operations. The design suggestions generated by these scenarios are considered assumptions in the IIICAD system.

The selection and use of object level scenarios can be illustrated using an example.

Suppose we have a problem called *rotationPair(r1,r2)* to be solved. In the object level scenario base, two scenarios can be used for making design suggestions for this problem, as shown in Figure 4.9. By selecting either of these two, a design suggestion named *beltGearDrive(r1,r2)* or *gearPair(r1,r2)* could be generated by abduction reasoning.

```
beltGearDriveDescription
BEGIN
IF beltGearDrive(X,Y)
   THEN rotationPair(X,Y),
IF beltGearDrive(X,Y)
   THEN sameDirection(X,Y),
... ...
END
```

(a) "beltGearDriveDescription" Scenario

```
gearPairDescription
BEGIN
IF gearPair(X,Y)
   THEN rotationPair(X,Y),
IF gearPair(X,Y)
   THEN ~sameDirection(X,Y),
... ...
END
```

(b) "gearPairDescription" Scenario

Figure 4.9: Two Object Level Scenarios for Suggestion

3. *Concept development* Details of the generated suggestions are further produced in the concept development subprocess. The two scenarios, shown in Figure 4.9, could also be used for concept development in deductive reasoning. For instance, these scenarios lead to the conclusion that *gearPair(r1,r2)* has the property of *~sameDirection(r1,r2)*. Generally speaking, the rules at the object level scenarios represent the design knowledge K_o (see Section 1.3.1); *gearPair(r1,r2)* represents the design solution D_s; *rotationPair(r1,r2)* and *~sameDirection(r1,r2)* represent the properties P of the design object.

4. *Candidate evaluation* The given design specifications in the form of constraints on design attributes can be used as criteria for design evaluation. The discussed aspect models and external models can also be used for design evaluation.

5. *Conclusion* If a design cannot successfully pass the evaluations, one can remove the design suggestion and relevant data using ATMS. When the fault design suggestion is removed, a new subproblem of the design will be introduced.

4.5.2 Representation of Design Objects

The representation of a design object is based on the metamodel mechanism [16]. As discussed earlier, the design object knowledge is represented by prototypes and physical features, and the evolutionary design object is described by a set of worlds.

In the design process, an empty world is first created. Given design specifications are described in the primary model of this world. The primary model, which represents the basic elements and configuration of the design object, is built up by the scenario-based reasoning.

Prototypes and physical features are used as the standard description for instantiating the specific design object. The metamodel, which consists of various information about the design object, is generated from the primary model using qualitative reasoning. Prototypes are used as the individual views and process views for reasoning. At each stage of the design, a user can work on a manageable aspect model, which is defined by part of the design data from a complete metamodel. The performance evaluations of the design object are conducted using these aspect models. Since all the data derived from qualitative reasoning are organized by ATMS, the consistency of these aspect models can be maintained. Both the primary model and the metamodel in a world are described by IDDL objects and facts.

4.5.3 An Example

In this section, an example is given to show how a design is performed in the prototype IIICAD system. The task is to design a weighing scale. The design knowledge used in this example is acquired from a design experiment. Figure 4.10 shows part of the protocol data obtained from this design experiment. Figure 4.11 shows the formalized knowledge written in production rules. In the IIICAD system, these rules are described using several object level scenarios and are selected when needed. For ease of presentation, each rule is presented in Figure 4.11 and numbered accordingly. A snapshot of this design performed in the present prototype IIICAD system is given in Figure 4.7.

In the initial stage of design, given design specifications, shown in Figure 4.12, are described in world *w0* (see Figure 4.7). By applying abduction three times with *rule (1)*, *rule (2-1)*, and *rule (2-2)*, the design object is evolved to world *w3*. In this world, it is suggested that the weight supported by the scale should be measured by the displacement caused by the weight. Furthermore, applying *rule (3)* or *rule (4)*, the world, *w3*, can be evolved to world *w41* or world *w42*, respectively (see Figure 4.13). In either of these two worlds, a spring mechanism is suggested to be used for measuring the weight by reading the displacement caused by the weight. However, because the mechanism described in world *w42* (see Figure 4.10B) is more easily realized than the mechanism described in world *w41* (see Figure 4.10A), world *w42* is selected for continuing the design. In world *w6*, a rack-pinion mechanism is used for transforming linear displacement to rotational displacement. Since the linear displacement is too big, a contradiction caused by *rule (2-2)* and *rule (6)* is found. This contradiction can be solved by circumscription (see Figure 4.14). In order to satisfy the revised rule, the original design specifications should also be changed. We give up the idea of changing the original specifications, and return to world *w5* to continue the design by adding a gear box. In world *w10*, another contradiction is caused by *rule (10)* and *rule (1)*, that is, the indicator cannot be seen from top. This contradiction could also be solved by circumscription. However, after the circumscription operation, some design specifications cannot be reached again. We thus return to world *w0* to restart the design. This time, because we have some experience about this design, the design evolves smoothly. By adding a helical gear pair to the design, one can now see the indicator from the top.

In world *w18*, the design of the gear box is performed. With given specifications and knowledge, a gear box with two gears, or a gear box with three gears, is selected and the

(1) What mechanism does a standard scale use?
(2) It measures the displacement caused by weight like this (Figure A).
(3) So we should use a spring.
(4) We can also use a spring this way (Figure B).
(5) If we use a rack and pinion (Figure C), we can measure the weight, because the displacement is in proportion to the weight.
(6) As it translates 100kg of the weight to 5mm of the displacement, it is impossible to realize it with this mechanism.
(7) If we do not mind the accuracy, it is possible to use many gears .
(8) But we cannot see the indictor upward.
(9) The indictor is difficult to see either.
(10) The indictor in a standard scale is easy to see.
(11) If the indictor can be seen upward, it is easy to see.
(12) How about using a helical gear?
... ...

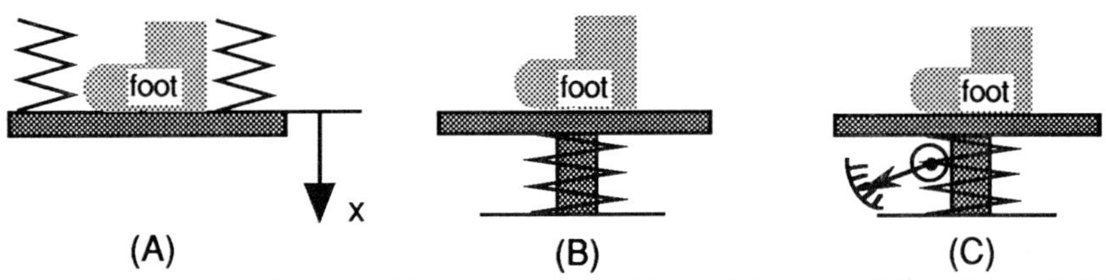

Figure 4.10: Design Experiment Protocol Data

(1) **IF** weight(W) & canMeasure(S,W) & support(S,W) **THEN** scale(S),
(2-1) **IF** displacement(D) & indictor(I) & has(S,I) & weight(W) & translate(S,W,D)
THEN canMeasure(S,W),
(2-2) **IF** proportion(S,W,D) & weight(W) & displacement(D)
THEN translate(S,W,D),
(3) **IF** spring(SP) & weight(W) & displacement(D) & has(S,SP) & pull(SP,W)
& move(SP,D) **THEN** proportion(S,W,D),
(4) **IF** spring(SP) & weight(W) & displacement(D) & has(S,SP) & push(SP,W)
& move(SP,D) **THEN** proportion(S,W,D),
(5) **IF** rackAndPin(RP) & has(S,RP) & weight(W) & displacement(D)
THEN proportion(S,W,D),
(6) **IF** proportion(S,W,D) & weight(W) & displacement(D)
& >(maxValue[W]/maxValue[D],10) **THEN** ~translate(S,W,D),
(7) **IF** proportion(S,W,D) & weight(W) & displacement(D)
& >(maxValue[W]/maxValue[D],10) & manyGears(MG)
& hasManyGears(S) **THEN** translate(S,W,D),
(8) **IF** manyGears(MG) & hasManyGears(S) **THEN** ~hasUpwardIndictor(S),
(9) **IF** ~hasUpwardIndictor(S) **THEN** ~easyToSee(S),
(10) **IF** ~easyToSee(S) **THEN** ~scale(S),
(11) **IF** hasUpwardIndictor(S) **THEN** easyToSee(S),
(12) **IF** hasManyGears(S) & helicalGear(HG) & has(S,HG)
THEN hasUpwardIndictor(S),
... ...

Figure 4.11: Formalized Knowledge

specifications:
scale(sc1)
=(maxSupportWeight[sc1],100)
=(maxDisplacement[sc1],5)
...

Figure 4.12: Design Specifications

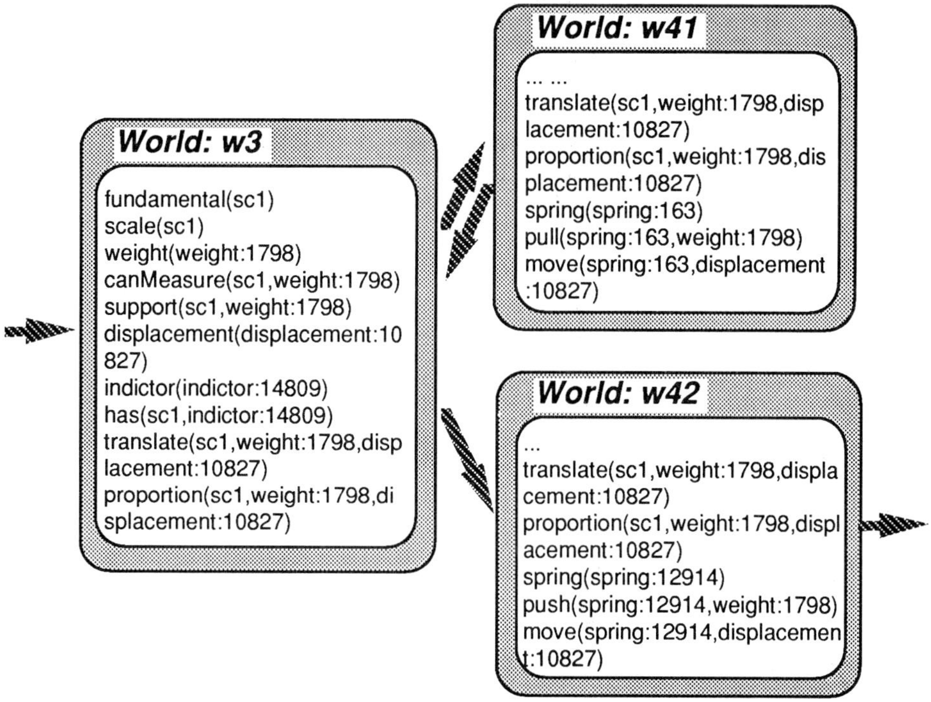

Figure 4.13: Two Possible Spring Mechanisms

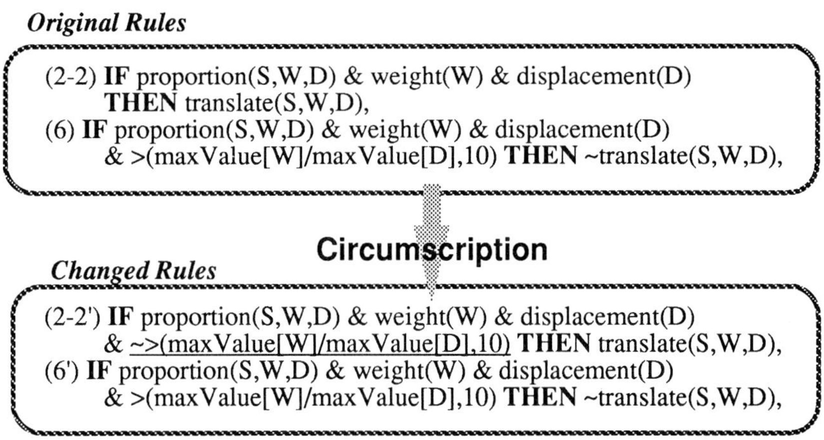

Figure 4.14: Revision of a Rule by Circumscription

design object evolves to world *w191* or world *w192*, respectively. However, the rotation direction of the indicator in world *w191* is counterclockwise. This direction is opposite from the rotation direction we are used to. We thus select the mechanism in world *w192* to continue our design. In world *w191*, the evaluation of the design object behavior is performed by qualitative reasoning (see Figure 4.15). First, from the primary model where the fundamental elements and configuration of the design object are described, a metamodel, consisting of various behavior descriptions of this design object, is generated. By focusing on part of the data related to the kinematic aspect in this metamodel, a kinematic aspect model is generated. The qualitative reasoning is performed in this kinematic aspect model. Before the reasoning is conducted, all parameters of individual views and process views are initiated with qualitative values. For example, in the process view *rackAndPinionMechanism* shown in Figure 4.15, the description *(pinion theta) direct (displacement value)* indicates that the parameter *theta* of an object *pinion* changes proportionally to the parameter *value* of an object *displacement*. If the parameter *value* of the object, *displacement:11377*, increases, the parameter *theta* of the object, *pinionGear:4169*, will also increase. So the parameter *theta* could change from the initial value *zero* to *plus*. In this way, it is concluded that the parameter *theta* of the indicator object, *displayHelicalGear:13767*, will change to *minus*, that is, the counterclockwise rotational displacement to the zero position.

Figure 4.16 shows part of the database preserved in world *w20*. An external geometric modeler can be used for representing the geometric information of this design object. A Smalltalk-80 interface to this geometric modeler for representing the gear pair, using a Boundary Representation model, is shown in Figure 4.17.

4.6 Conclusions

The theoretical studies and implementations of the third-generation intelligent CAD system, intelligent integrated interactive CAD (IIICAD), are discussed in this chapter. The research starts with theoretical study on design processes, design objects, and design knowledge. Based upon this study, the architecture of the IIICAD system is proposed. In the prototype IIICAD system, the design process is represented by a computable design process model, and the design object is modeled by a metamodel mechanism. This system has been implemented using Smalltalk-80. A design knowledge representation language, called Integrated Data Description Language (IDDL), is used as the kernel for implementing the IIICAD system.

The following conclusions are drawn from this research:

1. The computable design process model, which is transformed from a cognitive design process model, is useful for describing the design process. The metamodel mechanism is useful for maintaining the consistency of various design object models.

2. Design knowledge is divided into two categories, that is, knowledge regarding the experience of design and knowledge of the physical world. In the first category, the knowledge is used for building up the design object model and is appropriate for representation in production rules. In the second category, the knowledge is used

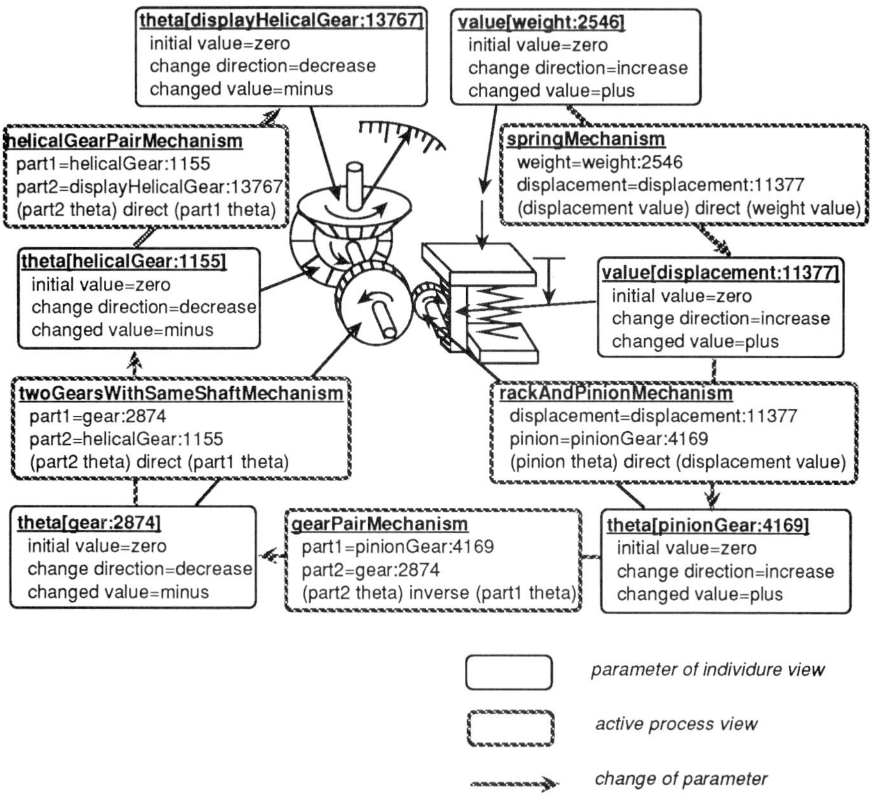

Figure 4.15: Qualitative Reasoning of a Gear Box Mechanism

> scale(sc1), easyToSee(sc1), weight(weight:2546),
> canMeasure(sc1,weight:2546), support(sc1,weight:2546),
> displacement(displacement:11377),
> translate(sc1,weight:2546,displacement:11377),
> proportion(sc1,weight:2546,displacement:11377),
> spring(spring:13414), push(spring:13414,weight:2546),
> rackAndPin(rackAndPin:5480),manyGears(manyGears:2599),
> hasManyGears(sc1), hasUpwardIndictor(sc1),
> helicalGear(helicalGear:1155), has(sc1,helicalGear:1155),
> threeGearsMechanism(threeGearsMechanism:13994),
> gear(gear:9600), gear(gear:16252), gearPair(gear:9600,gear:16252),
> ~sameDirection(gear:9600,gear:16252),
> ... ...

Figure 4.16: Part of the Database in a World

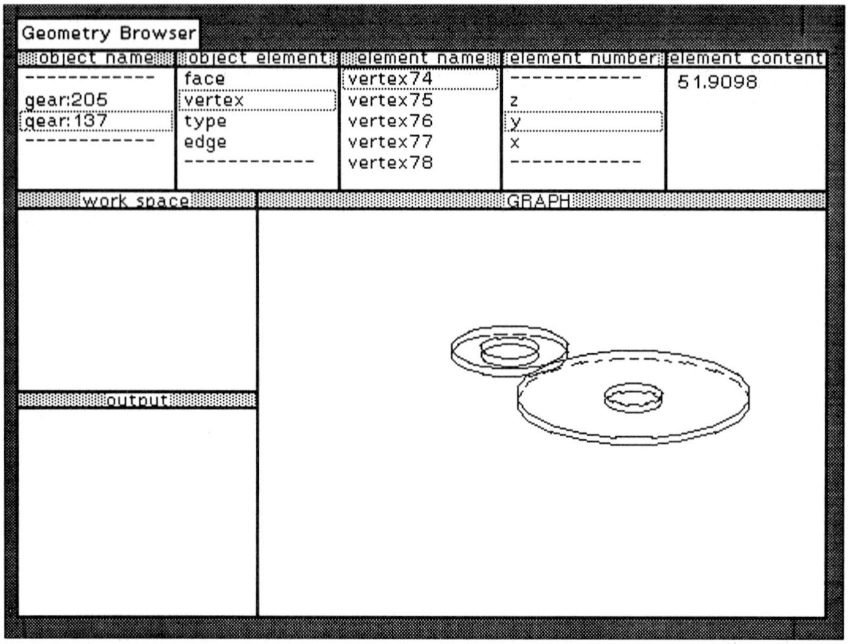

Figure 4.17: Interface for Representing Geometric Information

for evaluating the design object, such as analyzing the behavior of the design object. The qualitative process theory is appropriate for representing the knowledge of the physical world.

3. The IIICAD system is useful for organizing information of both the design process and design object in an integrated way. The IDDL kernel language is useful for describing both the design process knowledge and design object knowledge.

Future work includes the following research topics:

1. The IIICAD system (or more precisely the metamodel mechanism) should allow "plugging in" external modeling systems, such as the geometric modeler and the finite element analysis package, for various evaluation and computation purposes. To do so, the metamodel mechanism should incorporate concepts that are represented in those external systems. In the case of mechanical engineering design, this inevitably requests incorporating spatial and temporal knowledge into the framework of qualitative physics, which is a future research issue.

2. Incorporating functional knowledge into the framework of qualitative physics is also essential, since we are dealing with design. Special techniques should be developed for this purpose [29].

3. In the present implementation, the coupling of the design process handling system and the metamodel mechanism is "weak" in that object level operations of the metamodel mechanism are not recognized by the design process handling system or by scenarios. This must be solved by having more mutual coupling, which is also a research issue.

4. Scenarios for the current system must be manually "encoded." This should be automatically done by tracing back a design history and extracting useful knowledge from it, which largely depends on machine learning techniques.

Acknowledgement

We would like to thank the students of Yoshikawa-Tomiyama Laboratory of the Department of Precision Machinery Engineering of the University of Tokyo for their continuous help in developing the IIICAD system. We would especially like to thank Dr. T. Kiriyama and Dr. H. Takeda for their enormous contribution in the theoretical studies on both design processes and design objects. We also wish to thank Dr. Z. Dong at the University of Victoria, Victoria, B.C., Canada, for his important comments and suggestions in revising this chapter. Financial support from Nippon Steel Corporation, Japan, is gratefully acknowledged.

References

[1] Bobrow, D.G., eds. (1985). *Qualitative Reasoning about Physical Systems*, MIT Press, Cambridge, MA.

[2] Breige, A. (1992). "Towards a Formal Definition of IDDL," in *Intelligent CAD Systems IV: Added Value of Intelligence to CAD*, Barthes, J.P., and Dahshan, K.El., eds., Springer–Verlag, Berlin, in Press.

[3] Cutkosky, M.R., and Tenenbaum, J.M. (1990). "Research in Computational Design at Stanford," *Research in Engineering Design*, Vol. 2, No. 1, pp. 53–59.

[4] de Kleer, J. (1986). "An Assumption–based TMS," *Artificial Intelligence*, Vol. 28, pp. 127–162.

[5] Dixon, J.R., and Cunningham, J.J. (1989). "Research in Design with Features," in *Intelligent CAD, I*, Yoshikawa, H. and Gossard, D.C., eds., North-Holland, Amsterdam, pp. 137–148.

[6] Dong, X., and Wozny, M. (1988). "Feature Extraction for Computer-Aided Process Planning," in *Proc. Third International Conference on Computer-Aided Production Engineering*, June 1-3, University of Michigan, Ann Arbor, MI.

[7] Ericsson, K., and Simon, H. (1985). *Protocol Analysis: Verbal Reports as Data*, MIT Press, Cambridge, MA.

[8] Fann, K.T. (1970). *Peirce's Theory of Abduction*, Martinus Nijhoff, The Hague, The Netherlands.

[9] Forbus, K.D. (1984). "Qualitative Process Theory," *Artificial Intelligence*, Vol. 24, No. 3, pp. 85–168.

[10] Gero, J.S., ed. (1985). *Optimization in Computer-Aided Design, Proceedings of the IFIP W.G. 5.2 Working Conference 1983*, North-Holland, Amsterdam.

[11] Gero, J.S., ed. (1987). *Expert Systems in Computer Aided Design, Proceedings of the IFIP W.G. 5.2 Working Conference 1987 (Sydney)*, North-Holland, Amsterdam.

[12] Glegg, G.L. (1968). *The Design of Design*, Cambridge University Press, U.K.

[13] Goldberg, A., and Robson, D. (1983). *Smalltalk-80: The Language and Its Implementation*, Addison–Wesley, Reading, MA.

[14] Hubka, V. (1976). *Theorie der Konstrukutionsprozeße*, Springer–Verlag, Berlin.

[15] Inui, M., and Kimura, F. (1990). "Representation and Manipulation of Design and Manufacturing Processes by Data Dependency," In *Intelligent CAD, II*, Yoshikawa, H., and Holden, T. eds., North-Holland, Amsterdam, pp. 183-201.

[16] Kiriyama, T., Tomiyama, T., and Yoshikawa, H. (1990). "Qualitative Reasoning and Conceptual Design with Physical Features," in *Proceedings of the 4th International Workshop on Qualitative Physics*, Lugano, Switzerland, pp. 153–160.

[17] Lenat, D.B., and Guha, R.V. (1989). *Building Large Knowledge–base Systems: Representation and Inference in the Cyc Project*, Addison–Wesley, Reading, MA.

[18] McCarthy, J. (1980). "Circumscription—A Form of Non–Monotonic Reasoning," *Artificial Intelligence*, Vol. 13, pp. 27–39.

[19] Mistree, F., Karandikar, H., and Kamal, S. (1987). "Knowledge-based Mathematical Programming: A Hybrid Approach for Decision Making in Design," in *Design Theory for CAD, Proceedings of the IFIP Working Group 5.2 Working Conference 1985 (Tokyo)*, Yoshikawa, H., and Warman, E.A., eds., North-Holland, Amsterdam, pp. 201.

[20] Nakagawa, H., and Mori, T. (1987). "Computable Circumscription in Logic Programming," in *Transactions of Information Processing Society of Japan*, Vol. 28, No. 4, pp. 330–338, in Japanese.

[21] Nevins, J.L., and Whitney, D.E. (1989). *Concurrent Design of Products and Processes*, McGraw-Hill, New York.

[22] Roth, K. (1981). *Konstruieren mit Konstruktions Katalogen*, Springer–Verlag, Berlin.

[23] Sutherland, I.E. (1965). "SKETCHPAD: A Man-Machine Graphical Communication System," MIT Lincoln Laboratory Report No. 296, Cambridge, MA.

[24] Suzuki, H., Ando, H., and Kimura, F. (1990). "Synthesizing Product Shapes with Geometric Design Constraints and Reasoning," in *Intelligent CAD, II*, Yoshikawa, H., and Holden, T., eds., North-Holland, Amsterdam, pp. 309–324.

[25] Takeda, H., Veerkamp, P.J., Tomiyama, T., and Yoshikawa, H. (1990). "Modeling Design Processes," *AI Magazine*, Vol. 11, No. 4, pp. 37–48.

[26] Tomiyama, T., and ten Hagen, P.J.W. (1987). "The Concept of Intelligent Integrated Interactive CAD Systems," CWI Report No. CS-R8717, Centre for Mathematics and Computer Science, Amsterdam.

[27] Tomiyama, T., Xue, D., and Ishida, Y. (1989). "An Experience with Developing a Design Knowledge Representation Language," in *Intelligent CAD Systems III: Practical Experience and Evaluation*, ten Hagen, P.J.W., and Veerkamp, P.J., eds., Springer–Verlag, Berlin, pp. 131–154.

[28] Tomiyama, T. and Yoshikawa, H. (1987). "Extended General Design Theory," in *Design Theory for CAD, Proceedings of the IFIP Working Group 5.2 Working Conference 1985 (Tokyo)*, Yoshikawa, H., and Warman, E.A., eds., North-Holland, Amsterdam, pp. 95–130.

[29] Umeda, Y., Takeda, H., Tomiyama, T., and Yoshikawa, H. (1990). "Function, Behaviour, and Structure," in *Application of Artificial Intelligence in Engineering V, Vol 1: Design*, Gero, J.S., ed., Springer–Verlag, Berlin, pp. 177–193.

[30] Veerkamp, P.J., Pieters Kwiers, R., and ten Hagen, P.J.W. (1991). "Design Process Representation in ADDL," in *Intelligent CAD Systems III: Practical Experience and Evaluation*, ten Hagen, P.J.W., and Veerkamp, P.J., eds., Springer–Verlag, Berlin, pp. 155–167.

[31] Veth, B. (1987). "An Integrated Data Description Language for Coding Design Knowledge," in *Intelligent CAD Systems I: Theoretical and Methodological Aspects*, ten Hagen, P.J.W., and Tomiyama, T., eds., Springer–Verlag, Berlin, pp. 295–313.

[32] Weld, D.S., and de Kleer, J., eds. (1990). *Readings in Qualitative Reasoning about Physical Systems*, Morgan-Kaufmann Publishers, San Mateo, CA.

[33] Xue, D., Takeda, H., Kiriyama, T., Tomiyama, T., and Yoshikawa, H. (1992). "An Intelligent Integrated Interactive CAD—A Preliminary Report," in *Intelligent Computer Aided Design*, Brown, D.C., Waldron, M.B., and Yoshikawa, H., eds., North–Holland, Amsterdam, pp. 163–187.

[34] Yoshikawa, H. (1981). "General Design Theory and a CAD System," in *Man-Machine Communication in CAD/CAM, Proceedings of the IFIP Working Group 5.2 Working Conference 1980 (Tokyo)*, Sata, T., and Warman, E.A., eds., North-Holland, Amsterdam, pp. 241–253.

Part II

Product and Production Planning Using Intelligent Systems

Chapters 5 through 7 cover product planning, production sequence generation, and production scheduling using knowledge modeling, knowledge-based systems, numerical optimization, and computer automation.

Chapter 5, Product Planning Optimization Using Quality Function Deployment, by Biren Prasad, focuses on the customer-driven process for product design and development – quality function deployment (QFD). The definition and development of the QFD technique is first outlined. Extension of the QFD concept to the solution of a generalized class of multicriterion optimization problems, and application of the QFD concept for optimizing engineering resources and planning for product and process quality characteristics deployment are discussed.

Chapter 6, Automated Generation of Minimum Cost Production Sequence, by Zuomin Dong, presents a method for automatically generating production sequence using production knowledge modeling and process tolerance optimization. The approach combines a knowledge-based system with numerical optimization to generate the optimal production sequence for a given design. The knowledge-based system is used for generating all feasible production sequences based on the models of machine availability, feasible and alternative production operation, and empirical production cost-tolerance relations. The numerical optimization is applied to evaluate the generated candidate plans and to identify the most economical production sequence with least production cost and optimal process tolerances. By combining qualitative knowledge deduction and quantitative numerical optimization, the method can significantly improve present computer-automated process planning systems. A typical production sequence generation example is used to illustrate the approach.

Chapter 7, A Knowledge-based System for Building Scheduling Models, by Andrew Kusiak and Juite Wang, presents an intelligent system for selecting and building models of production planning problems. The system integrates an operation research and knowledge-based approach. A hybrid knowledge representation scheme that combines production rules and an object-oriented representation scheme is used in the system. The object-oriented model representation provides an efficient structure for model management as well as knowledge representation. The framework developed, which is based on the blackboard model and object-oriented approach, reduces the complexity of model management and increases the flexibility of the model-building process. Furthermore, the developed environment is capable of integration of symbolic and numerical computing and consistency maintenance. The approach is illustrated with a production planning application.

Chapter 5

Product Planning Optimization Using Quality Function Deployment

BIREN PRASAD
Electronic Data Systems, CPC Headquarters
Warren, Michigan

5.1 Introduction

5.1.1 Background

Mitsubishi Heavy Industry builds ocean transportation vessels at the Kobe Shipyards in Japan. In the early 1970s, the shipyard's management developed a form of upstream quality assurance to cope with the great deal of capital outlay. The market demanded quality products coupled with stringent government regulations. This management technique has come to be known as quality function deployment (QFD). The Kobe engineers in 1972 developed a matrix relating WHATs, such as government regulations, critical characteristics and customer requirements with HOWs to control how the company would incorporate WHATs in their design. The matrix also showed the relative importance of each item, which ensured proper prioritization and allocation of available resources to the production planning.

Other companies started to adopt the QFD methodology in the mid 1970s [20, 22]. Toyota began to use QFD in the late 1970s and reported a 20 percent reduction in start-up costs, when it introduced four new van-type vehicles, which increased progressively to a 61 percent cumulative total by 1984 [5]. Aisin Warner, which produced transmissions for Toyota, reported that QFD reduced the number of engineering changes in half resulting in significant improvement in quality [5]. QFD has also been used successfully by Japanese manufacturers of consumer electronics, home appliances, clothing, integrated circuits, synthetic rubber, construction equipment, and agricultural engines [17].

The application of QFD is a fairly recent innovation in the United States [22, 23]. Recent success can be attributed to industry interests in study missions to Japan and training in QFD offered by organizations such as the American Supplier Institute (ASI) [29] and the Growth Opportunity Alliance of Lawrence, Massachusetts/Quality Productivity Center (GOAL/QPC) [14, 17]. The first recorded case studies in QFD were in 1986 [29, 17, 21]. Kelsey Hayes used it to develop a coolant sensor which fulfilled critical customer requirements such as "easy-to-add coolant," "easy-to-identify unit," and "provide cap removal instructions" [22]. Now, several companies have begun using QFD [29, 23]. The companies applying the technique include Ford, General Motors, Chrysler, AT&T, Procteor & Gamble, Hewlett-Packard, Digital Equipment, ITT, and Baxter Healthcare.

5.1.2 Importance of QFD

Today many companies are interested in improving their competitive edge in the global marketplace. In order to achieve that they must know the wants (like-to-have), needs (must-have) and desires (wish-to-have) of their customers, the end users of products or services and their business. It is important for companies to bring product innovations and value-added services to the market in a timely fashion. This is because those companies that introduce new concepts at high-quality levels often wrestle away the largest share of the market. QFD is an aid to achieving such goals as quality, cost, and timeliness. This is schematically shown in Figure 5.1. Two objectives are shown crossing each other at an angle: (a) the Company Profits and (b) the Customer Satisfaction. In each case two parallel lines are shown separating the current level from the competition level. They cross in the middle creating four cross-points (A, B, C, D). Each point on the chart reflects the state of a company depending upon what choice of strategies they have resorted to. The bottom-most cross point A reflects a very poorly performing company. At this stage, as shown, minimum customer satisfaction and marginal profit exist. This is often the case when companies are desperately trying to win back the customer and not performing well (showing poor return on investment). When a product does not measure up to the customer's expectations, it does not sell well. This often forces companies to introduce incentives to move their product lines. Some of these techniques are quite expensive and cut into the company's shrinking profit margins. Companies which engage in such tactics reach point B on the chart. Some, by resorting to these temporary fixes, do improve customer satisfaction. However, it comes at the expense of their gross profit. There are other kinds of fixes which do the reverse – increase the profit margin temporarily which happen at the expense of customer satisfaction. This is shown by point C on the chart. In either case, at point B or C, the company is not able to achieve an increase in both customer satisfactions and profit.

Besides declining profits, companies today are facing a variety of new challenges. Persistent among these are global competition, increasing labor costs, rising customer expectations, shorter product life cycles, and increasing government regulation. The old techniques of coping with "short-term fixes" in "reactionary mode" are not enough to sustain the declining sales. Companies are slowly realizing an increasing need to focus on "precautionary measures" (problem prevention versus fighting fires most of the time). There is a need to plan ahead, combine the talents – marketing people, design engineers, and manufacturing staff – to work closely together, and plan a product for all important

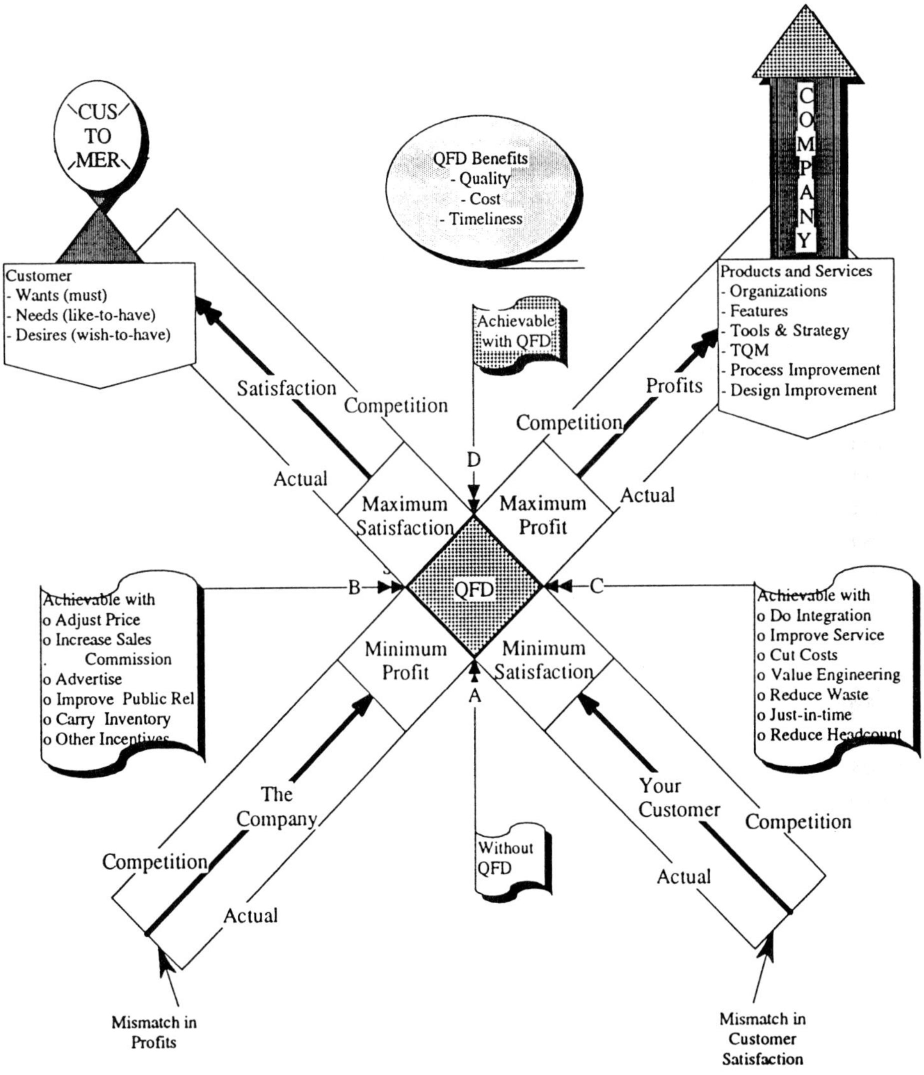

Figure 5.1: Importance of QFD

life-cycle values. By designing and manufacturing products to reflect the customer's desires and tastes, everybody wins. Customers see the benefits and are willing to buy the products. Manufacturers bag more profits. QFD is a technique that lets you achieve just that. This is shown by point D in Figure 5.1. At point D both the mismatches in customer satisfaction and profit are removed.

QFD is a simple and powerful tool that leads to longer-range thinking and better communications across functions. It enables teams of multidisciplinary groups to focus their planning operations (product planning, process planning and production planning) in concert with each other's needs and expectations. This is shown in Figure 5.2 through an ellipse chart. The process ensures that customers' wants (voice of customers) are first translated into key product characteristics (KPCs). Initial KPCs serve as product specifications for the planning phase. At the end of the product planning phase, key product characteristics are identified and selected. The chosen key product characteristics are transferred into the process planning phase as shown. Selection, specification, and the transfer process continue through the process and production planning phases. At the end of the production planning phase, the control parameters are then selected as shown in Figure 5.2. The three-phase QFD cycle is repeated as many times as needed. The cycle removes all inconsistencies arising from assumptions regarding the characteristics' parameters that were initially made. Key product characteristics are thus continuously enriched and updated in the cyclic process.

5.2 What is Quality Function Development

5.2.1 Definition

There are several ways to define QFD:

- QFD is a planning tool for translating customers' needs and expectations into products' requirements.
- QFD is a systematic disciplined method that assists users in prioritizing resources, such as when to focus time and where to concentrate on product improvements.
- QFD is a type of conceptual road map that provides the means for inter-functional planning and communications.
- QFD is a mechanism to bring new products and improved existing products to market sooner with lower cost and higher quality.
- QFD is a system that ensures that product and processes will be designed right the first time and everything will fall in place.

All these definitions taken together convey the main idea behind QFD.

The intent of QFD is to incorporate "voice of customers" into all phases of the product development cycle, through concept, engineering and analysis, design, prototyping,

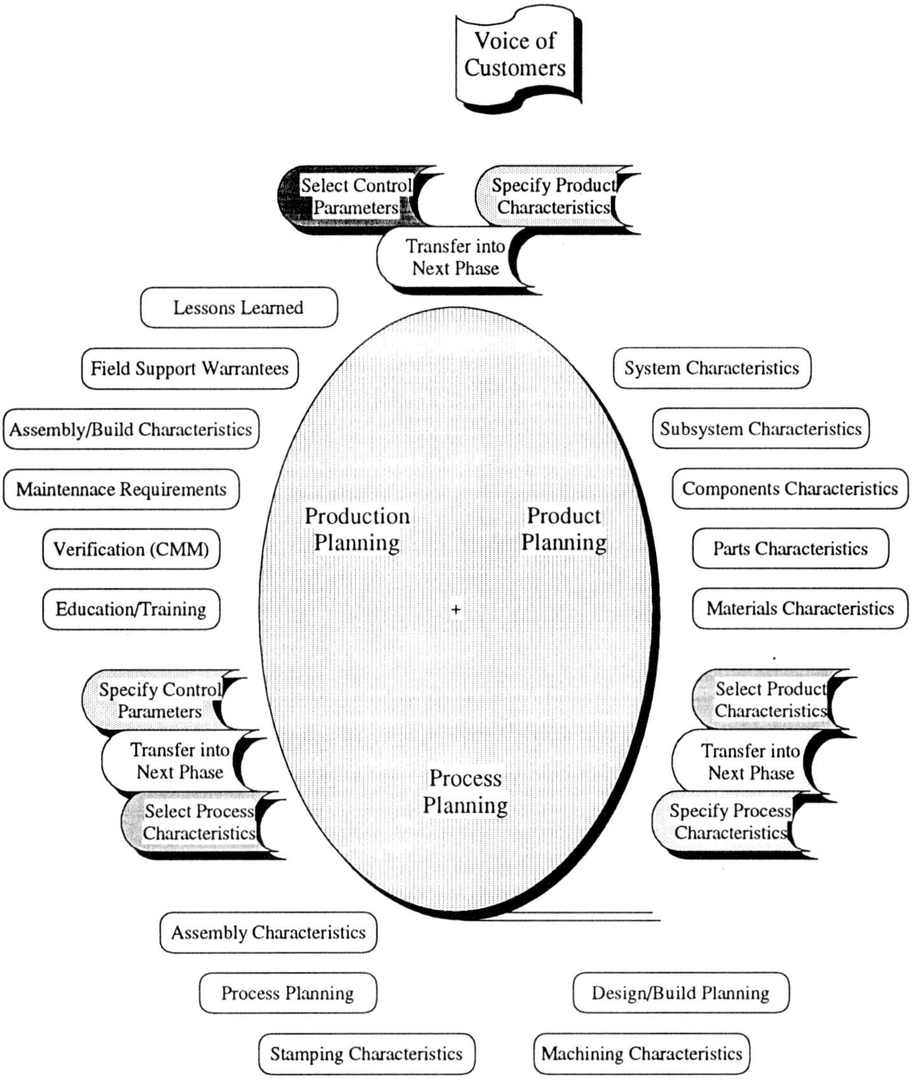

Figure 5.2: Planning Operations of Quality Function Development

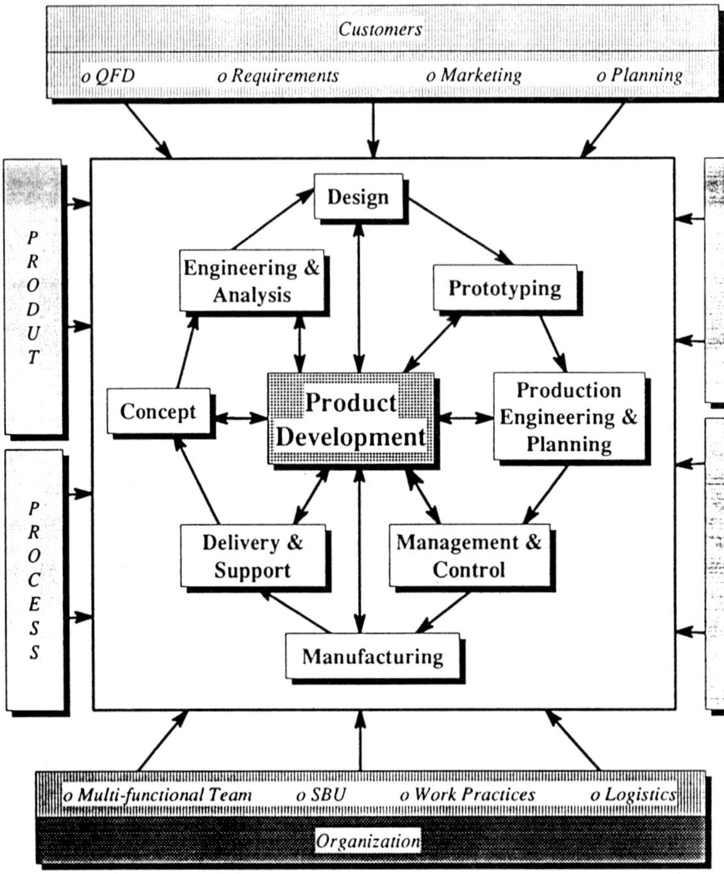

Figure 5.3: Life-cycle Phases of Product Development

production engineering and planning, management and control, manufacturing, and finally into delivery and support, as shown in Figure 5.3. In other words, QFD is customer driven product development methodology. The major goals for QFD are to incorporate the customers' voice during product improvement and reduce lead time of product development. The focus of QFD is on systematically capturing product information, such as competitive market analysis and customer satisfaction ratings, analyzing these ratings to improve product functionality, and then adding values which are important to the customers.

5.2.2 Overview

QFD is accomplished by multidisciplinary teams employing a series of matrix-based techniques to propagate critical customers' wants throughout the product life cycle. A three-legged structure of QFD – product planning, process planning and production planning

– is proposed. This is shown in Figure 5.2. This is different from the American Supplier Institute's structure which considers parts' deployment as a separate phase – independent of the product deployment phase. The present deployment is a hierarchical filtering technique; it is hierarchical because we have included part deployment as a subphase of product planning. The three phases are

- Phase I, product planning
- Phase II, process planning
- Phase III, production planning

Phase I is further broken down into five subphases:

- I.1, systems planning
- I.2, subsystems planning
- I.3, components planning
- I.4, parts planning
- I.5, materials planning

Figure 5.2 shows these phases and their subphases.

Filtering as the name implies is a method used in QFD to narrowdown the list of available options. In the beginning phases, all sorts of HOWs which can possibly affect any WHATs are usually included. A predetermined set of objectives or criteria can be used as a filtering technique for QFD. This concept is shown schematically in Figure 5.4 by a funnel pipeline. Some use criteria such as important, new, difficult and high risks [14, 4] for this purpose. Others filter the characteristics based upon cost, time, features, or most economical use of resources. Later in this chapter a filtering technique based upon a weighted multicriterion optimization formulation is proposed. Whatever the objectives may be, the end effects of all these filtering and hierarchical deployments are further reductions of the available HOWs. This reduced list of options then forms a new set of requirements for the following phase or subphases of the QFD process. This is a recursive process to obtain the key characteristics.

5.3 QFD Methodology

5.3.1 House of Quality (HOQ)

The basic tool of QFD is the "relational matrix" concept. Matrices are schemata to define generically and to relate directionally to multiple lists of identifiers, often referred to as vectors [20, 14]. The basic matrix of QFD is the "house of quality," so named because the triangular matrix which forms its top structure – the roof – makes the diagram resemble a house. The quality matrix in QFD translates the voice of the customer into the engineering,

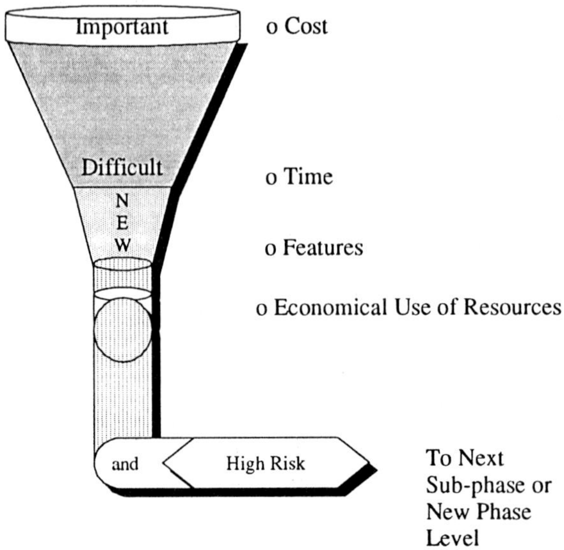

Figure 5.4: Filtering Method in QFD to Narrow Down the List of Available Options

design, process, and production stages. The original ASI version had only four rooms. Recent works have included WHYs in the matrix, as shown in Figure 5.5. In this way, the definitions and manipulations of generic matrices can be combined into an expanded house of quality. Expanded QFD enables support for all prevailing QFD approaches that have recently emerged. Figure 5.5 is a schematic view of an expanded house of quality (HOQ) [20, 15]. This expanded house of quality has eight rooms. Four of these rooms form the basic perimeters of the house. These are lists of two-row rooms – WHATs and HOW MUCHes – and two-column rooms: HOWs and WHYs. Expanded HOQ also encompasses relationships between these lists resulting in four relational matrices: HOWs versus HOWs; WHATs versus HOWs; HOWs versus HOW MUCHes; and WHATs versus WHYs. Figure 5.6 identifies all rooms in the QFD house of quality by list-vectors and matrices.

5.3.2 Components of QFD

This quality house consists of eight fundamental areas, all of which are not essential to apply QFD. Figure 5.6 depicts the scope and purpose of each area, giving us a glimpse of the full potential of the process. In the following section, we visit each room of HOQ and examine the essential features of the house of quality.

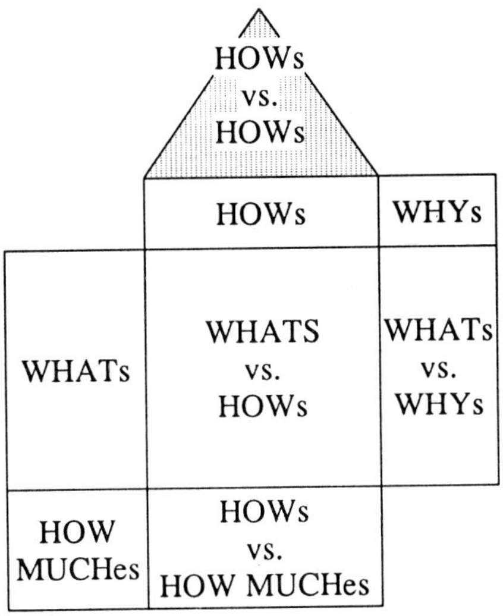

Figure 5.5: Expanded House of Quality

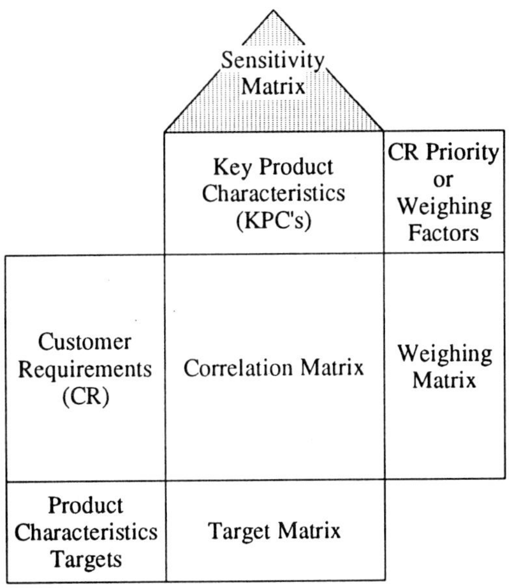

Figure 5.6: QFD Expanded House of Quality: List-vectors and Matrices

5.3.2.1 List-Vectors

WHATs: Market Quality Requirements Customers define what constitutes WHATs in QFD/HOQ. In simple terms, this is a list of customers' wants. In most consumer goods manufacturing companies, the voice of the customer (VOC) is considered a market requirement. Customers are listened to and a list of customer needs and expectations is created. This list is called WHATs. A typical WHAT might be "looks well built" or "provides good visibility," "opens and closes easily," and so on. They serve as the end-user requirements of what a consumer wants or would like to see in a future product. Noriaki Kano developed an expanded concept of quality. The Kano model of quality/features, as it is frequently called, defines three types of WHATs that he refers to as assurance to quality: basic, performance, and excitement. The Kano model relates customer satisfaction for each WHAT to its degree of achievement. An improved form of the Kano model is shown in Figure 5.7. Degree of achievement is plotted against customer satisfaction. The four quadrants are divided into three regions. The performance WHATs form the middle region that is enclosed in Figure 5.7 by a flashlight or a speaker-shaped boundary. The upper region of the flashlight model describes the areas for excitement WHATs. The lower region in the flashlight model shows the domain of unspoken basic WHATs.

- *Basic* WHATs

  These are the most basic sets of WHATs — the core quality feature or function that a customer normally expects in a product. This is schematically shown by a region below the speaker area in Figure 5.7. It is assumed that the customer does not have to ask for it. These characteristics are typically hidden or implied functions of the products under review. Such features exist as a natural part or function of the products in some form, unless it fails to work. The basic quality hardly ever increases customer satisfaction, but when left unattended, it adversely affects customer satisfaction. One example of basic quality is a car engine's starting and not stalling. A customer will not necessarily be happy just because the car started or it did not stall, but sure enough, will be frustrated and dissatisfied if it does not start or if it stalls frequently. Another example of a basic quality/feature is that a customer would normally assume that the oil pump works in the car and would not even think about it in a marketing survey unless, it gives trouble.

- *Performance* WHATs

  These are usually determined through market research and are one dimensional. They are often referred to as "spoken" WHATs. A performance quality describes how well the product measures up to the customer's wants. This is represented at the center of Figure 5.7 by a speaker-shaped area. Customer satisfaction increases with the degree of achievement. Examples of these WHATs for a car would include fuel economy, quietness, and a comfortable ride.

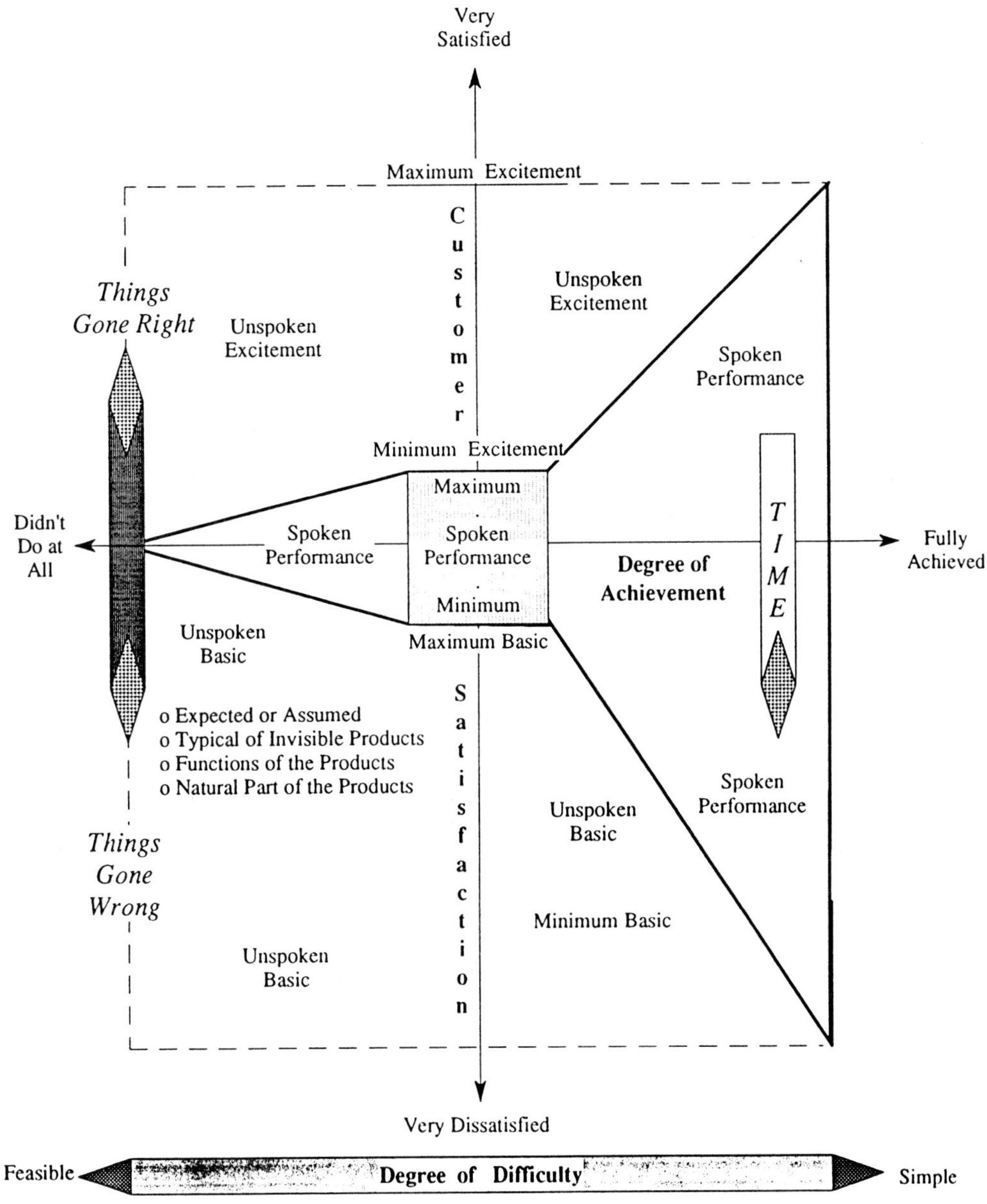

Figure 5.7: Three Types of WHATs (quality characteristics)

- *Excitement* WHATs

  These are "unspoken" WHATs. They involve new innovations and thoughtful engineering and provide pleasant surprises to the customer. These are shown in Figure 5.7 in a region above the speaker-shaped area. They increase customer satisfaction, even though the customer wouldn't necessarily be dissatisfied without them. The 3M corporation calls these WHATs "customer delights." Examples are "split sun visors," "cargo nets in the trunk to hold down plastic grocery bags," and "holes drilled for fixing toddler seats."

Over a period of time, a quality feature can shift character on the flashlight model. This is shown in Figure 5.7 by a TIME arrow pointing down. Over time, a feature which was an excitement character at one point (say part of the upper region) can take the form of basic quality (lower region) when products have matured. An example would be power steering, which was an excitement quality in the 1960s, performance quality in the 1970s and now is almost a basic quality. If a feature of the products is contained in the upper-half region, things are said to be going right for the products. The companies are expected to make a debut in the marketplace with the product introduction. On the other hand, if some features fall bellow the speaker region, products are considered to be noncompetitive in the marketplace. The normal cycle of change is from excitement to performance and into a basic quality. The WHATs can further be categorized into MUST (primary), MAY BE (secondary) and TERTIARY (like-to-have) categories.

HOWs: Quality Characteristics Items A list-vector in the quality house represents HOWs, as shown in Figure 5.8. HOWs define a set of quality characteristics through which WHATs can be realized. HOWs thus represent an array of design variables or solutions which may or may not be independent. Each of the HOWs provides a set of solutions for attacking one or more WHATs. We do not know what magnitudes of each of these HOWs will lead to the realization of the WHATs. These provide the operational definitions for the market quality requirements. This is a list which a company can measure and control in order to ensure the satisfaction of the WHATs'. The HOWs are the tools to translate the voice of the customer into design evaluation criteria. Typical entries on the HOWs vector-list are parameters for which means of measurements or a measurable target value can be established. For example, a customer's need of "good ride" (a WHAT) is achieved through "dampening," "shock isolation," "anti-roll," or "stability requirements" (the HOWs). The HOWs determine the set of alternate quality items to satisfy the customers' needs and expectations (WHATs). For this reason, these are also called quality characteristics. A typical HOW might be a length, a width, a height, a thickness, or a usable surface area. For every WHAT in the list, there should be at least one or more HOW to describe possible means of achieving customer satisfaction.

HOW MUCHes: Bounds on Quality Items This is a vector-list which identifies the bounds on the feasibility of HOWs. These are also called HOW MUCHes and represent the target values for each quality characteristic, as shown in Figure 5.8. In other words,

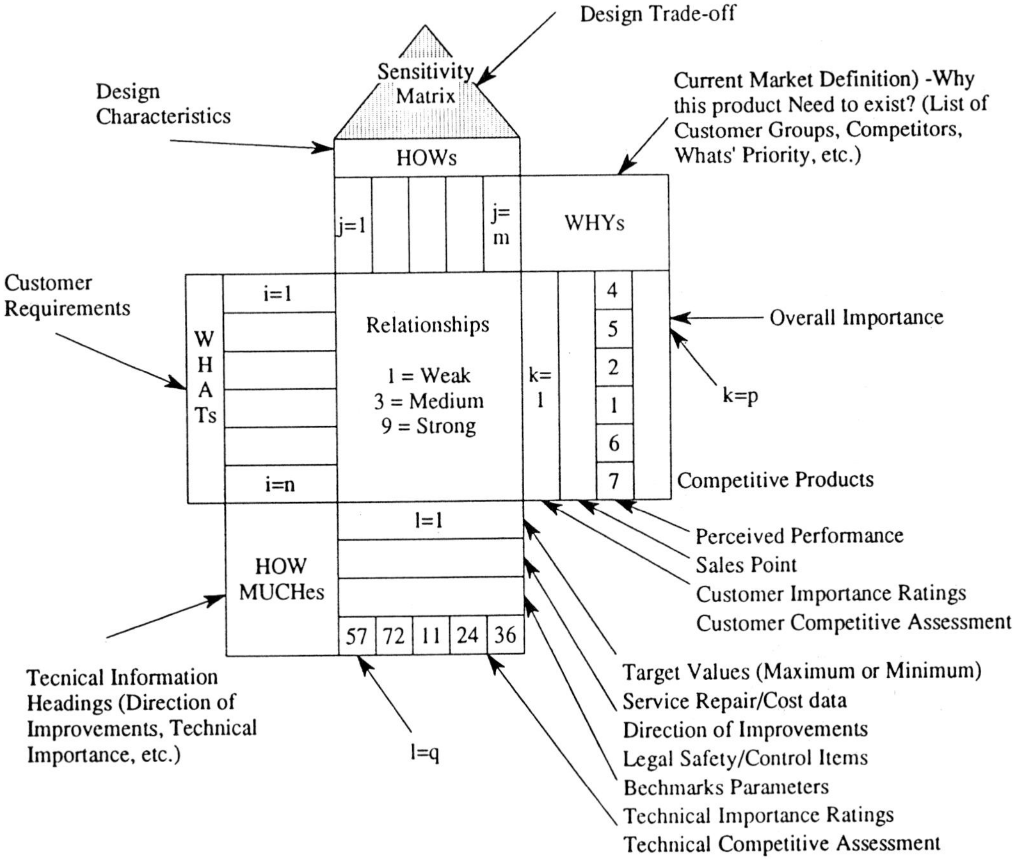

Figure 5.8: Expanded House of Quality - Terminology and Conventions

for each HOW on the list-vector, a HOW MUCH should be developed. The idea is to quantify the solution parameters into achievable ranges or a specification table, thereby creating a criterion for assessing success. This information is often obtained through market evaluation and research. HOW MUCHes capture the extremes — the permissible target values (positive or negative depending upon the sentence's construction or statements). A typical HOW MUCH might be a measure on "the importance of HOWs," "measured performance of product X," or "target values." In an optimization formulation discussed in later sections, a row of HOW MUCHes is used to collect bounds for the HOWs list.

WHYs: Weighting Factors on WHATs This is a vector-list which describes the relative importance of current competitive products referred to as "world class" or "best of the class" products that satisfy those WHATs in some way, as shown in Figure 5.8. It is the way of defining a relative priority for the WHAT's objectives. If a product solution (a HOW) exists today, HOWs can be looked upon as proportions in which customers' requirements (WHATs) are satisfied. It is the way to assess the feasibility of a product in the marketplace. It also allows defining the target projection in percentage to WHATs' requirements. Once these target values are multiplied with WHATs and summed, it can provide a single pseudo measurement index for "overall customer satisfaction." In an optimization sense it will represent a weighted sum of WHATs. Included in this category are the vector-lists of "world class products" that you would like to compete with and their relative importance with respect to customers' wants. If the product is targeted to multiple customer groups, such as American and Japanese, this list must include these customer groups and their relative wants. WHYs are names of competitors, competitive products, market segments, or other items that describe the current market conditions, criteria used to weigh decisions that a product must take into account. This usually translates into specifying weighting factors for WHATs — setting priority (what's significant in WHATs and what's not). A typical WHY might be "overall importance," "importance to the world purchaser," or "world-class achievable performance of product X."

5.3.2.2 House of Quality Relational Matrices

Conventions: HOQ relational matrices use either numbers or symbols depending upon the purpose of QFD and the context in which it is being used, as shown in Figure 5.8. Two possible rationales have commonly been proposed depending whether it is used for calculations or for visual aids.

- *Quantitative Reasoning*: Numbers are used. They facilitate comparison of relationship magnitudes through mathematical means.
- *Qualitative Reasoning*: Symbols are used. They provide better visual communication. Three symbols are often used to indicate the relationship between the WHATs and the HOWs: A solid circle (•) implies a strong relationship; an open circle (○), a medium relationship; and a triangle ($\triangle$), a weak or small relationship.

Table 5.1: **Standard Relationship Conventions (weights and symbols)**

Matrix	Quantitative	Qualitative
WHATs versus HOWs	Weight	Symbols
Strong	9	Double or Solid Circle
Moderate	3	Circle
Weak	1	Triangle
None	0	Blank
HOWs versus HOWs	Weight	Symbols
Strong Positive	9	Double or Solid Circle
Medium Positive	3	Solid Triangle
Positive	1	+
None	0	Blank
Negative	-1	-
Medium Negative	-3	Open Triangle
Strong Negative	-9	Open Circle

This process of establishing relationships gives the viewer a basic idea of comparing the strengths and weakness, importance, and interactions of these columns (WHATs, WHYs) or row-vectors (HOWs, HOW MUCHes). The symbols used here follow the convention adopted by the employees at the Kobe Shipyards, who incorporated the local horse racing symbols. By convention, each symbol in the relationship matrix receives a value. Table 5.1 shows a typical convention followed in defining different QFD relational matrices.

WHATs versus HOWs: Matrix Relationship between Market Requirements and Quality Items

To get a correlation, a matrix is created by placing the HOWs list in columns and the WHATs list in rows, as shown in Figure 5.8. The rectangular area between the rows and columns then depicts the relationships between the WHATs and the HOWs. The matrix thus developed is called the relationship matrix. It correlates what customers want in a product and how an enterprise can achieve those requirements. It is the core relational matrix of QFD. Relationships within this matrix are usually defined using a four-step procedure: strong, medium, weak, or none, as shown in Table 5.1. An example is shown in Figure 5.8. This matrix may be densely populated (more than one row or column affected). This results from the fact that some of the quality solutions affect more than one market requirement. For example, what a customer wants in "good ride" and "good handling" (WHATs) are both affected by quality characteristics such as dampening, anti-roll, and stability requirements (HOWs). The more densely populated and spread in ranks the matrix is the more valuable the information is likely to be. A diagonal matrix means there are very few interactions among the rows and the columns. HOWs also include some design evaluation criteria such as, "How do we know we can satisfy our customers?" "How can we test?" and "How can we control the quality?"

WHATs versus WHYs: Matrix of Influence Coefficients This is a relationship matrix that can be used to prioritize the WHATs based upon criteria for competitiveness. Usually, the data in the matrix (say a column) consist of one or more of the following, as shown in Figure 5.8:

- Marketing information ratings which identify the relative importance of each of the WHATs.
- Ratings on how important different customer groups perceive each of the WHATs (often referred as customer importance ratings or CIRs).
- Ratings of how well competitive products are perceived to meet each of the WHATs.
- Factors indicating where your product is perceived relative to competition.
- Factors that a company would like to consider in defining a specification for the design of a "world-class quality product."

The above data constitute a process of identifying the stated importance ratings and factoring in where your product is perceived relative to competition. These objectives help establish a realistic target that a company would like to shoot for in future product offerings.

Most importantly, the data help identify many ways of obtaining a weighted average of WHATs as a single performance index. The latter can serve as a quantitative measure of overall customer satisfaction for the product that the company is planning to launch.

HOWs versus HOW MUCHes: A Feasibility Matrix of Bounds on HOWs This is a feasibility matrix that lets you decide how much of each HOW you can possibly vary to meet customer wants. Typically, the data in this matrix (say a row) consist of one or more of the following, as shown in Figure 5.8:

- Factoring in HOW MUCHes of HOWs you perceive your product relative to competition.
- Ratings which identify the relative importance of each of the HOWs.
- How your competition performs relative to each of the HOWs.
- Estimates or calculations of the realistic upper limits that a HOWs can take.
- Estimates or calculations of realistic lower limits that a HOWs can take.

- Technical importance rating (TIRs) — a computed weighted sum of the characteristics with respect to customer importance ratings (CIRs).

This information can lead you to establish realistic target values (upper and lower bounds) for each HOW. Product values or target values identify engineering tolerances and specification limits on quality items. These can also be used to pinpoint where current technology is sufficient or where manufacturing research is necessary. Varying the appropriate combinations of HOWs in such a way that they lie within the bounds will ensure that you meet the customer requirements.

HOWs versus HOWs: Correlation Matrix Relationship among Quality Characteristics

This is described by means of a correlation matrix which forms the roof of the house of quality, as shown in Figure 5.8. The purpose of the roof is to identify the qualitative correlations among the characteristic items (HOWs). This is a very important feature of the quality house since, at times, the possible solutions could be redundant and could not add any value to the customers' wants. Other times they may be at cross purposes with each other. The correlation matrix helps identify such situations. If two HOWs help each other meet their target values (HOW MUCHes), they are rated as positive or strong positive. If meeting one HOW target value makes it harder or impossible to meet another HOW target (HOW MUCH), those two HOWs are rated as negative, or strongly negative relationships. One example of this is when "0-60 mph time" and "fuel economy" are two quality items. Efforts to decrease the "0-60 mph time" would have an adverse effect on the "fuel economy" item. In this case, the two HOWs have a negative correlation.

In actuality, correlations between solution parameters could be positive or negative and in varying degrees: strong, medium, or none. When one HOW adversely affects another HOW, a qualitative negative correlation results. On the other hand, if the first HOW favorably supports the second HOW, a positive correlation results. For example, "fuel economy" and "gross weight" are considered to have positive correlations because reducing gross weight will increase fuel economy keeping all other remaining parameters constant. Standard QFD practices use the weights as shown in Table 5.1. Symbols can be used to portray visually the different types of correlation.

After the relationship matrix is developed, care is taken in reviewing the matrix. Blank rows or columns call for closer scrutiny. A blank row implies a potential unsatisfied customer and emphasizes the need to develop one or more quality items (HOWs) for that particular market requirement (WHATs). A blank column implies that one of the quality items does not directly relate to any of the market requirements. It may in turn emphasize a reallocation of resources or might point to a new customer want that has not been identified.

5.3.3 QFD Deployment

Figure 5.9 illustrates one QFD concept of deploying the voice of the customer through QFD phases. In Figure 5.10 the phase I product planning level deployment is further spanned into a second-level QFD deployment. This breakdown corresponds to the subphases that were listed. In the example described herein only a two-level hierarchy is shown. This is the most common. However, such a hierarchical structure can have as many levels as needed. In these diagrams, the filtering process is shown by a solid pipeline connecting the "characteristic" HOWs room to the WHATs room, as shown in Figures 5.9 and 5.10. It ensures that characteristics that are critical to meeting the determined objectives are given proper attention in early phases. It also ensures that HOWs are further spanned into "root or key characteristics' factors" during the subsequent phases or subphases as the case may be.

During product planning, customer wants (product characteristics) or WHATs are related to key quality characteristics for which a list of WHYs and a list of HOW MUCHes are then identified. HOWs define the desired key product characteristics (KPCs) of a product to counter WHATs. WHYs are the overall evaluation criteria used within the organization to define acceptability of the product. Targets for the KPCs (HOW MUCHes) are established based upon competitive benchmarks and the customers' competitive assessment. The subphase of the product planning phase (subphases I.1 through I.5) deployment uses the KPCs defined in Phase I to evaluate alternatives and filter a design that meets most of the customer demands. At the end of Phase I (subphase I.5), a set of product KPCs is identified that represents best of the class. The process planning phase deals with selection of a process concept as an identification of critical operation parameters (called herein key process characteristics — process KPCs) that can cause the product KPCs identified in Phase I to be satisfied. Production planning (Phase III) identifies production KPCs (control requirements, maintenance requirements, quality control, education, training issues, and so on.) to make the process KPCs work.

To deploy QFD, certain assumptions are made and certain conditions are satisfied. Most problems are system related and not people related. Multidisciplinary teams are often recommended. Traditional processes are altered to accommodate QFD's heavy emphasis on the voice of the customer. More times are spent upstream to develop the matrix. Tools such as Statistical Process Control (SPC), Design of experiment (DOE), and other product improvement concepts are encouraged in the decision making process.

5.4 Applications of QFD

The four key relational matrices in the house of quality discussed previously are seen in every QFD chart. They are basic to QFD methodology. Many of these matrices are useful in drawing conclusions on the relative importance of WHATs, WHYs, HOWs, and so on. There are computer programs and software which allow you to enter these matrices interactively. They also provide a variety of sorting and matrix analysis algorithms such as "weighted average," "ranking," "technical importance," normalized ratings, "sum of WHYs or HOWs

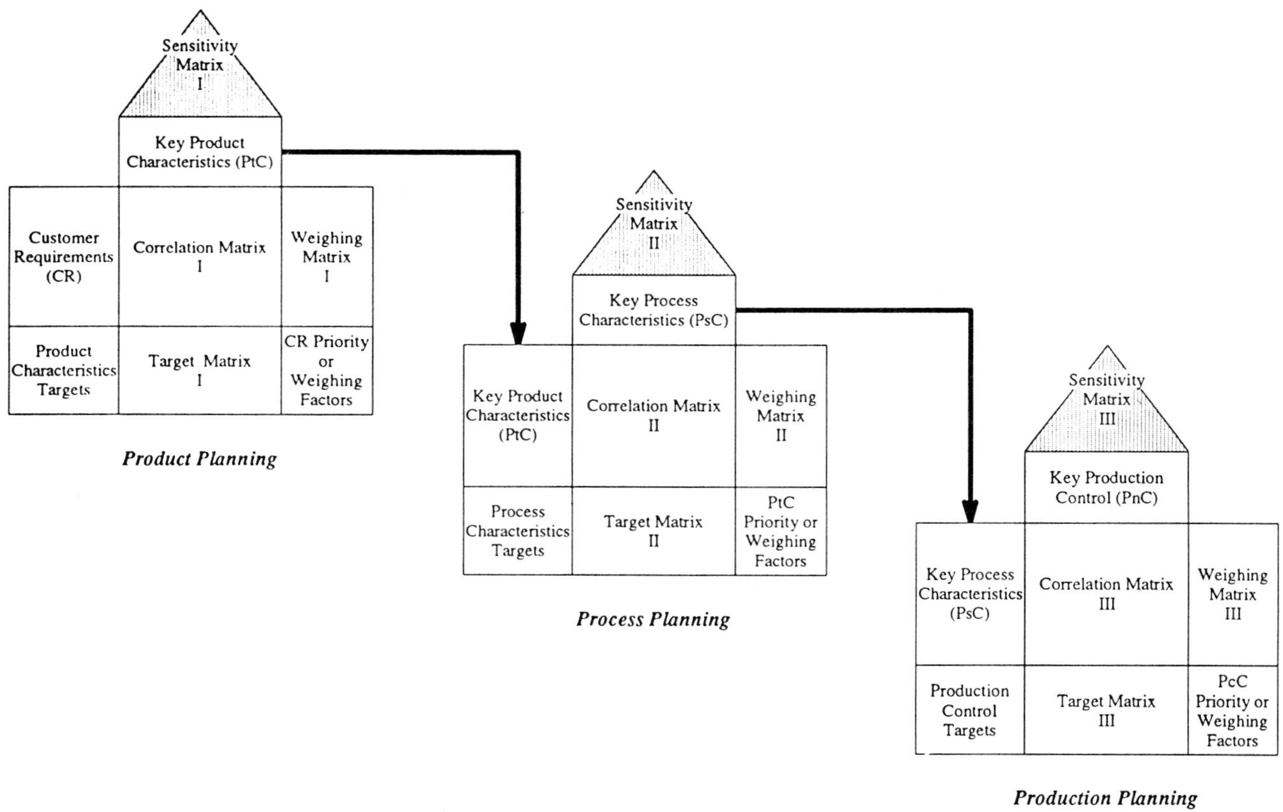

Figure 5.9: Linked QFD House of Quality: First Level Deployment

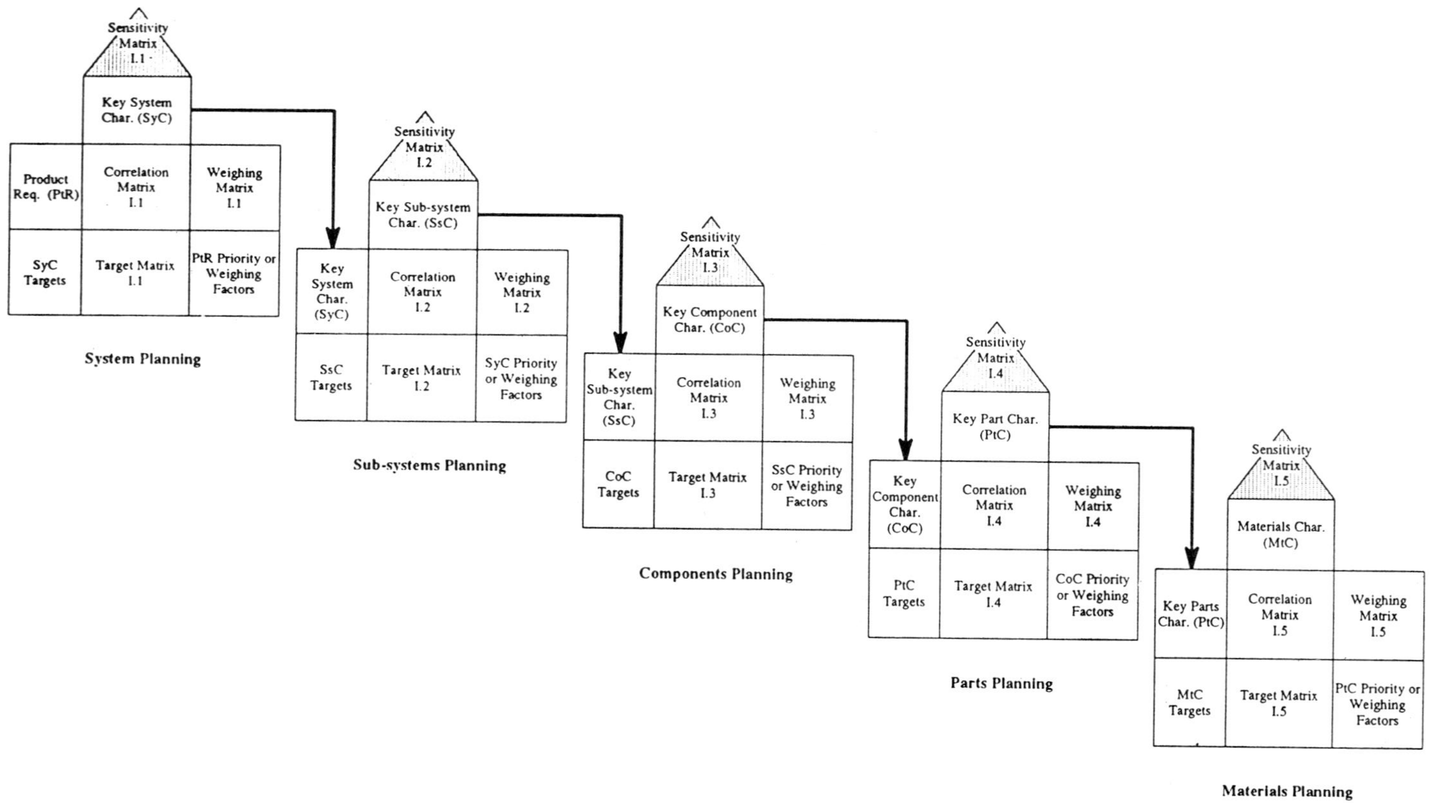

Figure 5.10: Linked QFD House of Quality: Second Level Deployment

matrix column," "weighting factors," "weighting average," graphics utilities (bar charts, line chart), etc. [20].

Before we show how QFD can be applied to solving a variety of problems, let us establish some nomenclatures for QFD list-vectors and relational matrices.

Let us denote the following:

- HH: A correlation matrix defined by HOWs versus HOWs
- WH: A sensitivity matrix defined by WHATs versus HOWs
- WW: An influence matrix defined by WHATs versus WHYs
- HM: A feasibility matrix defined by HOWs versus HOW MUCHes

In the next section, we consider some optional features of a QFD chart.

5.4.1 Competitive Assessment (CAs) Comparisons

Competitive assessment comparisons are used in a QFD chart to compare the competition's products with the company products. There are two types of competitive assessments.

- The customer competitive assessment (CCA) is developed by customer surveys to rate the customer's opinion of the various products in a particular marketplace. In the house of quality, this is listed in the WHATs versus WHYs relationship matrix. Customer assessments rate the WHATs (perceived response).
- In a similar fashion, the HOWs for the same competitive products are rated from a technical perspective. This second type of comparison is known as engineering competitive assessment (ECA). In the house of quality, this is found in the HOWs versus HOW MUCHes relationship matrix. Engineering assessment quantify the HOWs (engineered or measured outputs).

These two assessments are said to be in conflict when corresponding to a (WHAT, HOW) location (say wh_{ij}) where there exists a strong positive (WHAT, WHY) CCA value (say ww_{ik}), at the same time as there exists a strong negative (HOWs, HOW MUCHes) ECA value (say hm_{lj}). In mathematical terms, this can be represented as follows:

By definition:

$$CCA == ww_{ik} \tag{5.1}$$

where $i = 1, \ldots, n$; and $k = 1, \ldots, p$

and

$$ECA == hm_{lj} \tag{5.2}$$

where $j = 1, \ldots, m$; and $l = 1, \ldots, q$.

In general, a conflict occurs for an (i, j) combination, if there exists a row l in matrix HM and a column k in matrix WW, such that

$$Limit\ \ [ww_{ik}/hm_{lj}] \rightarrow -1. \tag{5.3}$$

corresponding to that (i, j) location

Note a conflict occurs only when the ratio is a negative unity. For any other values no conflict occurs. For example, if the ratio is a positive number, it represents a strong unity and the existence of a good supporting data. Any negative values other than minus 1 indicate the existence of a minor discrepancy for that location in the correlation matrix.

When the two CAs are in conflict, it often results from failure to understand the voice of the customer. In such a case, the HOWs list must be amended to reflect customer perception. This is most often resolved by letting engineers get directly involved in the process of comparing the in-house and the competitive products.

5.4.2 Ratings Comparison

Two types of ratings are commonly used in QFD:

- Customer importance ratings (CIRs), which are derived from customer surveys. These ratings are for each WHAT based on overall evaluation of the products in the field. It is posted as a column (WHATs versus WHYs) of the matrix. If that column is a kth column of matrix WW, then by definition,

  $$CIR == ww_{ik} \tag{5.4}$$

  where $i = 1, \ldots, n$.

- Technical importance ratings (TIRs), which are the results of calculations from the QFD matrix defined earlier. The steps used in the calculations of TIRs are as follows:

  1. Assigning a numerical value for each symbol used in the correlation matrix (WHATs versus HOWs). The conversion listed in Table 5.1 is most commonly employed.
  2. Multiplying a numerical value to each WHAT of the (WHATs, HOWs), which is equivalent to the correlation matrix (WHATs, WHYs) CIR value.
  3. Repeating the multiplications in step 2 for all WHATs and adding the results in each HOW column.
  4. Entering the results of step 3 into a HOWs versus HOW MUCHes row. The row of computed numbers stored in the HOWs versus HOW MUCHes matrix represents the TIR for each HOW. If that row is a lth row of an HH matrix, then, by definition

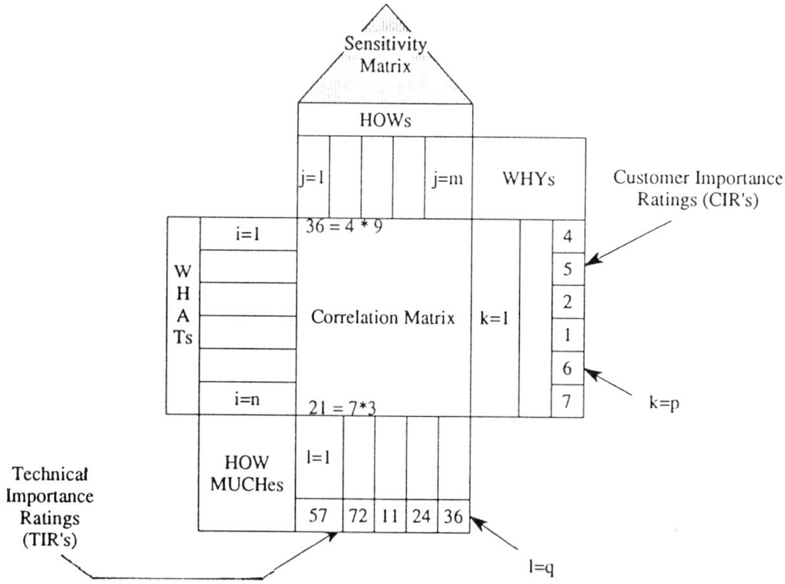

Figure 5.11: QFD Application to Determine Ratings

$$TIR == hm_{lj} == \sum_{i=1}^{i=n} [(wh_{ij}) * (ww_{ik})] \tag{5.5}$$

where, $i = 1, \ldots, n$, and $j = 1, \ldots, m$.
Ratings, stored in lth row of the HOWs versus HOWs MUCHes matrix, are a relative comparison of each jth element, provided the kth row contains the computed value of Eq. (5.4). Eq. (5.5) represents an overall measurement of the quality of the final product.

Figure 5.11 illustrates the application of this concept for computing technical importance ratings (TIRs). The CIR column in Figure 5.11 contains the customer importance ratings (CIRs) values obtained through surveys. A typical correlation matrix may have symbolic representation. If so, the matrix is first converted into a quantitative value matrix using the conversions shown in Table 5.1. These values are multiplied by the customer importance rating, resulting in an importance value for each location in the matrix. The technical importance rating for each HOW is then found by adding together the importance values in each column. As an example, in the first column of the matrix in Figure 5.11, the first relationship has a value of 36 (4×9), and the only other relationship has a value of 21 (7×3). The technical importance rating for this column is thus the sum of these two values, 57 ($36 + 21$). Technical importance ratings are used to determine what areas should receive the most resources and are particularly useful in trade-off decisions.

The technical importance rating is an example used to show the rating (prioritization) process. Other factors might include service repair/cost data, technical difficulty, safety control items, newness, reliability, timing, costs, and so on, as shown in Figure 5.8. Two significant outcomes and important benefits of a successful QFD application are communication and complete teamwork by all disciplines within an organization. It is impossible to fill-in and complete the house of quality effectively without teamwork. Additional tables do the same with suppliers and end users of the product.

5.5 Formulation of QFD as an Optimization Problem

QFD is a methodology of capturing the information necessary for making decisions on future product upgrades and/or analyzing competitors and their competing products. Though these applications are a useful resource in aiding the management and marketing personnel in decision making, QFD has never been applied to solving an optimization problem, although the basic methodology for QFD is quite general. In the following section, a way to extend the QFD methodology to solve a class of optimization problems is discussed. As stated before QFD captures four list-vectors — WHATs, HOWs, HOW MUCHes and WHYs — and four relational matrices — WHAT versus HOWs; WHAT Vs. WHYs; HOWs versus HOW MUCHes; and HOWs versus HOWs. A methodology has been developed to capture the optimization formulation as part of the QFD process. It employs the same structure for list-vectors and relational HOQ matrices to reformulate the original problem in a rigorous mathematical programming setting.

5.5.1 General Statement of the Mathematical Programming Problem

The general statement of the optimization problem can be posed as: Find a vector of design variables, v, that minimizes the objective functions, $F(v)$, while satisfying the constraint equation, $g_j(v)$. In mathematical notation

$$F(v) \rightarrow \text{minimum or maximum} \tag{5.6}$$

Subject to

$$g_j(v) > 0; \tag{5.7}$$

for j =1,2, ... m

$$V_{min} < v < V_{max} \tag{5.8}$$

In most cases, $F(v)$ is termed the objective function and $g_j(v)$ the constraints. V_{min} and V_{max} denote lower and upper bounds on design variables.

Note the question we are posing is how can QFD be used in this setting? What elements of QFD represent design variables, constraints and objective functions. Note that the rows of "WHATs in QFD" form customer wants. It is thus similar to specifying a set of objective functions which a company must strive for and pay attention to please the customers. They are not all independent. It is also safe to assume that one or more can be satisfied individually by choosing the right HOWs but it will be very difficult to satisfy all WHATs simultaneously by trial and error. Let us introduce some additional notations:

$$\{WHATs\} == \{f_i\} \tag{5.9}$$

where $i = 1, \ldots, n$.

Since HOWs are possible solution parameters that can be chosen to address WHATs, they are like the variables of the original problem. In an optimization setting vector HOWs of QFD will be like design variables in optimization. This means,

$$\{HOWs\} == \{v_j\} \tag{5.10}$$

where $j = 1, \ldots, m$.

If HOWs represent system variables, in order to achieve an optimal solution we must have these variables independent of each other. From QFD matrices we know that all HOWs are not independent. The correlation matrices, the "roof" of QFD, denotes the dependencies of HOWs versus HOWs. In an optimization formulation we can pose such relationships between HOWs versus HOWs as constraints of the problems. This situation can be handled through what we call a linking technique in optimization. The correlation matrix gives a linking matrix for the design variables. The latter matrix can be formulated as follows:

$$\{v_m\} = [HH]\{v_n\} \tag{5.11}$$

where HH is a square matrix of size (m, m) or

$$v_j = \{hh_{ji}\}_T * v_i \tag{5.12}$$

where HH is a triangular matrix whose diagonal terms are all zero. This is because in the QFD correlation matrix for HOWs versus HOWs,

$$H_{ij} = 0; \qquad for \quad \forall i = j; \tag{5.13}$$

where i and j are the rows and columns of the HOWs matrix.

The above relationship simply says that if we consider HOWs as a set of design vectors, all terms in a HOWs vector are not independent of each other. Eq. (5.12) thus serves as a linear set of equality constraints in the corresponding optimization formulation.

Let us define a term of an HM matrix as hm_{lj}. Since, in Eq. (5.10) we denote a row vector of HOWs as $\{v\}$, by definition of the HOWs versus HOW MUCHes matrix, a value stored in hm_{lj} would represent a limit on $\{v\}$. If we denote two vectors of an HM matrix ($l = L$ and $l = U$) as representing permissible upper and lower bounds on HOWs, the following statements can be made:

$$hm_{Lj} < v_j < hm_{Uj} \tag{5.14}$$

where $j = 1, \ldots, m$, and hm_{Lj} denotes a lower bound on v_j, and hm_{Uj} denotes an upper bound. It is interesting to note that Eq. (5.14) represents an equation of bounds on the system variable for an equivalent optimization formulation.

So far we have introduced the concept of design vectors, constraints, and a series of objectives defined by WHATs vectors. The only term left in completing an optimization formulation is a weighted objective function. Earlier we noted that a list of WHATs represents a series of objective functions. We also note from matrix WW that a column of this matrix contains the weighting factors associated with WHATs. If we denote a pth column of WW matrix as ww_{ip}, a cumulative or weighted function, say $F(v)$, can be formed using TIR formulation (Eq. (5.5)) as follows:

$$F_{pj}(v) == \sum_{i=1}^{i=n}[[wh_{ij}]^T * \{ww_{ip}\}] \tag{5.15}$$

or in a matrix form,

$$F_{pj}(v) == \sum_{i=1}^{i=n}[\{wh\}_j^T \{ww\}_p []] \tag{5.16}$$

Note in Eq. (5.15) or Eq. (5.16) a weighted multiobjective function has been formed, which is the weighted sum of all its component. The result can be stored as the qth row of the HM matrix. That is,

$$hm_{qj} == F_{pj}(v) \tag{5.17}$$

where $j = 1, \ldots, m$.

In the WHATs versus WHYs subsection, many ways of obtaining these weighting factors were described. Eq. (5.16) thus represents one of many cumulative forms that are possible. Now, if $F(v)$ is employed as an objective function, the general statement of optimization previously described can be transformed in QFD context as outlined in the following section.

5.5.2 QFD-based Optimization

The original optimization problem can now be transformed to: Find a vector of design variables v $\{D,\ where\ \ v == \{HOWs\}\}$; that minimizes the objective functions $F_{pj}(v)$ where

$$F_{pj}(v) == \sum_{i=1}^{i=n}[[wh_{ij}]^T * \{ww_i\}_p] \tag{5.18}$$

or

$$\{F(v)\}_p == [WH]^T \{ww\}_p \tag{5.19}$$

where $i = 1, \ldots, n$; $j = 1, \ldots, m$, and p is a column index of HW matrix.

While satisfying the following constraint equations

$$\{v_j\} = [HH]\{v_i\} \tag{5.20}$$

$$hm_{Lj} < v_j < hm_{Uj} \tag{5.21}$$

or

$$\{hm_L\} < \{v\} < \{hm_U\} \tag{5.22}$$

In Eqs. (5.18) to (5.22), the vectors $\{wh\}$ and $\{ww\}$ belong to QFD/HOQ relational matrices, WH and WW, respectively. Lower and upper bounds vectors $\{hm_L\}$, and $\{hm_U\}$ belong to the QFD/HOQ HM matrix.

5.6 Illustrative Example of Expanded QFD/HOQ

Accurate creation of the QFD house of quality vectors and matrices is not a simple task. The data have to be consistent and should represent real-world situations. It requires putting together a team of experts familiar with the products, process and the competition. Inputs are required from various sources including the management council, the engineering steering committee, quality control directors, marketing product managers, the manufacturing steering committee, and strategic business units. It takes a long time to come up with substantiated data that everyone is comfortable with. Each person must understand the mission of the product introduction, its goals, strategy, and key success factors. Assumptions and ground rules are carefully laid down to guide each team member in the same direction. Not many practical examples of QFD exist today. References [20, 15] list four such examples to illustrate the use of the CAPTURE program [20]. One of the examples called HATCH deals with a car Hatch Door. We have selected this example because it is simple and everyone can relate to it. In the following discussion several snapshots from HATCH examples are used to explain QFD and draw the basic features of expanded HOQ. The data are adapted from references [20, 15]. Figure 5.12 shows a complete expanded QFD/HOQ diagram with vectors and matrices all shown in one. For this example, there were 18 customers' wants (shown in WHATs list), 21 key quality characteristics (shown in the HOWs list), 14 customer and competitive data sets (shown in the WHYs list) and 11 competitive evaluations and assessments (shown in the HOW MUCHes list).

Figure 5.13 shows a WH(I, J) matrix with WHATs and HOWs list-vectors. The size (I, J) of the matrix is $(18, 21)$. The lists are organized in a bilevel hierarchical structure. The WHATs are organized under lead groups 1, 6, and 14. The HOWs are organized into lead groups 1, 7, and 15. The areas corresponding to these rows and columns are shaded since they represent main headings. The correlations are established in the lowest-level hierarchy. The values associated with the symbols are listed on the upper-left corner of the WH matrix.

Figure 5.14 shows a WW(I, K) matrix with WHATs and WHYs listed as rows and columns.The size (I, K) of the WW matrix is $(18, 14)$. Note that the column marked "overall importance" shown in the upper-left-hand corner of Figure 5.12 is listed as the last column of this matrix. There is a horizontal row marked "weighting" which quantifies the weighting factors associated with the age groups totaling 1. The weighted average of importance is calculated as the weighted sum of WW matrix columns 2 through 4.

$$WW(I, 5) = [WW(I, K) * Weighting(K)] \tag{5.23}$$

where I takes value from 1 to 18 and K goes from 1 through 14. Weighting (K) is a vector of factors which capture the relative weights that one can assign to a row or a column.

Product Planning Matrix for Car Hatch Door

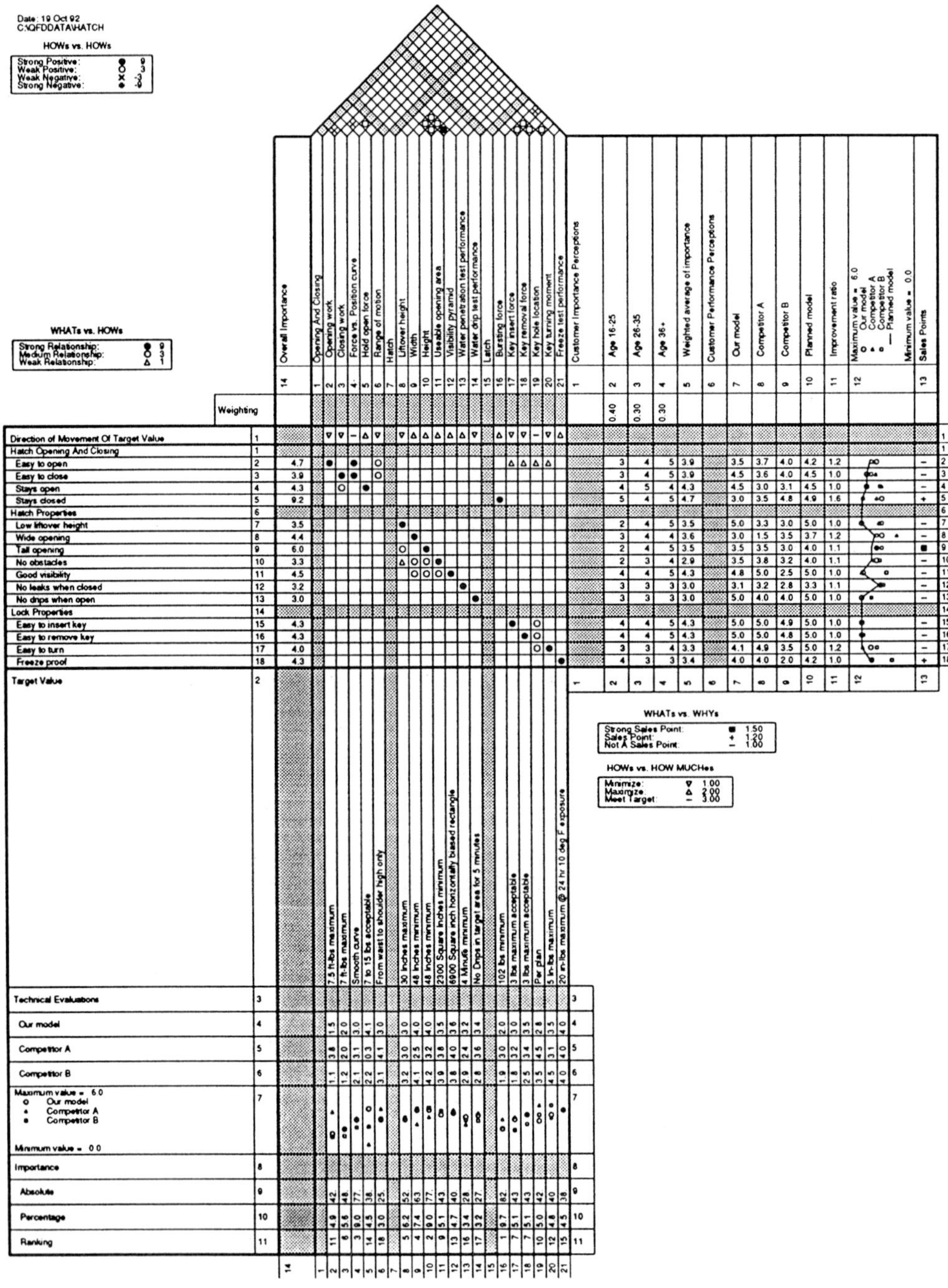

Figure 5.12: A Complete QFD/HOQ Lists and Matrices for the HATCH Example

Product Planning Matrix for Car Hatch Door

Date: 19 Oct 92
C:\QFDDATA\HATCH

WHATs vs. HOWs

Strong Relationship:	●	9
Medium Relationship:	○	3
Weak Relationship:	△	1

		1 Opening And Closing	2 Opening work	3 Closing work	4 Force vs. Position curve	5 Hold open force	6 Range of motion	7 Hatch	8 Liftover height	9 Width	10 Height	11 Useable opening area	12 Visibility pyramid	13 Water penetration test performance	14 Water drip test performance	15 Latch	16 Bursting force	17 Key insert force	18 Key removal force	19 Key hole location	20 Key turning moment	21 Freeze test performance	
Hatch Opening And Closing	1																						1
Easy to open	2		●		●		○											△	△	△	△		2
Easy to close	3			●	●		○																3
Stays open	4			○		●																	4
Stays closed	5																●						5
Hatch Properties	6																						6
Low liftover height	7								●														7
Wide opening	8									●													8
Tall opening	9								○		●												9
No obstacles	10								△	○	○	●											10
Good visibility	11									○	○	○	●										11
No leaks when closed	12													●									12
No drips when open	13														●								13
Lock Properties	14																						14
Easy to insert key	15																	●		○			15
Easy to remove key	16																		●	○			16
Easy to turn	17																			○	●		17
Freeze proof	18																					●	18
		1	2	3	4	5	6	7	8	9	10	11	12	13	14	15	16	17	18	19	20	21	

Figure 5.13: WHATs versus HOWs (WH) Matrix for the HATCH Example

The importance value is captured in the weighting column of the HOWs screen. These values should total 1 (i.e., 0.3 + 0.5 +0.2).

Column 12 plots WHATs data for the planned model with data for the current model. Competitor A and Competitor B plots are shown as scattered diagrams. The improvement ratio in column 11 is defined as follows:

$$Improvement\ Ratio = \frac{Planned\ Model\ Data}{Our\ Model\ Data}$$

$$WW(I,11) = \frac{WW(I,10)}{WW(I,7)}$$

The symbols used in column 12 are quantified in the upper-left-hand corner of Figure 5.12.

Figure 5.15 shows a HW(L, J) matrix with HOWs listed along the columns and HOW MUCHes along the row axes. The HW matrix is of size $(11, 21)$. The HOW MUCH columns are organized as two groups: Group 3 is a "technical evaluation" and Group 8 as "importance." Each group is further broken down into three parameters. Row 7 graphically plots the points under "technical evaluations" as a scattered diagram.

Figure 5.16 shows the HH(I, J) matrix — the roof of the QFD/HOQ. Both rows and columns are made out of HOW vectors listed along x and y axes. The HOWs are organized into three groups: 1, 8, and 16. No relationship is defined for these rows and columns, or the diagonal of the matrix HH(I, J) where $I = J$ is undefined. Symbols for off-diagonal terms are shown in Figure 5.16. The corresponding values are shown on the upper-left-hand corner of the matrix HH. As you may notice this is a square matrix with terms on the opposite sides of the diagonals being equal. The triangular matrix indicated by the roof of QFD are thus enough to define sufficiently the HH matrix.

5.7 Conclusion

QFD is a customer-driven process for product design and development. QFD is a system of identifying and prioritizing customer needs obtained from every available source. By using QFD, engineers can methodically analyze the details of design and process improvement to meet those needs. QFD is a highly versatile engineering tool translating the voice of the customer through all phases of the product life cycle. In this chapter we have shown that the concept of QFD can be extended to solve a generalized class of a multicriterion optimization problem. The theoretical development is presented herein. An experimental investigation is a topic of the future paper. The key feature of this approach is that it is based on generally accepted QFD practices, conventions and nomenclatures. It uses the same set of line-vectors and relationship matrices that were previously defined. The formulation is a straightforward extension of any QFD computer programs that might exist today. There is no need to define an extra set of vectors or matrices beyond what QFD provides. The techniques allow the users of QFD now only to optimize the design characteristics for a given set of WHATs, but

Product Planning Matrix for Car Hatch Door

Date: 19 Oct 92
C:\QFDDATA\HATCH

WHATs vs. WHYs

Strong Sales Point:	■	1.50
Sales Point:	+	1.20
Not A Sales Point:	–	1.00

		Customer Importance Perceptions	Age 16-25	Age 26-35	Age 36+	Weighted average of importance	Customer Performance Perceptions	Our model	Competitor A	Competitor B	Planned model	Improvement ratio	Maximum value = 6.0 ○ Our model ▲ Competitor A □ Competitor B — Planned model Minimum value = 0.0	Sales Points	Overall Importance	
		1	2	3	4	5	6	7	8	9	10	11	12	13	14	
Weighting			0.40	0.30	0.30											
Hatch Opening And Closing	1															1
Easy to open	2		3	4	5	3.9		3.5	3.7	4.0	4.2	1.2		–	4.7	2
Easy to close	3		3	4	5	3.9		4.5	3.6	4.0	4.5	1.0		–	3.9	3
Stays open	4		4	5	4	4.3		4.5	3.0	3.1	4.5	1.0		–	4.3	4
Stays closed	5		5	4	5	4.7		3.0	3.5	4.8	4.9	1.6		+	9.2	5
Hatch Properties	6															6
Low liftover height	7		2	4	5	3.5		5.0	3.3	3.0	5.0	1.0		–	3.5	7
Wide opening	8		3	4	4	3.6		3.0	1.5	3.5	3.7	1.2		–	4.4	8
Tall opening	9		2	4	5	3.5		3.5	3.5	3.0	4.0	1.1		■	6.0	9
No obstacles	10		2	3	4	2.9		3.5	3.8	3.2	4.0	1.1		–	3.3	10
Good visibility	11		4	4	5	4.3		4.8	5.0	2.5	5.0	1.0		–	4.5	11
No leaks when closed	12		3	3	3	3.0		3.1	3.2	2.8	3.3	1.1		–	3.2	12
No drips when open	13		3	3	3	3.0		5.0	4.0	4.0	5.0	1.0		–	3.0	13
Lock Properties	14															14
Easy to insert key	15		4	4	5	4.3		5.0	5.0	4.9	5.0	1.0		–	4.3	15
Easy to remove key	16		4	4	5	4.3		5.0	5.0	4.8	5.0	1.0		–	4.3	16
Easy to turn	17		3	3	4	3.3		4.1	4.9	3.5	5.0	1.2		–	4.0	17
Freeze proof	18		4	3	3	3.4		4.0	4.0	2.0	4.2	1.0		+	4.3	18
		1	2	3	4	5	6	7	8	9	10	11	12	13	14	

Figure 5.14: WHATs versus WHYs (HW) Matrix in the HATCH Example

Product Planning Matrix for Car Hatch Door

Date: 19 Oct 92
C:\QFDDATA\HATCH

HOWs vs. HOW MUCHes

Minimize:	▽	1.00
Maximize:	Δ	2.00
Meet Target:	–	3.00

Row 7 (technical evaluation graph): Maximum value = 6.0; ○ Our model; ▲ Competitor A; ▫ Competitor B; Minimum value = 0.0

#	HOWs	1 Direction of Movement Of Target Value	2 Target Value	3 Technical Evaluations: 4 Our model	5 Competitor A	6 Competitor B	8 Importance: 9 Absolute	10 Percentage	11 Ranking
1	Opening And Closing								
2	Opening work	▽	7.5 ft-lbs maximum	1.5	3.8	1.1	42.	4.9	11
3	Closing work	▽	7 ft-lbs maximum	2.0	2.0	1.2	48.	5.6	6
4	Force vs. Position curve	–	Smooth curve	3.0	3.1	2.1	77.	9.0	3
5	Hold open force	Δ	7 to 15 lbs acceptable	4.1	0.3	2.2	38.	4.5	14
6	Range of motion	▽	From waist to shoulder high only	3.0	4.1	3.1	25.	3.0	18
7	Hatch								
8	Liftover height	▽	30 Inches maximum	3.0	3.0	3.2	52.	6.2	5
9	Width	Δ	48 Inches minimum	4.0	2.5	4.1	63.	7.4	4
10	Height	Δ	48 Inches minimum	4.0	3.2	4.2	77.	9.0	2
11	Useable opening area	Δ	2300 Square Inches minimum	3.5	3.8	3.9	43.	5.1	9
12	Visibility pyramid	Δ	6900 Square inch horizontally biased rectangle	3.6	4.0	3.8	40.	4.7	13
13	Water penetration test performance	Δ	4 Minute minimum	3.2	2.4	2.9	28.	3.4	16
14	Water drip test performance	▽	No Drips in target area for 5 minutes	3.4	3.6	2.8	27.	3.2	17
15	Latch								
16	Bursting force	Δ	102 lbs minimum	2.0	3.0	1.9	82.	9.7	1
17	Key insert force	▽	3 lbs maximum acceptable	3.0	3.2	1.8	43.	5.1	7
18	Key removal force	▽	3 lbs maximum acceptable	3.5	3.4	2.5	43.	5.1	7
19	Key hole location	–	Per plan	2.8	4.5	3.5	42.	5.0	10
20	Key turning moment	▽	5 In-lbs maximum	3.5	3.1	4.5	40.	4.8	12
21	Freeze test performance	Δ	20 in-lbs maximum @ 24 hr 10 deg F exposure	4.0	4.0	4.0	38.	4.5	15

Figure 5.15: HOWs versus HOW MUCHes (HM) Matrix in the HATCH Example

Product Planning Matrix for Car Hatch Door

Date: 19 Oct 92
C:\QFDDATA\HATCH

HOWs vs. HOWs

Strong Positive:	●	9
Weak Positive:	O	3
Weak Negative:	X	-3
Strong Negative:	✱	-9

		1 Opening And Closing	2 Opening work	3 Closing work	4 Force vs. Position curve	5 Hold open force	6 Range of motion	7 Hatch	8 Liftover height	9 Width	10 Height	11 Useable opening area	12 Visibility pyramid	13 Water penetration test performance	14 Water drip test performance	15 Latch	16 Bursting force	17 Key insert force	18 Key removal force	19 Key hole location	20 Key turning moment	21 Freeze test performance	
Opening And Closing	1																						1
Opening work	2			X																			2
Closing work	3		X			X																	3
Force vs. Position curve	4						O																4
Hold open force	5			X																			5
Range of motion	6				O																		6
Hatch	7																						7
Liftover height	8																						8
Width	9											O	O										9
Height	10											O	O										10
Useable opening area	11									O	O		●										11
Visibility pyramid	12									O	O	●											12
Water penetration test performance	13																						13
Water drip test performance	14																						14
Latch	15																						15
Bursting force	16																						16
Key insert force	17																		O	O		X	17
Key removal force	18																	O		O			18
Key hole location	19																	O	O		O		19
Key turning moment	20																			O			20
Freeze test performance	21																	X					21
		1	2	3	4	5	6	7	8	9	10	11	12	13	14	15	16	17	18	19	20	21	

Figure 5.16: HOWs versus HOWs (HH) Matrix in the HATCH Example

also to include the heuristics-based relationships in optimization formulation. The approach is capable of optimizing engineering resources and planning for product and process quality characteristics' deployment.

References

[1] Akao, Y. (1990). "Quality Function Deployment — Integrating Customer Requirements into Product Design," Japanese Standards Association.

[2] Aswad, A. (1989). "Quality Function Deployment — A Tool or Philosophy," SAE-890163.

[3] Bevan, N., Jebb, A., and Wynn, H.P. (1989). "Design to Cost," SAE-894120, *Engineering Design Conference*, MEP-333, Institution of Mechanical Engineers, London.

[4] Bossert, J.L. (1991). "Quality Function Deployment — A Practitioner's Approach," *SERIES: Quality and Reliability*, Vol. 21, American Society for Quality Control.

[5] Carey, W.R. (1992). *Tools for Today's Engineer "Strategy for Achieving Engineering Excellence", Section 1: Quality Function Deployment*, SP-913, SAE International Congress and Exposition, Detroit.

[6] Crosby, P.B. (1979). *Quality is Free: The Art of Making Quality Certain*, McGraw Hill.

[7] Davis, C.R. "Using the 'Voice of the Customer' to Establish Product Specifications," SAE-880333.

[8] Deming, W.E. (1986). *Out of Crisis*, 2nd ed., MIT Center for Advanced Engineering Study, Cambridge, MA.

[9] Eureka, W.E. and Ryan, N.E. (1988). "Customer-Driven Company — Managerial Perspectives on QFD," A.S.I. Press, Dearborn, MI.

[10] Ezop, E., Jacoby, T., and Leach, L. (1989). "Continuous Improvement with Quality Function Development, *SME Technical Papers*, SME-IQ890405.

[11] Freeze, D.E., and Aaron, H.B. (1990). "Customer Requirements Planning Process [CRPII] (Beyond QFD)," *SME Technical Papers*, SME-MS90003.

[12] Ginder, D.A., and Quinn, W.B. (1988). "Quality Function Deployment — Customer Driven Engineering," SAE-881334.

[13] Ginder, D.A. (1989). "Case Study — Quality Function Deployment," SAE-890652.

[14] *GOAL/QPC, Transactions from the Fourth Symposium on Quality Function Deployment (QFD)*, American Society for Quality Control, American Supplier Institute, 1992, Novi.

[15] Hales, R., Lyman, D., and Norman, R. (1990). "Quality Function Deployment and the Expanded House of Quality," International TechneGroup Inc., Technical Report, pp. 1-12.

[16] Hauser, J.R., and Clausing, D. (1988). "The House of Quality," *Harvard Business Review*, (May-June), pp. 63-73.

[17] King, B. (1987). "Better Designs in Half the Time - Implementing QFD Quality Function Deployment in America," *GOAL/QPC*, Methnen.

[18] Morrell, N.E. *Quality Function Deployment*, SAE-870272.

[19] Oakland, J.S. (1989). *Total Quality Management — Proceedings of 2nd International Conference on Total Quality Management*, London.

[20] *QFD/CAPTURE: Users Manual*, International TechneGroup Inc. Version 2.2, 1990.

[21] Sadri, H.A. (1991). "Quality Function Deployment Results in Less Engineering Changes," *SME Technical Papers*, SME-IQ910432.

[22] Sullivan, L.P. (1988). "Quality Function Deployment," *Quality Progress*, Vol. 21, No. 6.

[23] Taguchi, G., and Clausing, D. (1990). "Robust Quality," *Harvard Business Review*, (January-February), pp. 65-75.

[24] Thompson-Colon, J.A. (1989). "Quality Function Deployment as Applied to the Formulation of Polyu-Rethane Foams for Instrument Panels," SAE-890767.

[25] Vannoy, E.H., and Davis, J.A. "Test Development Using the QFD Approach," SAE-890807.

[26] Vigier, M.G. (1989). "Quality Function Deployment," *Ingenieurs de l'Automobile, Section 7, Value Analysis*, SAE-903001, pp. 59-61.

[27] Vigier, M.G. (1990). "ELAQ — A New Approach to Implement the Logical Chain of Quality Function Deployment," *Ingenieurs de l'Automobile, Section 11, Reliability and Maintenance*, SAE-903014, pp. 80-82.

[28] Vora, L.S., Veres, R.E., and Jackson, P.C. (1989). "Technical Information Engineering System (TIES)," *SME Technical Papers*, SME-MS890736.

[29] Taguchi, S. *Taguchi Methods and QFD: Hows and Whys for Management*, American Supplier Institute, A.S.I. Press, Dearborn, MI.

[30] Wu, Y. and Taguchi, S. "Orthogonal Arrays and Linear Graphs," A.S.I. Press, Dearborn, MI.

Chapter 6

Automated Generation of Minimum Cost Production Sequence

ZUOMIN DONG
Department of Mechanical Engineering, University of Victoria

6.1 Introduction

6.1.1 Background

When a product design has been completed, the production sequence for manufacturing the product is planned through process planning. As a traditional, but critical component of design and manufacturing, process planning has being automated today to meet the needs for higher productivity and lower production costs. Computer-automated process planning (CAPP) has been recognized as a crucial link to computer-aided design (CAD) and computer-aided manufacturing (CAM). Improvements in CAPP can essentially change the art of mechanical design and manufacturing [1, 2, 3, 20].

A production sequence is a selected group of production operations for changing the designed part from a raw material form into the designed specifications of dimension, tolerance, and surface finish. For example, a typical production sequence for producing a rotational (part) surface consists of four production operations: casting/forging, rough turning, semifinish turning, and finish grinding.

As part of the efforts for developing new generative CAPP systems, research on automated production process generation focuses on the development of knowledge-based systems that synthesize part geometry and generate process plans as expert process planners [2, 11, 13, 20]. Heuristics for process planning and models of machinability data are used for determining the production operations and machining parameters. The generative CAPP system, acting as an experienced production engineer, creates the production sequence for a part based on the stored knowledge on production processes and their capabilities, production codes, tooling and machining parameters, as well as constraints imposed on part manufacturing. This research has significantly contributed to the advance of computer-automated process planning. On the other hand, CAPP systems using a pure knowledge-based approach are qualitative in nature. Quantitative studies in process planning are currently limited to the computerized tolerance charting [19] and the optimal cutting parameter determination [1]. These both emerge at the later stage of the process plan generation, after the process sequence has been determined. The quantitative analysis aspect of process sequence generation has rarely been addressed. Since no quantitative comparison is made among the candidate production sequences and among the different process tolerance combinations of each production sequence, an optimal process plan cannot be obtained.

6.1.2 Quantitative Intelligent System Approach

Mechanical design and production planning are dominated by decision-making processes that require both qualitative knowledge reasoning and quantitative numerical analysis.

The ability to quickly make appropriate qualitative decisions and the quality of these decisions are often proportional to one's manufacturing knowledge and experience. Accumulation of this experience and knowledge normally takes a relatively long period of time. The new developments of artificial intelligence (AI), particularly the knowledge-based system approach, have made it possible to automate the knowledge acquisition and knowledge-reasoning processes and to make a computer system function as a human expert, to some extent. In addition, a major fraction of engineering decisions are made by referencing standards documented in handbooks and existing practice of previous cases. The symbolic manipulation and database management capabilities of a knowledge-based system suit these applications well. The limitation of traditional artificial intelligence techniques lies in their limited capability for handling complex knowledge-reasoning tasks involving quantitative decision making. This limitation is primarily due to their primitive first-order logic structure.

Numerical analysis is a traditional domain for applying computer technology to engineering applications. The numerical analysis tools frequently used in design and manufacturing include finite elements analysis, tolerance analysis, and numerical optimization. With the tremendous computing power of digital computers, one can conduct complex design performance and production cost analyses, which are normally too difficult to be performed manually. Through the quantitative design performance and production cost evaluation using the engineering analysis tool and the numerical optimization, an optimal design or an optimal production plan can be generated. A major obstacle to the wide application of engineering analysis and optimization methods today is its dependence to manual problem formulation. This formulation can be carried out only by an expert in the specific area, with special knowledge and experience acquired over a long period of time. Due to this high labor-cost overhead, engineering analysis can hardly be justified in many applications even though the direct benefit is significant.

The combination of artificial intelligence and numerical optimization can fundamentally improve the present practices of engineering design, analysis, and planning. The AI techniques such as an expert system can be used to formulate the engineering analysis and optimization problem and to generate all feasible designs or production plans [7, 8]. The problem formulation and candidate generation are performed using knowledge reasoning based on the domain knowledge extracted from human experts. On the other hand, numerical optimization can be used for identifying the optimal design or production plan, based upon the design performance or production cost evaluation. The approach can eliminate the bottleneck of engineering analysis — manual problem formulation and need for hard-to-find human experts. This methodology makes both artificial intelligence and numerical optimization more valuable to engineering design, analysis and planning.

In this chapter, the combination of one of the AI techniques, an expert system, and the numerical optimization for automatically generating the optimal production sequence and optimal process tolerances with least manufacturing costs is illustrated. Through this case study one can see the potential benefits of the quantitative intelligent system approach to engineering design and planning problems.

6.1.3 Overview of the Approach

The new approach for production sequence generation by combining a knowledge-based system with numerical optimization is introduced in our recent work [8, 9]. The knowledge based system is used for generating all feasible production sequences. The numerical optimization is applied to evaluate these generated candidates and to identify the optimal production sequence with least production cost. The approach is illustrated by Figure 6.1.

From the part design specified by a feature-based CAD system, various kinds of design information relevant to production sequence generation, including form and size of geometric features, part accuracy specified by tolerances and surface finishes, material, and batch size, are passed to the optimal production sequence generation system.

In this approach, multiple candidates of production sequence for a given feature, rather than a single solution, are introduced. The generation of all feasible production sequences is accomplished by acquiring knowledge of production processes, modeling machine availability, identifying feasible production operation, analyzing successive/alternative production operations, and reasoning using a knowledge-based system.

In addition to the qualitative reasoning, production cost models of various production operations and production sequences are introduced. The feasible production sequences generated by the knowledge-based system are quantitatively evaluated based upon their minimum achievable production costs using nonlinear optimization. The objective function of the cost minimization is constructed automatically by the knowledge-based system based upon the feature-tolerance-process relations. The optimal process tolerances for each sequence are also obtained through this minimization. As a final result, the optimal production sequence with minimum manufacturing costs is identified.

6.2 Generation of Feasible Production Sequence

6.2.1 Model of Feature-to-Production Operation Connection

Frequently used production operations can be classified into five major groups: casting, forging, machining, non-cutting processing, and special machining. These productions can

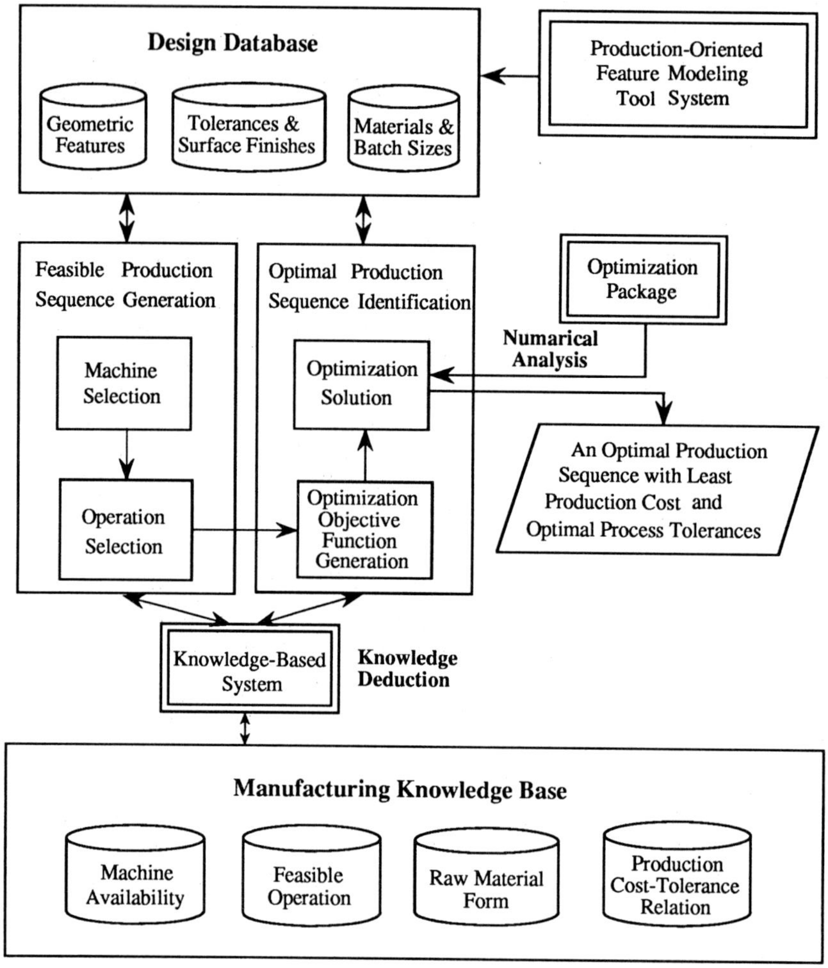

Figure 6.1: Optimal Production Sequence Generation

be regrouped into five categories based on the mechanical features that they can process. These categories are cylindrical surfaces, plane surfaces, holes, slots/groovers/keyways, and curved surfaces. Production operations corresponding to the these feature categories are listed in Figure 6.2. This list provides feasible production operations for a given design feature without considering the availabilities and capabilities of the machine tools involved.

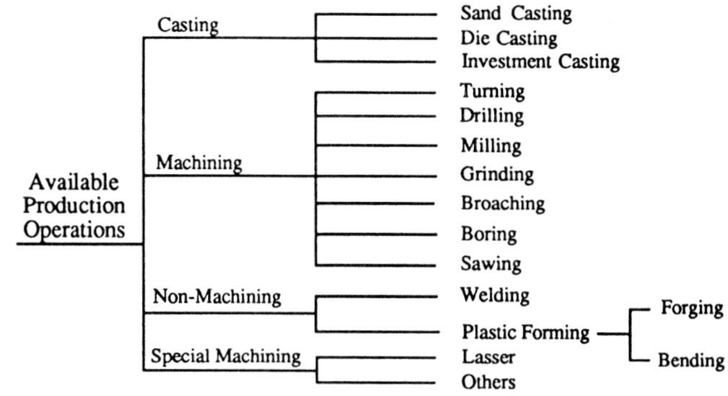

(a) Available Production Operations

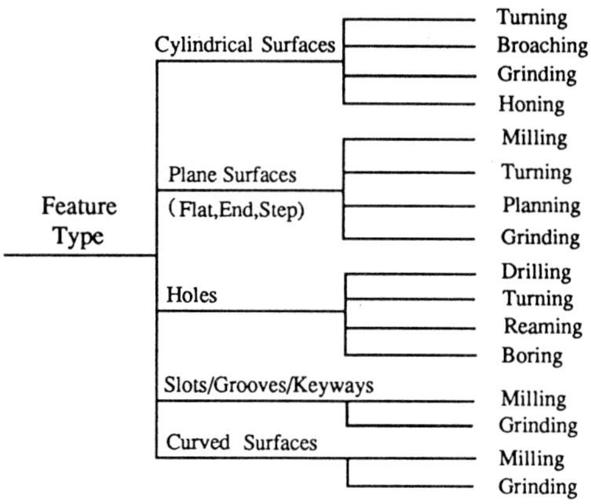

(b) Feasible Production Operation for a Feature

Figure 6.2: Feasible Production Operations for A Feature

6.2.2 Model of Machine Availability

In practice, availability and capability of the involved machine tools must be considered. For example, both turning and milling operations may be feasible for producing a feature, but the number of lathes and number of milling machines are likely to be different in a shop. The work loads of these machine tools are also different. These differences contribute to the availability of various machine tools. They are measured by a machine tool weighting factor, α, $(0 \leq \alpha \leq 1)$, as illustrated in Figure 6.3a.

In addition, the suitability and productivity of a production operation to part machining at different stages also need to be considered. For example, rough machining of a rotational surface can be conducted using both turning or grinding, but in general, turning is more suitable due to lower production cost and less wear on the machine tool. An operation suitability factor, β $(0 \leq \beta \leq 1)$, is introduced for each production operation to compensate the potential equipment wear, as shown in Figure 6.3b. The cost consideration will be discussed later.

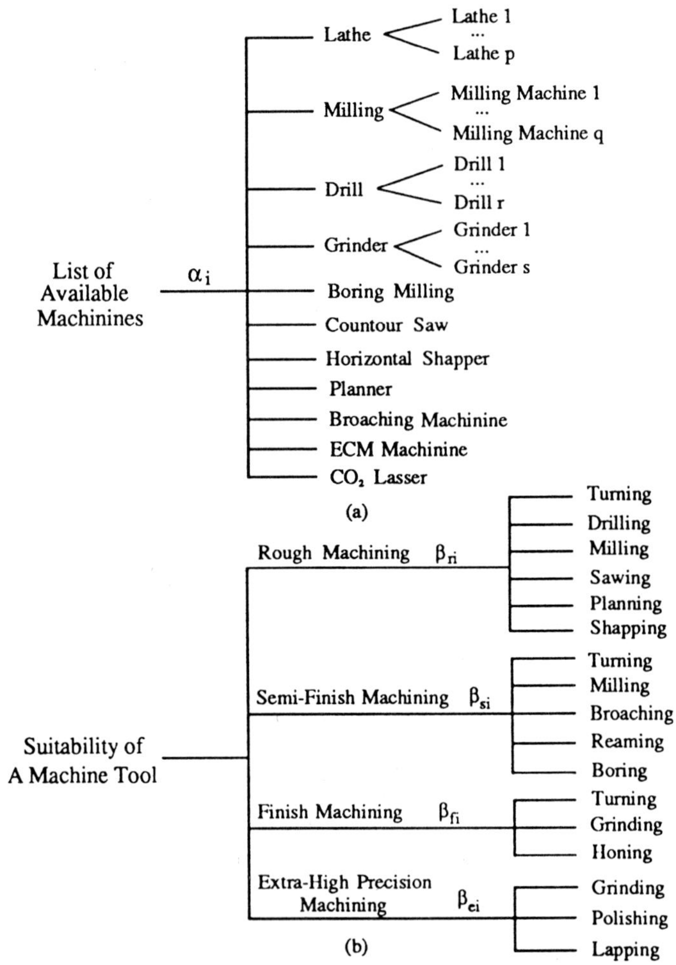

Figure 6.3: Machine Availability Model

Table 6.1: Tolerance Range of Machining Operations (± mm)

Operation	Range of Nominal Dimensions (mm)							
	0–14.9	15–24.9	25–37.9	38–69.9	70–114.9	115–199.9	200–344.9	345–533
Drilling	0.050 ~ 0.127	0.064 ~ 0.152	0.076 ~ 0.203	0.102 ~ 0.254	0.127 ~ 0.305	0.152 ~ 0.381	0.203 ~ 0.508	0.254 ~ 0.635
Milling	0.030 ~ 0.127	0.038 ~ 0.152	0.050 ~ 0.203	0.064 ~ 0.254	0.076 ~ 0.305	0.102 ~ 0.381	0.127 ~ 0.508	0.152 ~ 0.635
Turning, Boring, Planing	0.020 ~ 0.127	0.025 ~ 0.152	0.030 ~ 0.203	0.038 ~ 0.254	0.050 ~ 0.305	0.054 ~ 0.381	0.076 ~ 0.508	0.102 ~ 0.635
Reaming	0.008 ~ 0.050	0.010 ~ 0.064	0.013 ~ 0.076	0.015 ~ 0.102	0.020 ~ 0.127	0.025 ~ 0.152	0.030 ~ 0.203	0.040 ~ 0.254
Broaching	0.005 ~ 0.020	0.006 ~ 0.025	0.008 ~ 0.030	0.010 ~ 0.038	0.013 ~ 0.050	0.015 ~ 0.064	0.020 ~ 0.076	0.025 ~ 0.102
Grinding, Diamond Turning	0.004 ~ 0.013	0.004 ~ 0.015	0.005 ~ 0.020	0.006 ~ 0.025	0.008 ~ 0.030	0.010 ~ 0.038	0.013 ~ 0.050	0.015 ~ 0.064
Lapping and Honing	0.004 ~ 0.008	0.004 ~ 0.010	0.005 ~ 0.013	0.006 ~ 0.015	0.008 ~ 0.020	0.010 ~ 0.025	0.013 ~ 0.030	0.015 ~ 0.040

Combining the machine tool weighting factor, α_i, and the operation suitability factor, β_{ki}, a quantitative measurement of the machine availability for a production operation, a_k, can be obtained

$$a_k = \alpha_i \times \beta_{ki} \tag{6.1}$$

where α_i and β_{ki} are obtained based on analysis of machine shop data.

6.2.3 Model of Successive and Alternative Production Operations

In manufacturing, a part is machined from raw material to designed geometry through many production operations of a production sequence. Each production operation is characterized by the geometric shape and the shape accuracy that it can produce. The final design geometry is accomplished by applying a group of sequentially arranged production operations with increasing accuracy improving capability.

In reality, the accuracy improving capability of an operation is limited by a lower tolerance bound, δ_{min}, and an upper tolerance bound, δ_{max}, as shown in Table 6.1 [18]. This valid tolerance range of a production operation can be represented as $[\delta_{min}, \delta_{max}]$. In Figure

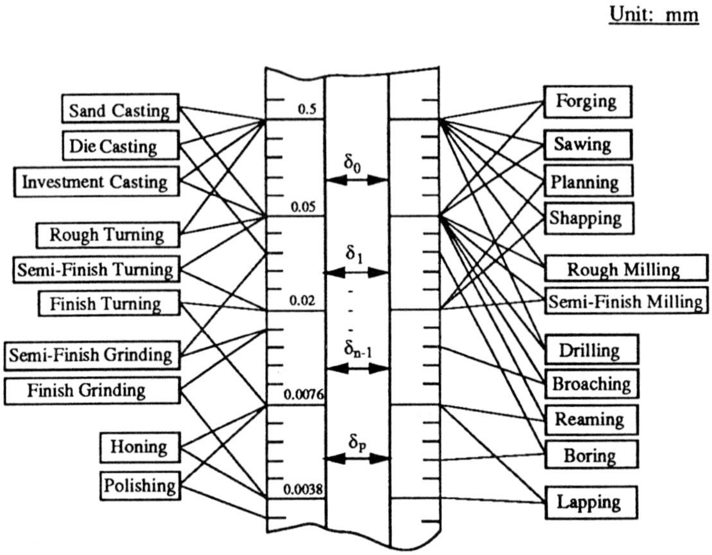

Figure 6.4: Successive and Alternative Production Operations

6.4, frequently used production operations are arranged according to their valid tolerance ranges. For a mechanical feature with a given raw material tolerance, δ_o, and a design tolerance, $\delta = \delta_p$, this chart illustrates the available production operations that can be used sequentially to improve the accuracy of the feature. The relations among the sequentially arranged production operations can be modeled by the AND logic. On the other hand, those production operations sharing common valid tolerance range form alternative production operations. Their relations can be represented by the OR logic. Frequently used production operations for producing mechanical features of a specific type can thus be represented as an AND/OR tree. As an example, the AND/OR tree for rotational surface machining is illustrated in Figure 6.5.

6.2.4 Generating Feasible Production Sequences by an AND/OR Tree Search

Feasible production sequences are formed by the combination of feasible production operations that span their valid tolerance ranges between the given raw material tolerance, δ_o, and the design tolerance, $\delta = \delta_p$. Since each path on the AND/OR tree between a given δ_o and δ_p represents a feasible production sequence, all feasible production sequences can be identified by an exhaustive search on the AND/OR tree. For example, given the design tolerance of a rotational surface $\delta = 0.013$ (mm) and free choice of blank part form, all feasible

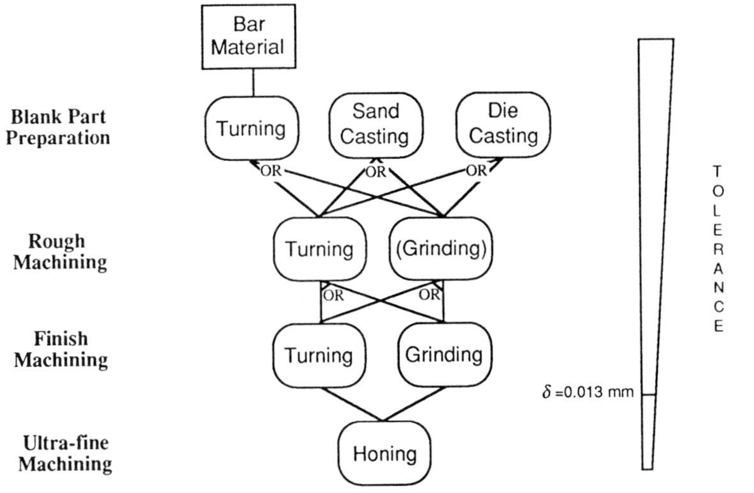

Figure 6.5: Production Operations for a Rotational Surface

production sequences for producing the rotational surface can be identified by an exhaustive search on the AND/OR tree shown in Figure 6.5, starting at the blank part preparation and ending at the finish machining. The 12 feasible production sequences obtained from the search are listed in Table 6.2.

Table 6.2: Feasible Production Sequences for Machining the Rotational Surface

Feasible Seq.	1	2	3	4	5	...	12
Blank Part Prep.	turning	turning	turning	turning	sand casting	...	die casting
Rough Machining	turning	turning	grinding	grinding	turning	...	grinding
Finish Machining	turning	grinding	turning	grinding	turning	...	grinding

6.2.5 Implementations of the Knowledge-based System

The discussed knowledge models have been represented using frames and production rules in a prototype knowledge-based system, shown in Figure 6.6. The expert system development software EXSYS Professional was used in the implementation due to its capability of interfacing numerical routines. The current implementation is limited to rotational parts.

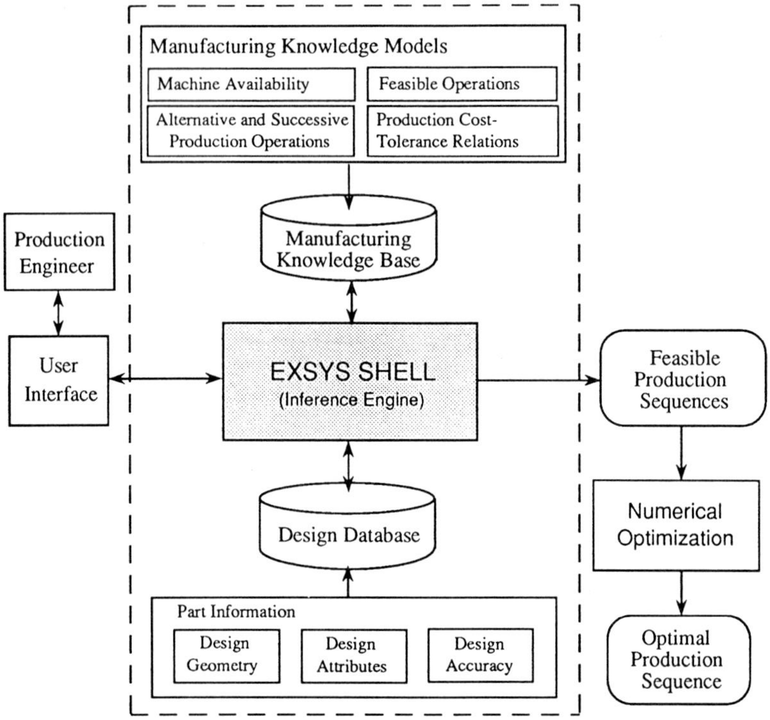

Figure 6.6: Structure of the Prototype Knowledge-based System

The developed system produces a list of all feasible production sequences and constructs the objective function of optimization.

6.3 Identification of Optimal Production Sequence and Process Tolerances

6.3.1 Empirical Production Cost-tolerance Relations

The valid tolerance range of a production operation, $[\delta_{min}, \delta_{max}]$, represents only the accuracy improving capability of a production operation. Within this valid range, a tighter tolerance (or a higher accuracy demand) leads to higher production costs, and a looser tolerance (or a lower accuracy demand) leads to lower production costs. Through a production sequence, a mechanical feature is modified from its raw material form to the designed shape and accuracy. The mechanical tolerance and surface finish of a part are continually improved by the selected production operations.

Empirical production cost-tolerance data curves for various production operations and processes have been obtained in earlier research [18]. One of these curves, the empirical relation of production cost versus the true position tolerance in hole making, is illustrated in Figure 6.7, where δ denotes a tolerance that describes the hole location accuracy and $c(\delta)$ represents the production costs for producing a hole to the location accuracy of δ. The curve covers production operations of drilling, jig boring, and machining using special equipment. This is the empirical curve representing the cost-tolerance relation of a hole-making production sequence that consists of production operations of rough machining, semifinish machining, finish machining, and accurate forming using special equipment.

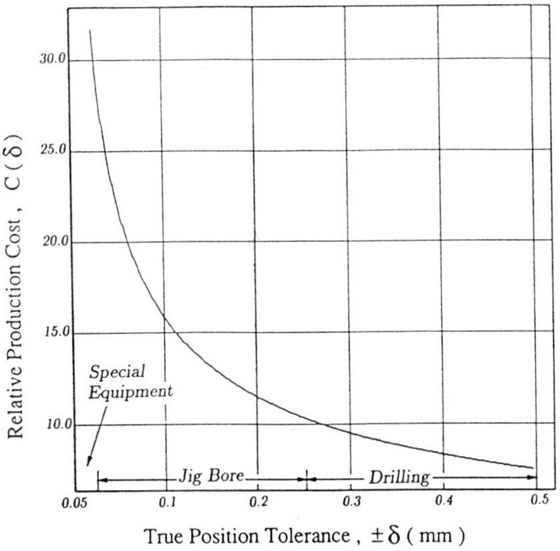

Figure 6.7: Production Cost versus True Position Tolerance

The illustrated cost-tolerance relation of a production sequence also applies to each production operation in the sequence. The tolerance produced through a production operation (such as drilling or boring) and the machining cost of the operation present a similar curve over a smaller valid tolerance region. From the empirical production data of different production operations, one can find that the cost-tolerance curve of a production sequence is actually the embracing curve for the cost-tolerance curves of its production operations. This relation is illustrated in Figure 6.8.

6.3.2 Mathematical Model of Empirical Cost-tolerance Relations

To incorporate the collected empirical cost-tolerance data of various production operations into the calculation of production costs of a production sequence, the appropriate mathematical model for these data curves needs to be introduced. Many researchers worked

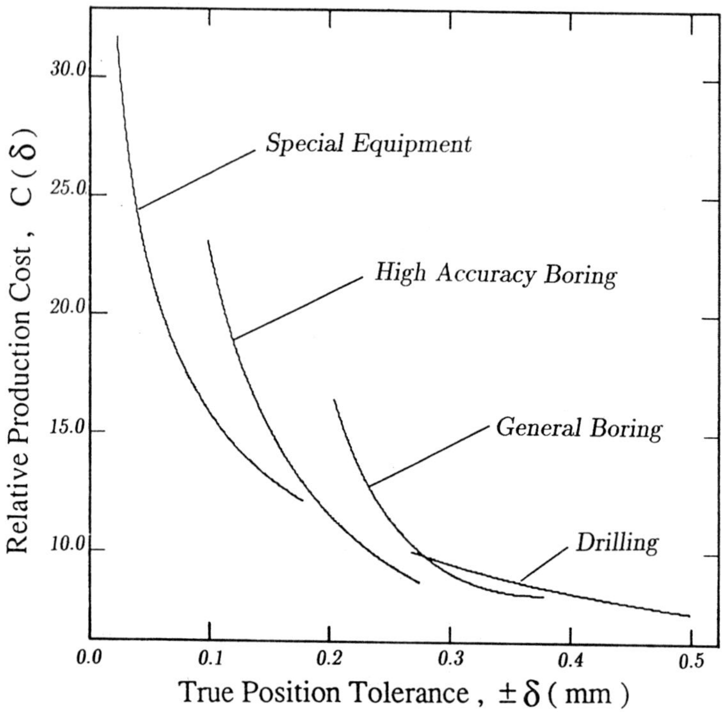

Figure 6.8: Production Cost-tolerance Curves for Hole Machining Sequence/Operations

on this subject and introduced close to a dozen cost-tolerance models. These models are listed in Table 6.3.

The earlier mathematical models for empirical cost-tolerance data were introduced over the tolerance range of $[0, +\infty)$, and were obtained using one specific cost-tolerance curve. It has been recognized that these models have relatively large modeling errors [21]. In the author's recent work, a systematic study on a wide variety of model forms was conducted [8]. Various published empirical production cost-tolerance data curves were collected and used for testing the existing and new models. The collected cost-tolerance data cover a wide variety of production processes, including hole producing (true position), face milling, turning, rotational surface grinding, internal grinding, die casting, and investment casting, compiled by Trucks [18], and the general cost-tolerance relation published by Dieter [5]. These empirical production cost-tolerance data curves are plotted in Figure 6.9.

Since each production operation has only limited accuracy improving capability, curve fitting of the new cost-tolerance models was conducted over the valid tolerance range of the operation. The new models showed significant modeling accuracy improvements over the existing models. This can be illustrated using Dieter's empirical data as a test curve, as shown in Figure 6.10.

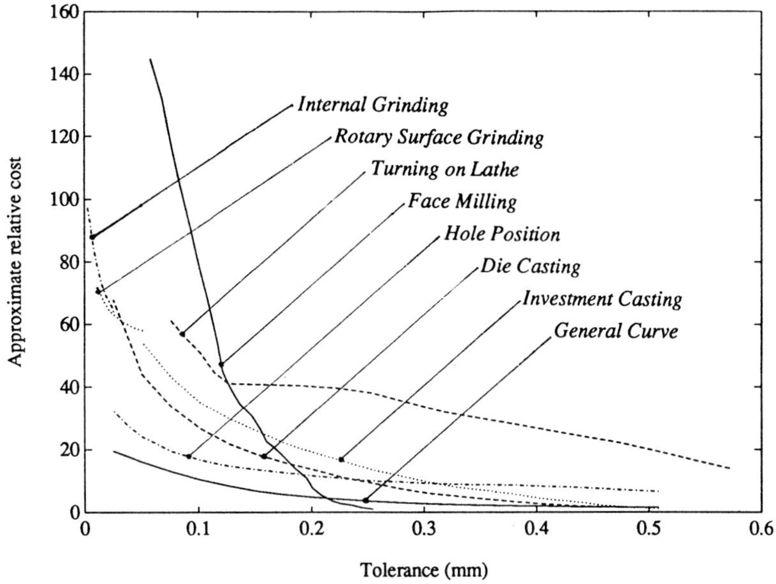

Figure 6.9: Empirical Production Cost-tolerance Curves

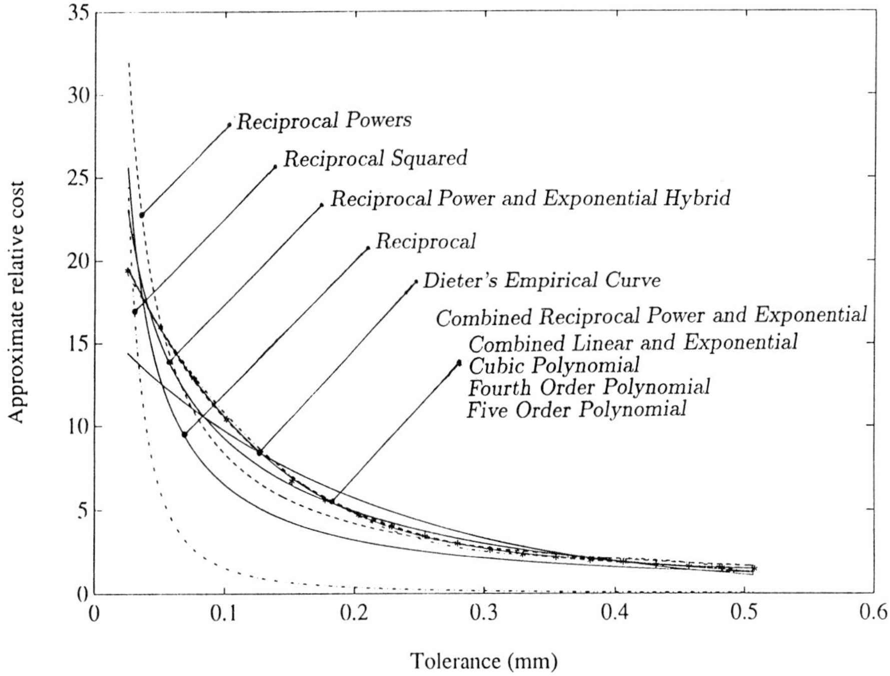

Figure 6.10: Model Comparisons Using Dieter's Cost-tolerance Curve

Table 6.3: Empirical Production Cost-tolerance Curves

No Bounds for δ	
Exponential [15]	$c(\delta) = a_0\ e^{-a_1\delta}$
Reciprocal Squared [16]	$c(\delta) = \frac{a_0}{\delta^2}$
Reciprocal Powers [17]	$c(\delta) = a_0\ \delta^{-a_1}$
Reciprocal Powers/Exponential [14]	$c(\delta) = a_0\ \delta^{-a_1}\ e^{-a_2\delta}$
Reciprocal [4]	$c(\delta) = \frac{a_0}{\delta}$
Discrete [12]	$\delta_i,\ c(\delta_i)\quad (1 \leq i \leq N)$
With A Valid Range $\delta \in [\delta_{min},\ \delta_{max}]$	
Modified Exponential [7, 6]	$c(\delta) = a_o\ e^{-a_1(\delta - a_2)} + a_3$
Reciprocal Powers/Exponential II [8]	$c(\delta) = a_o + a_1\ \delta^{-a_2} + a_3\ e^{-a_4\delta}$
Combined Linear and Exponential [8]	$c(\delta) = a_o + a_1\delta + a_2\ e^{-a_3\delta}$
Cubic Polynomial [8]	$c(\delta) = a_o + a_1\delta + a_2\delta^2 + a_3\delta^3$
Fourth Polynomial [8]	$c(\delta) = a_o + a_1\delta + a_2\delta^2 + a_3\delta^3 + a_4\delta^4$
Fifth Polynomial [8]	$c(\delta) = a_o + a_1\delta + a_2\delta^2 + a_3\delta^3 + a_4\delta^4 + a_5\delta^5$

6.3.3 Cost-tolerance Model for a Production Operation

The production cost reflects the required manufacturing efforts for changing a mechanical feature from the raw material state with considerably larger errors (very loose tolerance) to the tolerance corresponding to the cost. A reasonable tight tolerance can be obtained with high-quality producing and inefficient cutting parameter settings, such as small cutting depth, slow feed rate, and fine turning of the machine tool. The infinite initial tolerance can be considered to be corresponding to zero cost. A mechanical feature with small original errors (tight blank part tolerance) requires less manufacturing effort in machining and leads to low production costs. The relative production cost for improving the feature tolerance from δ_1 to δ_2 can thus be calculated by

$$\Delta C_{\delta_1 - \delta_2} = c(\delta_2) - c(\delta_1) \tag{6.2}$$

where δ_1 and δ_2 represent the tolerances of a mechanical feature before and after the production operation. The $c(\delta_1)$ and $c(\delta_2)$ are the corresponding relative production costs to change the tolerance of a feature from an infinite tolerance (very large error) to the tolerances of δ_1 and δ_2. The result $\Delta C_{\delta_1 - \delta_2}$ is the relative production costs for improving the tolerance of a feature from δ_1 to δ_2.

In the traditional process tolerance assignment practice, a recommended process tolerance δ is suggested for a given machining operation by a machining data handbook [10]. The recommended tolerance represents a point on the cost-tolerance curve. Since this fixed tolerance was generated for a general case without considering the interactions among different production operations, the optimal process tolerances and the minimum achievable production cost of the production sequence cannot be obtained.

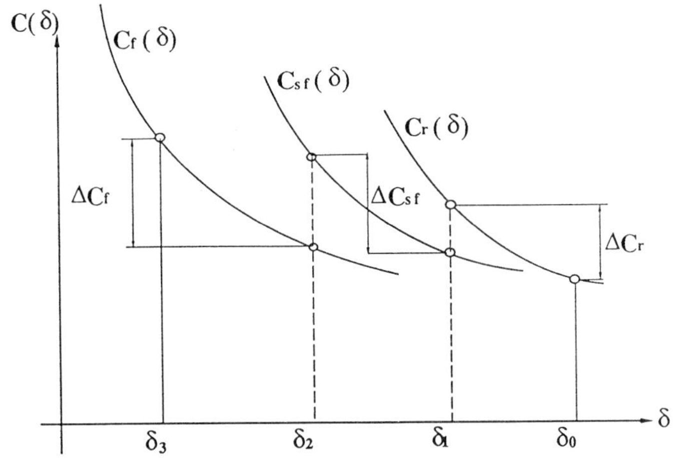

Figure 6.11: Cost-tolerance Model for a Production Sequence

6.3.4 Cost-tolerance Model for a Production Sequence

From the above analysis, the process for changing a mechanical feature from its raw material state with considerably larger error to the finished part state with the designed tolerance can be modeled as shown in Figure 6.11, provided the process sequence consists of three production operations: rough machining, semifinish machining, and finish machining. The production cost of this process sequence can be calculated by

$$\begin{aligned} C_{sequence} &= \Delta C_r + \Delta C_{sf} + \Delta C_f \\ &= [c_r(\delta_1) - c_r(\delta_o)] + [c_{sf}(\delta_2) - c_{sf}(\delta_1)] + [c_f(\delta_3) - c_f(\delta_2)] \end{aligned} \tag{6.3}$$

where δ_o is the tolerance of the blank part. δ_3 is the tolerance accomplished by the production process; $c_r(\delta)$, $c_{sf}(\delta)$ and $c_f(\delta)$ are the production cost-tolerance models for the rough machining, semifinish machining, and finish machining operations; and ΔC_i represents the tolerance improvement from the corresponding production operation.

A general formulation of the production cost for a process sequence with (19) "p" production operations becomes

$$C(\delta_1, \ldots, \delta_{p-1})_{sequence} = \sum_{i=1}^{p} \Delta C_i = \sum_{i=1}^{p} [c_i(\delta_i) - c_i(\delta_{i-1})] \tag{6.4}$$

where δ_o is the tolerance of the blank part, and δ_p is the tolerance accomplished by the production sequence. The value of δ_p is equal to (or smaller than) the designed tolerance δ of the feature.

In practice, a part is often unloaded from a machine tool after one production operation and loaded onto another machine tool to proceed to the next production operation. The set-up error for the ith production operation, $e_{setup,i}$, will be introduced. When a part is reloaded, the errors that the production operation has to handle include the error of the part that has been inherited from the previous production operation, and the error that has been introduced in the reloading. In other words, the existing tolerance for the ith production operation is $\delta'_{i-1} = \delta_{i-1} + e_{setup,i}$, rather than δ_{i-1}. The general formulation of the production cost for a process sequence with (19) "p" production operations thus becomes

$$C(\delta_1, \ldots, \delta_{p-1})_{sequence} = \sum_{i=1}^{p} \Delta C'_i = \sum_{i=1}^{p} [c_i(\delta_i) - c_i(\delta'_{i-1})] \tag{6.5}$$

$$\delta'_{i-1} = \delta_{i-1} + e_{setup,i}$$

The set-up error, $e_{setup,i}$, is production process dependent. Its values for various machining operations can be obtained by experiments. If a part is not reloaded before the ith production operation, $e_{setup,i}$ becomes zero.

6.3.5 Identifying the Minimum Cost Production Sequence

The introduced feasible production sequences are competitive candidates, each represented by their production cost, $C(\delta_1, \ldots, \delta_{p-1})_k$. By minimizing the formulated production cost of each competitive sequence, the minimum achievable production costs of a production sequence, C_k^*, and the optimal process tolerances of the sequence, $\delta_1^{k*}, \ldots, \delta_{p-1}^{k*}$, can be obtained. The cost minimization has the form

$$\min_{w.r.t.\delta_i} C_k(\delta_1, \ldots, \delta_{p-1}) = \min_{w.r.t.\delta_i} \sum_{i=1}^{p} [c_i^k(\delta_i) - c_i^k(\delta'_{i-1})] \tag{6.6}$$

$$\delta'_{i-1} = \delta_{i-1} + e_{setup,i}$$

subject to

$$\delta_{i,min} \leq \delta_i \leq \delta_{i,max}$$

The optimal production sequence is the production sequence, L, with the (19) "least" minimum achievable production costs, $C_L^* = \min(C_k^*,\ \ k \in \text{all production sequences})$.

6.4 An Example

The optimal machining sequence generation for a rotational surface feature is used to illustrate the method presented in this chapter. The expected size tolerance of the rotational surface δ is 0.013 (mm). To achieve this accuracy, the ultra-fine machining operation is not necessary as illustrated in Figure 6.5. Suppose the grinding machine for semifinish machining is tied up by other machine jobs and is not available. The feasible production sequences will be reduced from the original twelve (as shown in Table 6.2) to six. These six feasible production sequences are generated from an exhaustive depth-first search on the AND/OR

Table 6.4: The Calculated Relative Production Costs

Production Sequence	δ_0	δ_1^*	δ_2^*	δ	C_{min}^*	Relative C
turn, turn, turn	0.500	0.076	0.064	0.013	206.9	0.929
turn, turn, grind	0.500	0.076	0.064	0.013	116.8	0.524
die cast, turn, turn	0.500	0.076	0.064	0.013	237.8	–
die cast, turn, grind	0.500	0.076	0.064	0.013	147.7	–
sand cast, turn, turn	0.500	0.127	0.064	0.013	312.4	–
sand cast, turn, grind	0.500	0.127	0.064	0.013	222.3	–
handbook suggested	0.500	0.218	0.076	0.013	(223)	1.000

tree of production operations for the rotational surface feature. The search is conducted by the knowledge-based system for generating feasible production sequences.

Each of the six alternative production sequences has three production operations. The production costs of these sequences can be calculated by

$$C_k(\delta_1, \delta_2) = [g_f(\delta) - g_f(\delta_2)] + [g_{sf}(\delta_2) - g_{sf}(\delta_1)] + [g_r(\delta_1) - g_r(\delta_0)]$$

This cost function is constructed by the knowledge-based system based upon the generated feasible production sequences and the cost-tolerance models. The optimization results are listed in Table 6.4. The suggested handbook production sequence, (19) "turning-turning-turning," with the recommended process tolerances is also listed at the bottom of the table for comparison. The calculation results indicate that (a) the optimal process tolerance assignment presents a 7.1 percent production cost reduction for the identical production sequence, rough turning, semifinish turning and finish turning and (b) the optimal production sequence with optimal process tolerances leads to a 47.6 percent production cost reduction over the manual planned approach. The advantages of the optimal production sequence identification and optimal process tolerance assignment are demonstrated.

6.5 Summary

A method for automatically generating the optimal production sequence with minimum production costs is presented. The method follows the quantitative intelligent system approach combining artificial intelligent techniques and numerical optimization to solve the optimal production sequence generation problem. Through the modeling of both qualitative and quantitative types of manufacturing knowledge, the approach automatically generates all feasible production sequences, forms the objective functions for cost minimization, and identifies the optimal production sequence.

The qualitative manufacturing knowledge modeled in this work includes the connections between a design feature and its production operations, the machine availability, and the successive and alternative production operations for a given feature. These knowledge models have been incorporated into a prototype knowledge-based system developed using the EXSYS expert system shell. This system automatically generates all feasible production sequences and creates the cost models for minimizing production costs of the generated production sequences.

The quantitative manufacturing knowledge modeling has been focused on the introduction of new mathematical models for empirical production cost-tolerance curves. These newly introduced mathematical models with significantly improved model accuracy are used as building blocks for the objective functions of production sequence cost minimization. The optimal process tolerances of each feasible production sequence are calculated in the minimization, and the optimal production sequence with the least minimum production costs is identified.

The method can considerably improve the generative CAPP system and reduce production costs for machining a part. The method is illustrated using a practical production planning example.

References

[1] Chang, T.C., and Wysk, R.A, (1981). "An Integrated CAD/Automated Process Planning System," *AIIE Transactions*, Vol. 13, No. 3, pp. 223-233.

[2] Chang, T.C., and Wysk, R.A, (1983). "CAD/Generative Process Planning with TIPPS," *Journal of Manufacturing Systems*, Vol. 2, No. 2, pp. 127-135.

[3] Chang, T.C., and Wysk, R.A, (1985). *An Introduction to Automated Process Planning System*, Prentice Hall Inc.

[4] Chase, K.W., and Greenwood, W.H., (1988). "Design Issues in Mechanical Tolerance Analysis," ASME *Manufacturing Review*, Vol. 1, No. 1, pp. 50-59.

[5] Dieter, G.E. (1983). *Engineering Design: A Materials and Processing Approach*, McGraw-Hill.

[6] Dong, Z., and Soom, A. (1990). "Automatic Optimal Tolerance Design For Related Dimension Chains," *Manufacturing Review*, Vol. 3, No. 4, pp. 262-271.

[7] Dong, Z., and Soom, A. (1991). "Some Applications of Artificial Intelligence Techniques to Automatic Tolerance Analysis and Synthesis," Chapter 5 of *Artificial Intelligence in Design*, D.T. Pham, ed., Springer-Verlag, pp. 101-124.

[8] Dong, Z., and Hu, W. (1991). "Optimal Process Sequence Identification and Optimal Process Tolerance Assignment in Computer-Aided Process Planning," *Computers in Industry*, Vol. 17, pp. 19-32.

[9] Dong, Z., and Hu, W. (1991). "Candidate Machining Sequence Generation for Optimal Process Planning Using A Knowledge-Based System," *Proceedings of IEEE Pacific Rim Conference on Communication, Computers and Signal Processing*, May 9-10, Victoria, Canada, pp. 569-572.

[10] Machinability Data Center. (1980). *Machining Data Handbook*, Metcut Research Associates Inc., Cincinnati, OH.

[11] Iwata, K., and Sugimura, N. (1984). "A Knowledge Based Computer-Aided Process Planning System for Machine Parts," *16th CIRP International Seminar on Manufacturing Systems*, Tokyo.

[12] Lee, W.J., and Woo, T.C. (1989). "Optimum Selection of Discrete Tolerances," *Journal of Mechanisms, Transmissions, and Automation in Design*, Transactions of ASME, Vol. 111, pp. 243-251.

[13] Liu, R.C., and Srinivasan, R. (1984). "Generative Process Planning Using Syntactic Pattern Recognition," *Computer in Mechanical Engineering*, (March), pp. 63-66.

[14] Michael, W., and Siddall, J. N. (1981). "Optimization Problem with Tolerance Assignment and Full Acceptance," *Journal of Mechanical Design*, Transactions of the ASME, Vol. 103, pp. 842-848.

[15] Speckhart, F.H. (1972). "Calculation of Tolerance Based on a Minimum Cost Approach," *Journal of Engineering for Industry*, Transactions of ASME, Vol. 94, No. 2, pp. 447-453.

[16] Spotts, M.F. (1973). "Allocation of Tolerances to Minimize Cost of Assembly", *ASME Journal of Engineering for Industry*, (August), pp. 762-764.

[17] Sutherland, G.H., and Roth, B. (1975). "Mechanism Design: Accounting for Manufacturing Tolerances and Costs in Function Generating Problems," *Journal of Engineering for Industry*, Transactions of ASME, Vol. 98, pp. 283-286.

[18] Trucks, H. E. (1976). *Designing for Economical Production*, H.B. Smith, (Ed.), Society of Manufacturing Engineers, Dearborn, MI.

[19] Wade, O.R. (1967). *Tolerance Control in Design and Manufacturing*, Industrial Press Inc.

[20] Wang, H.P., and Wysk, R. A. (1987). "Turbo-CAPP: A Knowledge-Based Computer-Aided Process Planning System," *19th CIRP International Seminar on Manufacturing Systems*, pp. 161-167.

[21] Wu, Z., Elmaraghy, W.H., and Elmaraghy, H.A. (1988). "Evaluation of Cost-Tolerance Algorithms for Design Tolerance Analysis and Synthesis," ASME *Manufacturing Review*, Vol. 1, No. 3, pp. 168-179.

Chapter 7

A Knowledge-based System for Building Scheduling Models

ANDREW KUSIAK and JUITE WANG
Intelligent Systems Laboratory
Department of Industrial Engineering, University of Iowa

7.1 Introduction

Operations research (OR) techniques have been widely used to support decision making in many industries, including manufacturing, petroleum, steel, forestry, communications, and banking. However, difficulties have been encountered with formulation of models, management of data, and interpretation of results. Typically, the OR software (for example, LINDO, MPSX) operates in the "stand alone" mode under an implicit assumption that a user knows enough about the problem and is able to formulate appropriate models and select the software required. Therefore, the use of operations research models has been limited to professionals who are knowledgeable in operations research. In practice, however, a user may not know how to build or select a model required for the problem considered. For example, a manager who determines a production or inventory schedule may not know what model is appropriate, or may not be aware of the best solution approaches (integer programming [IP], linear programming [LP], nonlinear programming, and so on). An intelligent system is needed to build a model for problem solving. The purpose in this chapter is to develop a system capable of selecting and building a candidate model for a problem considered by the user. The problem domain selected is production scheduling.

A knowledge-based approach [2, 15, 30] is suitable for handling ill-defined and ill-structured domains. This approach is easily accessible and explanations for the solutions are usually provided. However, it tends to be ineffective when problems become large and complex [32]. Moreover, in contrast to the operations research techniques, the solution obtained from the knowledge-based approach is only satisficing [13].

The operations research and knowledge-based approaches share the common goal of developing practical methods for solving intractable problems, but their emphases differ. Neither knowledge-based nor optimization approach in its pure form is suitable for solving these problems. It is possible to have a system that takes advantage of the strength of the two approaches and such a system is called a tandem system [18, 19]. The numerical part of a problem is efficiently handled by the algorithm, while the knowledge-based approach is used to deal with the qualitative aspect of the problem. Depending on the nature of the interaction between a knowledge-based system and an operations research tool, Kusiak [19] classified the tandem-based system into four types: data-modifying system, model-based system, model-building system, and algorithm-building system.

Similar research for combining the optimization approach with the knowledge-based system approach has also been developed for applications pertaining to decision support systems (DSSs). DSS is proposed as an interactive problem-solving vehicle through which models and analytical techniques are made available to the decision maker. The need for managing models requires an information system called a model-based management system (MMS), which supports all aspects of the modeling life cycle. Most research attempts to develop a modeling language to define the problem encountered. Dolk [6] developed a prototype model management system called GXMP based on the concept of model abstraction [5]. An algebraic modeling language was developed. Geoffrion [9] proposed a conceptual framework (structured modeling) for model building and management. A modeling language called SML has developed based on the framework of structural modeling [10]. Liang [20] developed a knowledge-based model management system that is capable of model integration and selection. The domain selected is the economic order quantity (EOQ) and demand forecasting models. Murphy and Stohr [23] developed a prototype system known as LPFORM to assist a knowledgeable user in formulating linear programs. A modeling language called "block language" has been developed for problem definition. Murphy [24] presented a systematic view of semantic relationships in a linear programming model. Binbasioglu and Jarke [3] proposed an approach to formulate models that combines the knowledge-based classification techniques with interactive construction of specialized submodels for production planning.

The purpose of this chapter is to develop a system capable of selecting or building an operations research model for a problem considered by the user. The tandem architecture developed is based on the blackboard architecture [16, 27] as well as an object-oriented approach [11, 29]. The concept of a "knowledge object" is introduced for model representation; that is, the knowledge for model building in terms of production rules is incorporated into the model representation scheme. A model represented as a "knowledge object" provides a hierarchical, modular, and understandable knowledge structure that increases the reusability and flexibility of the model-building process. The problem domain selected is production scheduling [1, 8, 12].

The chapter is organized as follows. Section 7.2 describes production rules and objects that represent the knowledge in the system. An object-oriented approach for model representation is also discussed. Section 7.3 presents the framework of the system which employs the blackboard architecture and object-oriented paradigm. Section 7.4 discusses three levels of the problem-solving methodology: model selection, data acquisition, and model instantiation. Section 7.5 describes the implementation of the system. Finally, concluding remarks are presented in Section 7.6.

7.2 Knowledge Representation

Much of the artificial intelligence (AI) research has focused on determining appropriate knowledge representation schemes to achieve high performance of knowledge-based systems. Mylopoulos and Levesque [25] classified most of the existing knowledge representation schemes into four categories: logic, network, procedural, and frame-based schemes. Each representation scheme has its advantages and drawbacks. No single scheme appears to be suitable for all applications.

The knowledge representation scheme used in this chapter combines a rule-based approach [15, 25, 30] and object-oriented programming [4, 7, 11, 29]. The hybrid representation scheme leads to a totally integrated system for knowledge representation and inferencing. Moreover, in a tandem system, it is important to have a model representation scheme that captures all features of a model and supports the model management [6]. The concept of "knowledge object" is introduced for this reason.

Section 7.2.1 presents the hybrid knowledge representation scheme used in the prototype system. A model representation scheme based on "knowledge object" is described in Section 7.2.2.

7.2.1 Hybrid Knowledge Representation

Production rules are a subset of predicate calculus with an added prescriptive component indicating how the information in the rules is to be used during reasoning. They have significant expressive power for a range of domain-dependent inference rules and behavior specifications. However, they are inadequate for defining terms and describing domain objects as well as hierarchical relationships among objects.

An object is an individually defined, self-contained component representing a specific item of knowledge in terms of data and procedures. Objects in a knowledge base interact by sending messages conforming to the predefined protocol. The receipt of a message by an object invokes its internal procedures to generate the information requested, which is sent to the originator of the message. Another feature of an object-oriented representation is the support of inheritance that enables descriptive information to be shared among multiple objects. The major advantage of object-oriented representation is that it provides a concise structural representation of useful relations, and supports a definition-by-specialization technique that is easy to use by domain experts.

By combining production rules and objects, one can take advantage of both schemes. In the tandem system developed, production rules are used to control reasoning (for example, the knowledge for model selection) and objects are employed as an intelligent database that maintains the knowledge required for reasoning (for example, production information) [7]. The daemons that automatically take some actions when a slot value changes (similar to forward-chaining reasoning), are used to maintain the integrity of knowledge and allow for dynamic information management [22]. Moreover, in the tandem system it is important to access the external software to analyze or execute a model generated. The message passing mechanism provides the capability to integrate external software with the system developed. For example, in the prototype system, LINDO is invoked by the message passing mechanism.

Although objects have internal data structures representing their states and methods that take actions, they do not have reasoning capability to determine which action should be taken or what kind of message can be passed. Thus, objects need knowledge to guide their behavior. In this chapter, a knowledge object, which is a conventional object with attached rules (see Figure 7.1), is used for model representation [28].

Using the concept of a "knowledge object," a mathematical model can be represented with the model object that includes data, procedures, and rules. The model representation based on an object-oriented concept enhances the flexibility and efficiency of model representation and management. The concept of the object-oriented model representation scheme is presented in the next section.

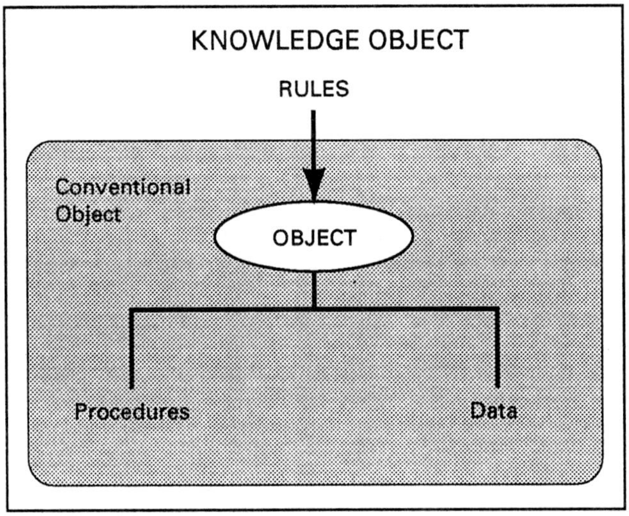

Figure 7.1: Knowledge Object versus Conventional Object

7.2.2 Object-oriented Model Representation

The object-oriented model representation developed in this chapter is based on the concept of a "knowledge object," which is an extension of the conventional object (see Figure 7.2). The "knowledge object" not only includes the state and behavior, which are described by the data and procedure sections respectively, but also possesses knowledge to guide its behavior depending on the state of the object.

The data section enumerates the data used in the model. The procedure section lists procedures (numerical programs) required for the model formulation. For example, the procedure "Solve(Lindo)" (see the appendix) is an interface that creates a process for calling the software "LINDO." The rule section includes the knowledge to formulate a model. Backward reasoning is employed for model building.

<u>**KNOWLEDGE OBJECT**</u>

DATA SECTION
data 1 # comment
... ...
data i

PROCEDURE SECTION
procedure 1
... ...
procedure j

RULE SECTION
rule 1
... ...
rule k

Figure 7.2: Structure of a Knowledge Object

A large model may consists of several simple models as building blocks, such as a product-mix model, transportation model, or blending model. Similarly, a model may be built in terms of several types of constraints, such as capacity constraint, requirement constraint, or precedence constraint.

In order to increase the reusability of a model, the well-known models as well as criteria and constraints are defined as template objects and used as building blocks to formulate a model. A template model object or template criterion or constraint object is a knowledge object which possesses the knowledge required for building a model, criterion, or constraint.

A specific model object is built in terms of the template model or criterion and constraint objects. To utilize the template objects, a metamodel object is defined. A metamodel object is a model of a model (Figure ??). Each model is defined as a metamodel object, which retrieves, stores, and reasons in order to construct the required model object. Depending upon the requirement, the metamodel object retrieves the needed template objects from the model base and attaches them as the subobjects of the model object.

A metamodel object possesses the necessary knowledge for building several formulations of a model. For example, a mixed integer linear programming model (see the appendix) for production scheduling may have several possible formulations depending on the type of scheduling criterion or constraints, for example, due date or precedence constraints. The

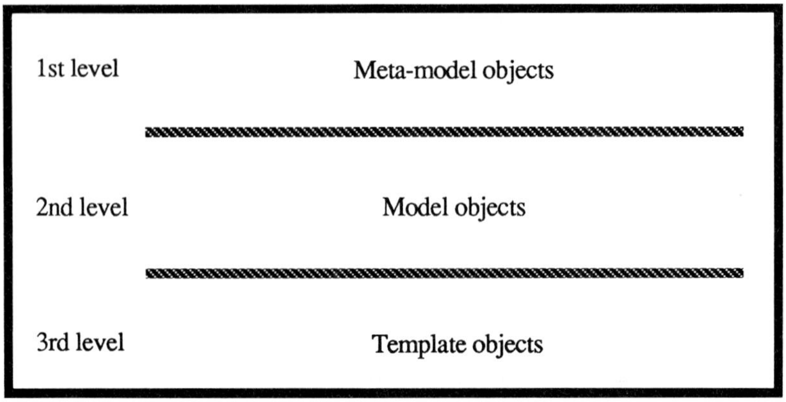

Figure 7.3: Three Levels of Objects for Model-Building

rules defined in the rule section of a metamodel object are called metarules. The metarules contain the knowledge for acquiring the data, knowledge for constructing a model object, and knowledge for the model formulation process.

The model representation based on "knowledge object" provides the hierarchical, modular, and understandable knowledge structure. The advantages are shown as follows:

- Increased reusability of a model in the model base.
- Easy maintainance and organization of the model base.
- Increased flexibility of the model-building process; that is, the model object can be constructed incrementally.

7.3 The Object-oriented Blackboard Architecture

The framework of the system (see Figure 7.4) developed is based on the blackboard architecture and an object-oriented approach [17, 27, 29, 31]. The object-oriented blackboard architecture organizes the knowledge in a modular way, which simplifies model management in the tandem system. The object-oriented feature, message passing, offers a capability for combining numerical programs with the knowledge-based approach, that is, integration of symbolic and numerical computing. Furthermore, the entire system environment that is based on the object-oriented blackboard architecture is incremental and additive. In other words, a new model is easily incorporated into the model base in terms of a model object without changing the rest of the system.

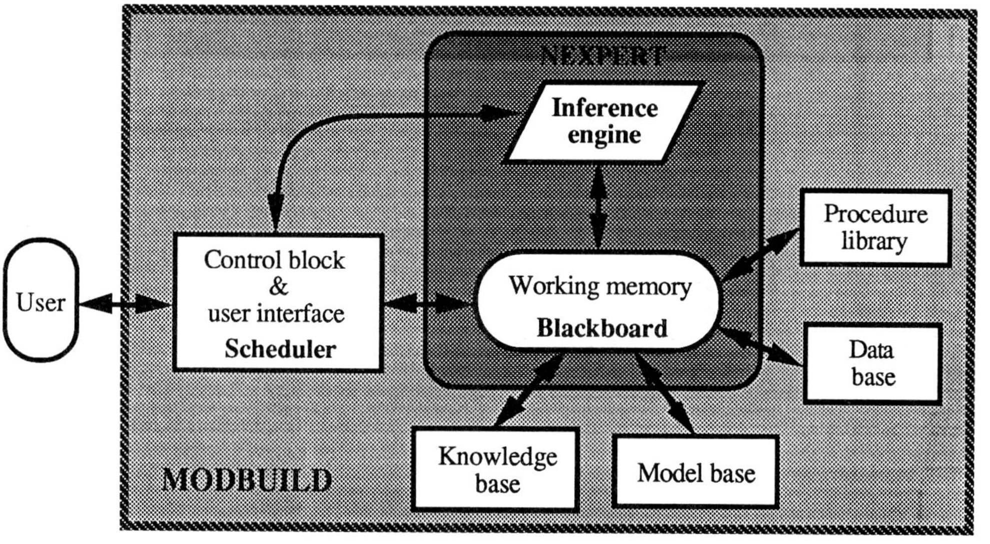

Figure 7.4: Framework of the System

The system consists of seven components: a scheduler, blackboard, inference engine, model base, knowledge base, database, and procedure library. The scheduler consists of a user interface and a control block. The control block monitors the changes on the blackboard and decides what actions to take next. Various kinds of information are made global and are available to the control block. As a knowledge object (knowledge source) is activated, it transfers control to the knowledge object for problem solving.

The purpose of the blackboard is to store the information produced during the model-building process. The blackboard records intermediate hypotheses and inferences made from the knowledge object. Each knowledge object can store and retrieve current information from the blackboard. In other words, the blackboard in the system is considered a global database (working memory) during the model-building process.

The model base contains the required information for building models. The metamodel objects and template objects are stored in the model base. They are retrieved from the model base when required during the model-building process.

The knowledge base stores the knowledge for model selection and domain-specific knowledge. The procedure library stores numerical programs that are called during the model-building process. The database stores the data required for building models. It is desirable that models and data be independent so that changes to one do not affect the other.

7.4 Knowledge-based Modeling

The process of model building is divided into three levels: model selection, data acquisition, and model instantiation. The model selection function classifies the problem by identifying an appropriate specification queried from the user and searches for an existing model (algorithm) in the model base. After a model has been selected, the required information for building the model selected is retrieved from the existing database or from the user. Then the model object can be built by reasoning the metamodel object corresponding to the model selected.

The following sections explains each of these levels in detail. The concepts are illustrated with a scheduling problem.

7.4.1 Knowledge-based Model Selection

The process of model selection is performed in two phases: problem recognition and problem mapping. The problem recognition phase is mostly performed in the user environment. To solve the problem encountered effectively, the user should specify important characteristics of the problem. In the problem mapping phase, the knowledge-based system selects an appropriate model according to the problem specification obtained from the user. Currently, the system does not have the capability for model modification when a suitable model cannot be found.

Due to the diversity of scheduling problems, it is difficult to build an abstract problem that adequately matches all the relevant features in the scheduling domain. In this chapter, a classification scheme proposed by Graves [12] is adapted to the function of model selection, where the scheduling problems are considered in four dimensions: shop type, production environment, processing complexity, and scheduling criteria. The production requirement is usually generated by either customers (open shop) or inventory replenishment decisions (closed shop). The production environment is basically divided into two types: static and dynamic. The processing complexity of the scheduling problem is determined primarily by the number of operations associated with each part (job). Scheduling criteria are expressed in terms of job-related variables, for example, makespan, flow-time, lateness, and tardiness. In addition, more detailed specifications are included, such as number of machines, due date, and operation preemption. Figure 7.5 shows the classification scheme of the scheduling problem used in this chapter. Having developed a classification scheme, the user can describe an abstract problem based on the information related to the existing problem. Production rules are used to select an appropriate model according to the abstract problem described by the user (see Figure 7.6). Considering rule 1 in Figure 7.6, if the left-hand side is matched with the abstract problem, then the shortest-processing-time (SPT) rule is used for solving the problem encountered.

7.4.2 Data Acquisition

After a candidate model has been selected, the system attempts to get the data required to formulate the model. There are two ways to obtain the data. One is to directly query

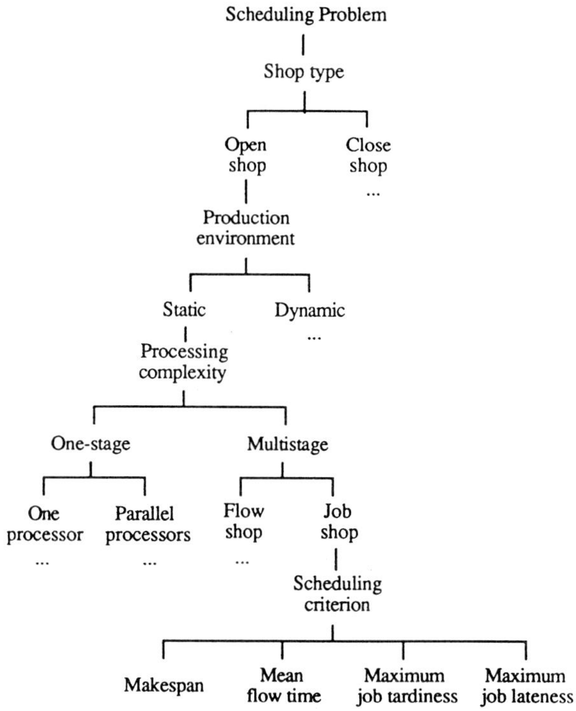

Figure 7.5: Classification Scheme for Scheduling Problems

it from the user and the other is to retrieve it from the existing database, which may be generated by another application. Storing the data required in the external database has several benefits

- The size of the knowledge base is reasonable because only the rules, objects, classes, and procedures are stored. The data (facts) are retrieved from the databases dynamically when needed.
- A large database can be managed by a database management system.
- The data can be accessed or produced by other applications.

After the data required has been retrieved from the existing database, they are converted into objects which inherit properties from their superclasses, for example, classes "JOB," "OPERATION," defined in the system.

7.4.3 Knowledge-based Model Instantiation

Traditionally the knowledge about models and model structure was embedded within a "black box" and was accessible only to programmers. By contrast, the knowledge-based system for model formulation attempts to explicitly code the modeling knowledge in a

RULE 1:	**IF**	job type is open
	AND	job environment is static
	AND	processing complexity is one-stage one-processor
	AND	there is no due-date requirement
	AND	scheduling criterion is to minimize the mean flow time
	THEN	SPT rule is selected
RULE 2:	**IF**	job type is open
	AND	job environment is static
	AND	processing complexity is one-stage one-processor
	AND	there is a due-date requirement
	AND	scheduling criterion is to minimize the maximum job lateness
	THEN	EDD rule is selected
RULE 3:	**IF**	job type is open
	AND	job environment is static
	AND	processing complexity is multistage flow-shop
	AND	number of machine is 2
	AND	scheduling criterion is to minimize the makespan
	THEN	Johnson's algorithm is selected
RULE 4:	**IF**	job type is open
	AND	job environment is static
	AND	processing complexity is multistage flow-shop
	AND	number of machines is greater than 2
	AND	scheduling criterion is to minimize the makespan
	THEN	a branch-and-bound algorithm (Gupta 1971) is selected
RULE 5:	**IF**	job type is open
	AND	job environment is static
	AND	processing complexity is multistage job-shop
	AND	number of machines is greater than 2
	AND	there is a due-date requirement
	AND	scheduling criterion is to minimize the mean flow time
	THEN	a mix integer programming model is selected
	...	... ...

Figure 7.6: Example of Production Rules for Model Selection

knowledge base. Thus, the model can be formulated automatically by reasoning with rules. The result of the model instantiation phase can be turned over to any of the commercially available software, for example, LINDO or MPSX.

The model instantiation process is divided into four steps:

Step 1: Construct a model object according to the problem specification.

Step 2: Combine the sets of rules under the hierarchical structure of a model object into the blackboard.

Step 3: Build the model by reasoning with rules in the blackboard.

Step 4: Invoke the required software (for example, LINDO) for solving the model.

A model object is constructed by reasoning the candidate metamodel object, which has been selected according to the problem specification defined by the user. The required objective function object and constraint objects, corresponding to the problem specification, are attached to the model object. The attached objects are called subobjects.

The metamodel object contains the metarules used for the model instantiation process. The template objects include the rules for formulating objective functions, constraints, or simple models. After the model object, which includes all the required model components, is constructed, the next step is to load the metarules stored in the model object and the corresponding rule sets stored in the subobjects into the working memory (blackboard). The above might be viewed as if the knowledge to formulate a specific model were distributed among several knowledge bases. By combining and reasoning the distributed knowledge bases, a required model is built. Furthermore, the required software is invoked to solve the model created. The existing software is considered an object in the system. Thus, an existing software is invoked by sending appropriate messages.

Combining the knowledge-based approach and object-oriented concept for model instantiation has the following benefits:

- It reduces the size of rules and objects in the working memory, because the rules and objects are loaded as required.
- It increases the flexibility of the model-building process. It allows the system to create a complicated model by integration of the existing models in the model base.
- The environment for model building is incremental and additive. New models or software are incorporated into the system without changing the remainder of the system.
- It provides the capability for integration of symbolic and numerical computing and increases the flexibility of linking with the existing software, for example, LINDO.

7.4.4 Example of the Model-building Process

In this section, an example is provided to illustrate some of the concepts used in the process of model-building. Assume that three parts with the eight operations shown in Figure 7.7 must be scheduled on three machines. The relevant data for each part are shown in Table 7.1.

Table 7.1: Relevant Data for Each Job

Part number	1		2			3		
Operation number	1	2	3	4	5	6	7	8
Machine number	1	2	3	1	2	2	1	3
Processing time	8	4	3	5	6	9	2	7

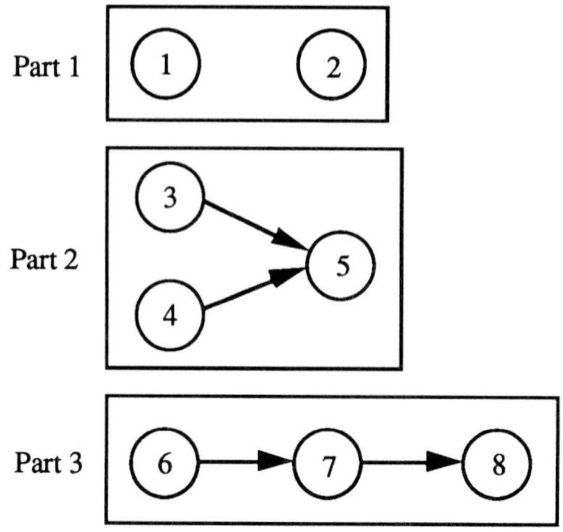

Figure 7.7: Parts with Operations and Precedence Constraints

First, the system interacts with the user to obtain the important characteristics for the scheduling problem at hand. An example of the model selection process is shown in Figure 7.8. As a result, the mixed integer programming model (see the appendix) is built for the scheduling problem. After the model has been selected, the required data have to be retrieved from the existing database or from the user. The data for each job in Table 7.1 are converted into objects based on the classes stored in the system and loaded into the blackboard.

In the model instantiation phase, the model object is constructed by reasoning the metarules in the metamodel object. The representation of the semantic net of the selected model "MixedIntegerLP" is shown in Figure 7.9. Only the required objective function and constraints are instantiated. The mechanism based on the "knowledge object" relaxes the inheritance property of the object-oriented programming. Figure 7.10 shows the hierarchical structure of the instantiated model object "MixedIntegerLP," as the job type is open shop, scheduling environment is static, due date is not considered, and the criterion is to maximize the makespan.

After the "MixedIntegerLP" model object is constructed, the rule sets corresponding to each subobject are loaded into the blackboard. The set of rules for building the mixed integer programming model is listed in the appendix. The system employs backward chaining as the inference strategy during the model instantiation process. The goal is to build the required model.

The inference process for model instantiation is to select a correct pattern, including indexes and parameters, to instantiate the model to be built. A pattern is a conditional description which can be created only by a rule. For example, the rule for formulation of the precedence constraint is shown below:

What is the type of a shop (open_shop / close_shop)?
: open_shop

What is the scheduling environment (static / dynamic)?
: Static

Is the problem single stage or multistage (single_stage / multistage)?
: multistage

Is the problem job shop or flow shop type (job_shop / flow_shop)?
: job_shop

How many machines do you need to perform the job?
: 3

Does the problem have precedence relationship between any two operations (true / false)?
: true

Does the problem have due-date constraint for each job (true / false)?
: false

Which scheduling criteria do you want to suggest (min_mean_flowtime / min_makespan_time / min_max_mean_tardiness / min_max_mean_lateness / min_weighted_mean_flowtime)?
: min_makespan_time

***** Mixed integer programming model is suggested for this problem *****

Figure 7.8: An Example Illustrating the Model Selection Process

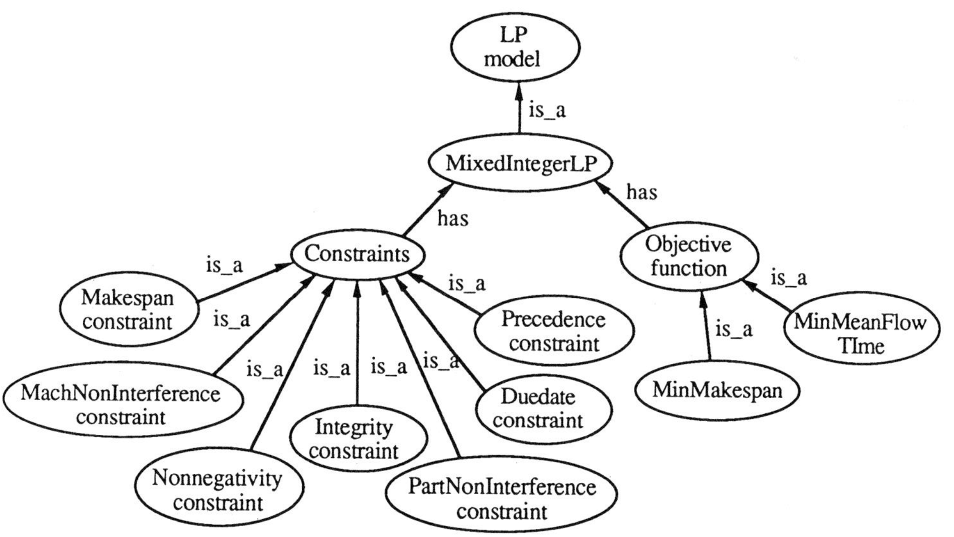

Figure 7.9: Representation of the Semantic Net of the Scheduling Model "MixedIntegerLP"

IF part i exists

AND operation k precedes operation l of part i

THEN precedence_constraint_is_built

AND EXECUTE "Build_constraint (PrecedenceConstraint, IndexSet, Parameter)"

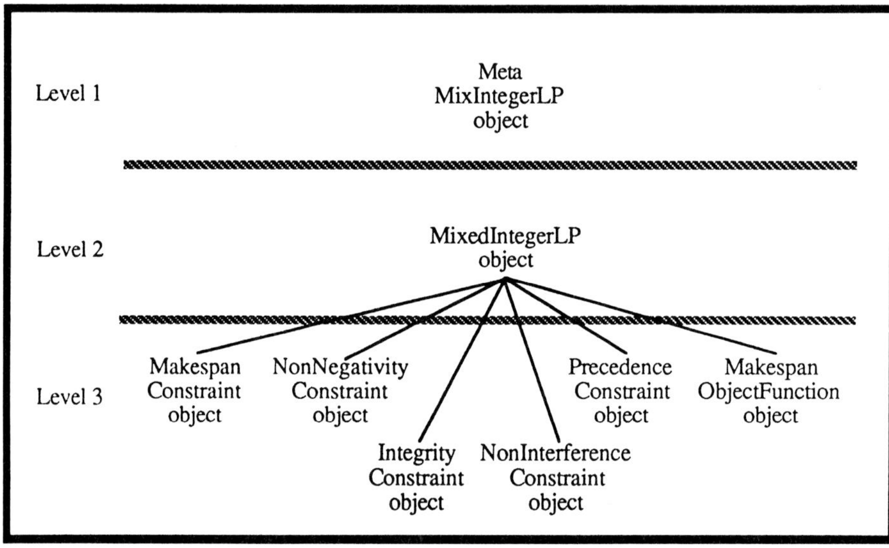

Figure 7.10: Hierarchical Structure of the Metamodel Object, Model Object, and Template Objects

If there are some operations in the blackboard with precedence relationships, then this rule is fired, that is, the hypothesis "precedence_constraint_is_built" is true and procedure "Build_constraint (PrecedenceConstraint, IndexSet, Parameter)" is executed. The satisfied operations resulting from reasoning with this rule are stored in the list "IndexSet." The satisfied patterns of each inference are shown below:

IndexSet for first inference $\rightarrow$ $< operation3, operation5 >$
Parameter for first inference $\rightarrow$ $< 6 >$
IndexSet for second inference $\rightarrow$ $< operation4, operation5 >$
Parameter for first inference $\rightarrow$ $< 6 >$
IndexSet for third inference $\rightarrow$ $< operation6, operation7 >$
Parameter for first inference $\rightarrow$ $< 2 >$
IndexSet for fourth inference $\rightarrow$ $< operation7, operation8 >$
Parameter for first inference $\rightarrow$ $< 7 >$

To instantiate a constraint, the <IndexSet> and algebraic equation of a precedence con-

```
MIN T
SUBJECT TO
x24-x26>=6
x24-x25>=6
x12-x13>=2
x11-x12>=7
x38>=8
x37>=4
x26>=3
x25>=5
x24>=0
x13>=9
x12>=0
x11>=0
x38-x25-999z85>=-991
x25-x38+999z85>=5
x38-x12-999z82>=-991
x12-x38+999z82>=2
x37-x24-999z74>=-995
x24-x37+999z74>=6
x37-x13-999z73>=-995
x13-x37+999z73>=9
x26-x11-999z61>=-996
x11-x26+999z61>=7
x25-x12-999z52>=-994
x12-x25+999z52>=2
x24-x13-999z43>=-993
x13-x24+999z43>=9
x38-T<=0
x37-T<=0
x24-T<=0
x11-T<=0
x38-x37-999y87>=-991
x37-x38+999y87>=4
END
INTEGER y87
INTEGER z85
INTEGER z82
INTEGER z74
INTEGER z73
INTEGER z61
INTEGER z52
INTEGER z43
```

Figure 7.11: The Instantiated Mixed Integer Linear Programming Model

straint are passed to the procedure "Build_constraint," which converts these objects into decision variables with indices and searches the required parameters corresponding to the indices. The instantiated precedence constraints are as follows:

$x25 - x23 >= 6$
$x24 - x23 >= 6$
$x37 - x36 >= 2$
$x38 - x37 >= 7$

where

x23: completion time of operation 3 of part 2
x24: completion time of operation 4 of part 2
x25: completion time of operation 5 of part 2
x36: completion time of operation 6 of part 3
x37: completion time of operation 7 of part 3
x38: completion time of operation 8 of part 3

Similarly, the objective function and the other constraints are instantiated and listed in Figure 7.11.

7.5 System Implementation

The system for model building involves the following modules:

- main program, algorithm, and procedures used during the inference written in C
- Nexpert (expert system shell, [Nexpert 1987])
- LINDO (optimization software)

The main program calls Nexpert which builds a model and generates the data required. The linear programming model is solved by LINDO. The system was implemented on an Apollo 3500 workstation in an X-window environment. The prototype system has 63 rules and the model base currently includes the SPT rule, the EDD rule, the Johnson algorithm, a branch-and-bound algorithm, and a mixed integer linear programming model [1, 19]. Two sample windows are shown in Figure 7.12.

7.6 Conclusion

The research presented in this chapter was concerned with the development of a tandem system that combines an optimization and knowledge-based system approach for problem solving. A domain-specific approach to model selection and model formulation was proposed which uses the structural knowledge for building a mathematical model along with the application-specific knowledge about the domain. A model representation scheme based on the "knowledge object" is developed. The hybrid knowledge representation scheme leads to a totally integrated system for model representation and inferencing. Moreover, the framework of the system, based on the object-oriented approach and blackboard model, not only integrates the qualitative and quantitative aspects of problem solving but also enhances the model management capability .

The approach presented in this chapter has some limitations:

- The prototype system is limited on the open-shop and static production environment.
- The model validation and interpretation are not included in the prototype system.
- The current prototype system does not have the capability for model modification, if the model required cannot be built.

There are a number of issues for future development. The most important of these are listed below.

- Build an application front-end. Since this study is not a consultation system in a real environment that includes a variety of models or algorithms for problem solving, an application front-end can be built to help a decision maker select an appropriate model or algorithm within a specific domain, such as production planning or scheduling.

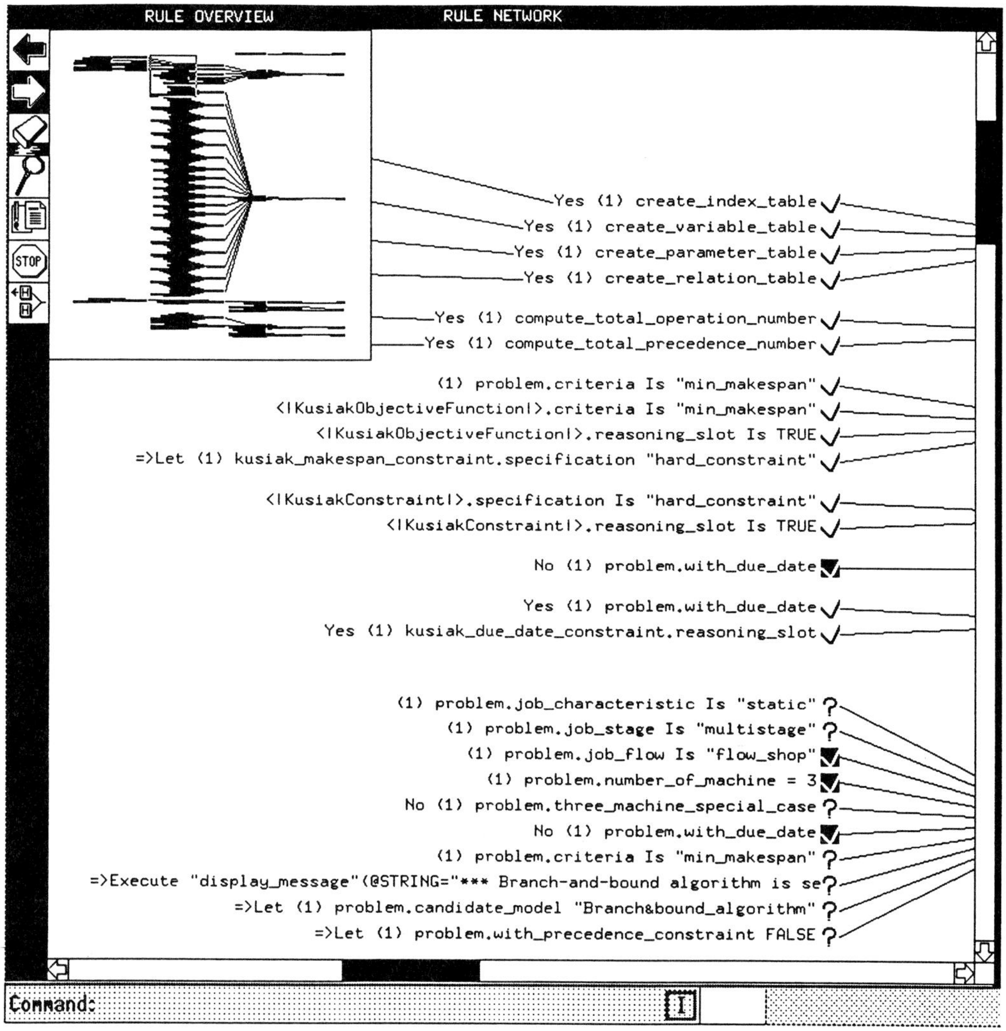

Figure 7.12: Portion of the Rule Base

- Develop a graphic, user-friendly interface. A hierarchical structure of the model object can be represented as a network of icons. Since a model or constraint is represented as a "knowledge object," adding a constraint can be considered as attaching a constraint object to the model object. Similarly, removing a constraint is like moving a constraint object into a trash box.

- Develop a set of rules that can help the decision-maker to integrate a variety of models for problem solving. The causal relationship between the models should be captured.

- Incorporate explanation capability into the reasoning process so that the decision maker will know how the model is built.

The study offers a computerized modeling environment for a novice user. However, much work remains to be done on incorporating more intelligence into other aspects of the problem-solving strategies, for example, model validation and aiding the analysis of the results generated by models. Research is also needed to handle the uncertainty involved in the modeling process.

References

[1] Baker, K. (1974). *Introduction to Sequencing and Scheduling*, Wiley, New York.

[2] Barr, A., and Feigenbaum, E. A. (1981). *The Handbook of Artificial Intelligence*, William Kaufman, Los Altos, CA.

[3] Binbasioglu, M., and Jarke, M. (1987). "Knowledge-Based Formulation of Linear Planning Models," in *Expert Systems and Artificial Intelligence in Decision Support Systems*, Sol, H. G., Takkenberg, C. A. T. and De Vries Robbe, P. F. eds., pp. 113-136.

[4] Coad, P., and Yourdon, E. (1990). *Object-Oriented Analysis*, Prentice-Hall, Englewood Cliffs, NJ.

[5] Dolk, D. R., and Konsynski, B. R. (1984). "Knowledge Representation for Model Management Systems," *IEEE Transactions on Software Engineering*, Vol. SE-10, No. 6, pp. 619-628.

[6] Dolk, D. R. (1986). "A Generalized Model Management System for Mathematical Programming," *ACM Transactions on Mathematical Software*, Vol. 12, No. 2, pp. 92-126.

[7] Fike, R., and Kehler, T. (1985). "The Role of Frame-Based Representation in Reasoning," *Communications of the ACM*, Vol. 28, No. 9, pp. 904-920.

[8] French, S. (1982). *Sequencing and Scheduling: An Introduction to the Mathematics of the Job-Shop*, John Wiley, New York.

[9] Geoffrion, A. M. (1987). "Introduction to Structured Modeling," *Management Science*, Vol. 33, No. 5, pp. 547-588.

[10] Geoffrion, A. M. (1991). "FW/SM: A Prototype Structured Modeling Environment," *Management Science*, Vol. 37, No. 12, pp. 1513-1538.

[11] Goldberg, A., and Robson, D. (1983). *SmallTalk 80: The Language and Its Implementation*, Addison-Wesley, Reading, MA.

[12] Graves, S. C. (1981). "A Review of Production Scheduling," *Operations Research*, Vol. 29, No. 4, pp. 646-675.

[13] Grant, T. J. (1986). "Lessons for O.R. from A.I.: A Scheduling Case Study," *Journal of the Operational Research Society*, Vol. 37, No. 1, pp. 41-57.

[14] Gupta, J. N. D. (1971). "A Branch-and-Bound Solution to the General Scheduling Problem," *Operations Research*, Vol. 16, pp. 353-361.

[15] Hayes-Roth, B., Waterman, D. A. and Lenat, D. eds. (1983). *Building Expert Systems*, Addison-Wesley, Reading, MA.

[16] Hayes-Roth, B. (1985). "A Blackboard Architecture for Control," *Artificial Intelligence*, Vol. 26, pp. 251-321.

[17] Kitzmiller, C. T., and Kowalik, J. S. (1986). "Symbolic and Numerical Computing in Knowledge-Based Systems," in *Coupling Symbolic and Numerical Computing in Expert Systems*, Kowalik, J. S., ed., Elsevier Science Publishers, North-Holland, Amsterdam, pp. 3-17.

[18] Kusiak, A. (1987). "Artificial Intelligence and Operations Research in Flexible Manufacturing Systems," *Information Systems and Operational Research*, Vol. 25, No. 1, pp. 2-12.

[19] Kusiak, A. (1990). *Intelligent Manufacturing Systems*, Prentice-Hall, Englewood Cliffs, NJ.

[20] Liang, Ting-Peng (1988). "Development of a Knowledge-Based Model Management System," *Operations Research*, Vol. 36, No. 6, pp. 849-863.

[21] Manne, A. S. (1960). "On the Job-Shop Scheduling Problem," *Operations Research*, Vol. 8, No. 1, pp. 218-223.

[22] Murata, M. and Kusumoto, K. (1989). "Daemon: Another Way of Invoking Methods," *Journal of Object-Oriented Programming*, Vol. 8, No. 6, pp. 8-12.

[23] Murphy, F. H., and Stohr, E. A. (1986). "An Intelligent System for Formulating Linear Programs," *Decision Support Systems*, Vol. 2, No. 1, pp. 39-47.

[24] Murphy, F. H. (1988). "A Knowledge-Base for Formulating Linear Programs," In *Mathematical Models for Decision Support*, Mitra, G., ed., NATO ASI Series, Vol. F48, Springer-Verlag, Berlin, pp. 451-470.

[25] Mylopoulos, J., and Levesque, H. J. (1984). "An Overview of Knowledge Representation," in *On Conceptual Modeling*, Brodie, M. L., Mylopoulos, J. and Schmidt, J. W., eds., Springer-Verlag, New York, pp. 3-17.

[26] Nexpert, (1987). *Nexpert Object Fundamental*, User Manual, Neuron Data Inc., CA.

[27] Nii, H. P. (1986). "Blackboard System: The Blackboard Model of Problem Solving and the Evolution of Blackboard Architectures," *AI Magazine*, Vol. 7, No. 2, pp. 38-53.

[28] Rozenshtein, D. and Minsky, N. H. (1989). "A Law-Governed Object-Oriented System," *Journal of Object-Oriented Programming*, Vol. 1, No. 6, pp. 14-28.

[29] Stefik, M. and Bobrow, D. G. (1986). "Object-Oriented Programming: Themes and Variations," *The AI Magazine*, Vol. 6, No. 4, pp. 40-62.

[30] Waterman, D. A., ed. (1986). *A Guide to Expert Systems*, Addison-Wesley, Reading, MA.

[31] Winkelbauer, L. (1988). "A Structure for the Incremental Construction of Coupled Systems: A Bridge Between Shallow and Deep Coupling," in *Coupling Symbolic and Numerical Computing II*, Kowalik, J. S., ed., Elsevier Science Publisher, North-Holland, Amsterdam, pp. 145-156.

[32] Yu, D. H. (1987). "A System for Utilizing Operations Research Techniques in Knowledge-Based Planning Systems," *Technique Report M87-1*, The Mitre Corporation, Bedford, MA.

Appendix: The Model Object of a Mixed Integer Program

The appendix shows an integer programming model for the job shop scheduling problem. The kernel of this model was formulated by Manne [21]. Baker [1] modified it by introducing an indicator variable and noninterference constraint. Kusiak [19] adopted the model to a manufacturing environment by incorporating due date and multiple resource constraints. To present the model, the following notation is introduced:

n number of parts

m number of machines

R_i set of pairs of operations $[k, l]$ for part p_i, where k precedes l, $i = 1, \cdots, n$

Q_i set of pairs of operations $[k, l]$ for part p_i, where k and l can be performed in any order, $i = 1, \cdots, n$

I_i set of operations without precedence constraints, $i = 1, \cdots, n$

N_p set of operations to be performed on machine p, $p = 1, \cdots, m$

n_i number of operations in part p_i, $i = 1, \cdots, n$

t_{ik} processing time of operation k of part p_i, $k = 1, \cdots, n_i$, $i = 1, \cdots, n$

M an arbitrary large positive number

x_{ik} completion time of operation k of part p_i, $k = 1, \cdots, n_i$, $i = 1, \cdots, n$

$x_{(j)}$ completion time of the last operation of part p_i, $i = 1, \cdots, n$

$d_{(j)}$ due date of part p_i, $i = 1, \cdots, n$

y_{kl} 1, if operation k precedes operation l,
0, otherwise $k = 1, \cdots, n_i$, $l = 1, \cdots, n_i$, $i = 1, \cdots, n$

z_{kl} 1, if operation k precedes operation l,
0, otherwise $k = 1, \cdots, n_i$, $l = 1, \cdots, n_i$, $i = 1, \cdots, n$

T the maximum makespan of all of parts

The model is formulated as follows:

$$MIN \qquad Z = T \tag{A7.1}$$

subject to:

$$x_{il} - x_{ik} \geq t_{il}$$
for all $[k, l]$ belong to R_i, for all i (A7.2)

$$x_{il} - x_{ik} + M(1 - y_{kl}) \geq t_{il}$$
$$x_{ik} - x_{il} + My_{kl} \geq t_{ik}$$
for all $[k, l]$ belong to Q_i, for all i (A7.3)

$$x_{jl} - x_{ik} + M(1 - z_{kl}) \geq t_{jl}$$
$$x_{ik} - x_{jl} + Mz_{kl} \geq t_{ik}$$
for all $[k, l]$ belong to N_p, for all p, $i \neq p$ (A7.4)

$$x_{(j)} \leq d_{(j)}$$
for all i (A7.5)

$$T \geq x_{(j)}$$
for all i (A7.6)

$$x_{ik} \geq t_{ik}$$
for i, k belong to I_i (A7.7)

$$x_{ik} \geq 0$$
for all other i, k (A7.8)

$$y_{kl} = 0, 1$$
for all k, l

$$z_{kl} = 0, 1$$
for all k, l (A7.9)

Constraint (A7.2) imposes that the operations of each part are processed according to the precedences required. Constraint (A7.3) ensures that any two operations belonging to the same part cannot be processed at the same time. Analogously, constraint (A7.4) ensures that any resource cannot process more than one part at the same time. Constraint (A7.5) imposes due dates. Constraint (A7.6) imposes that completion time for each part cannot exceed the maximal makespan time. Constraints (A7.7), (A7.8) and (A7.9) impose nonnegativity and integrality, respectively.

Another formulation is to minimize the total completion time of all parts. The formulation is similar to the above, except that the constraint (A7.6) is removed and the objective function is replaced by the following one:

$$MIN \quad \sum_{i=1}^{n} x_{[i]} \tag{A7.10}$$

The model class "SchedulingModel" is shown as follows:

MODEL OBJECT SchedulingModel
DATA SECTION
ObjectiveFunction1(objective function (A7.1))

ObjectiveFunction2(objective function (A7.10))
Constraints
    HardConstraints
        PrecedenceConstraint(constraint (A7.2))
        PartNonInterferenceConstraint(constraint (A7.3))
        MachineNonInterferenceConstraint(constraint (A7.4))
        MakespanConstraint(constraint (A7.5))
        NonNegativityConstraint1(constraint (A7.7))
        NonNegativityConstraint2(constraint (A7.8))
        IntegrityConstraint(constraint (A7.9))
    SoftConstraints
        DueDateConstraint(constraint (A7.6))
IndexSet(i: job)
IndexSet(j: job)
IndexSet(k: operation)
IndexSet(l: operation)
DecisionVar(x)
    Link(job, operation)
DecisionVar(T)
DecisionVar(y)
    Link(operation, operation)
    Type(integer)
DecisionVar(z)
    Link(operation, operation)
    Type(integer)
Parameter(t)
    Link(job, operation)
Parameter(d)
    Link(job)

PROCEDURE SECTION

Solve(Lindo)
Create_index_table(IndexSet)
Create_parameter_table(Parameter)
Create_variable_table(DecisionVar)
Create_relation_table(Link)
Add_objective_function(LinearEquation)
Delete_objective_function(LinearEquation)
Add_constraint(LinearEquation)
Delete_constraint(LinearEquation)
Build_objective_function(ObjectiveFunction, IndexSet, Parameter)
Build_constraint(PrecedenceConstraint, IndexSet, Parameter)

Build_constraint(NonInterferenceConstraint, IndexSet, Parameter)
Build_constraint(MakespanConstraint, IndexSet, Parameter)
Build_constraint(NonNegativityConstraint1, IndexSet, Parameter)
Build_constraint(NonNegativityConstraint2, IndexSet, Parameter)
Build_constraint(IntegrityConstraint, IndexSet, Parameter)
Build_constraint(DueDateConstraint, IndexSet, Parameter)

RULE SECTION

```
IF      model_is_built = "TRUE"
THEN    solve_by_Lindo
IF      relevant_tables_are_built = "TRUE"
AND     objective_function_is_built = "TRUE"
AND     all_constraints_are_built = "TRUE"
THEN    model_is_built
IF      create_index_table = "TRUE"
AND     create_parameter_table = "TRUE"
AND     create_variable_table = "TRUE"
AND     create_relation_table = "TRUE"
THEN    relevant_tables_are_built
IF      build_objective_function = "TRUE"
THEN    objective_function_is_built
IF      hard_constraints_are_built = "TRUE"
AND     soft_constraints_are_built = "TRUE"
THEN    all_constraints_are_built
IF      precedence_constraint_is_built = "TRUE"
AND     part_noninterference_constraint_is_built = "TRUE"
AND     machine_noninterference_constraint_is_built = "TRUE"
AND     makespan_constraint_is_built = "TRUE"
AND     nonnegativity_constraint_is_built = "TRUE"
AND     integrity_constraint_is_built = "TRUE"
THEN    hard_constraints_are_built
IF      due_date_constraint_is_built = "TRUE"
THEN    soft_constraints_are_built
IF      part i exists
AND     operation k precedes operation l for part i
THEN    precedence_constraint_is_built
AND     EXECUTE "Build_constraint(PrecedenceConstraint, IndexSet,
        Parameter)"
IF      part i exists
AND     operation l and operation k belong to part i
AND     operation l and operation k do not have precedence relationship
AND     operation k is not equal to operation l
```

THEN	part_noninterference_constraint_is_built
AND	EXECUTE "Build_constraint(NonInterferenceConstraint,IndexSet, Parameter)"
IF	operation k of part i is processed on machine m_1
AND	operation l of part j is processed on machine m_2
AND	machine $m1$ is equal to machine $m2$
AND	operation k of part i is not identical to operation l of part j
THEN	machine_noninterference_constraint_is_built
AND	EXECUTE "Build_constraint(NonInferenceConstraint, IndexSet, Parameter)"
IF	operation k of part i exists
AND	operation k does not belong to the parent operation in the precedence relationship
OR	operation k does not have precedence relationship
THEN	makespan_constraint_is_built
AND	EXECUTE "Build_constraint(MakespanConstraint, IndexSet,Parameter)"
IF	nonnegativity_constraint1_is_built
AND	nonnegativity_constraint2_is_built
THEN	nonnegativity_constraint_is_built
IF	operation k of part i exists
AND	operation k does not have precedence relationship
THEN	nonnegativity_constraint1_is_built
AND	EXECUTE "Build_constraint(NonNegativityConstraint1,IndexSet, Parameter)"
IF	operation k of part i exists
AND	operation k has precedence relationship
THEN	nonnegativity_constraint2_is_built
AND	EXECUTE "Build_constraint(NonNegativityConstraint2,IndexSet, Parameter)"
IF	type of DecisionVar is integer
THEN	integrity_constraint_is_built
AND	EXECUTE "Build_constraint(IntegrityConstraint, IndexSet,Parameter)"
IF	the problem has due date constraint
AND	operation k of part i exists
AND	operation k is not a parent operation in the precedence relationship
OR	operation k is not involved in a precedence relationship
THEN	due_date_constraint_is_built
AND	EXECUTE "Build_constraint(DueDateConstraint, IndexSet,Parameter)"

Part III

Intelligent Manufacturing Systems – Planning and Control

Chapter 8 through 10 cover the automated process planning of sculptured part machining, intelligent rough machining of sculptured parts, and neural networks application to welding process control.

Chapter 8, Automated Process Planning and Machining of Sculptured Parts, by Yuan-Shin Lee, presents methodologies of automatic process planning and cutter path generation to machine sculptured parts. These methodologies include analytical surface evaluation for collecting machining information, heuristic-based determination of machining strategy, and optimal tool selection. The approach takes a surface defined in a Non-Uniform Rational B-spline (NURB) model to generate an operation plan and the cutter paths for rough and finish machining. Both protrusion and cavity surfaces, with or without interior islands, are handled. Examples are given to illustrate the developed systems. The current trend and future work in sculptured surface machining are also addressed.

Chapter 9, Intelligent Rough Machining of Sculptured Parts – Planning and Programming, by Hui Li, Zuomin Dong, and Geoffrey W. Vickers, focuses on the productivity improvement of automated sculptured part rough machining. New methodologies of sculptured part rough machining, based on the quantitative intelligent system approach, which combines geometric modeling, artificial intelligence, and numerical optimization, is presented. Starting from the geometric models of a sculptured part and a parallelepiped stock, the approach decomposes a complex 3-D rough machining problem into 2-D contour machining problems of several slicing layers. The efficiency of various tool path patterns for given cutting layer shapes is studied. A method for automatically identifying the optimal tool path pattern for a cutting layer using fuzzy pattern classification and identification is developed. In addition to the optimal tool paths, the approach also optimizes the machining parameters, including feed rate, depth of cut, number of cutting layers and distribution of cutting depths for each cutting layer. The method can automate the optimal computer numerical control (CNC) tool path generation and significantly reduce the total machining time. Several examples are used to illustrate the approach.

Chapter 10, Artmap Neural Networks for Closed-loop Welding Process Control, by H.-H. Tom Huang, H.-P. Ben Wang, and Gary W. Fischer, presents and discusses an artificial neural network approach for closed-loop arc welding process control. Two modified Artmap neural networks, arranged in tandem, are developed and trained. One classifies signals collected from a manufacturing process being monitored; the other works as a feedback controller. This tandem network system has been trained and tested with real welding data and has shown satisfactory accuracy. In addition, the performance of the Artmap networks is compared with those of back propagation, simulated annealing, and tabu search. The Artmap neural network approach demonstrates great potential in performing real-world process control.

Chapter 8

Automated Process Planning and Machining of Sculptured Parts

YUAN-SHIN LEE
School of Industrial Engineering, Purdue University

8.1 Introduction

Complex surfaces, often called sculptured surfaces, are common in molds, stamping dies, forging tools, rolling shapes, and so on. In general, these products consist of a combination of complex surfaces, and their manufacturing is a time-consuming task. Growing industrial demand initiated the development of computer-aided systems for design and manufacture of free-form surfaces. Application of the advanced computer-aided design and manufacturing (CAD/CAM) techniques to sculptured surface design and manufacturing is desperately needed.

CAD/CAM is becoming increasingly important in the manufacturing industry to reduce lead time and production costs. The product development process can be divided into three stages: design, planning, and manufacturing. In the last few decades, tremendous efforts have been given to the development of these areas. Computer-aided design (CAD) has made the design task faster and easier. Computer-aided process planning (CAPP) has been developed to automate the planning process. The CAPP systems automatically determine the machining processes, conditions, setups, tools, and so on, required to convert a stock of raw material to the finished part that was specified by the design engineer [6]. More precise, more powerful, and faster machines have been developed. With these advances, more attention is being focused on the full automation of manufacturing systems through integrating CAD, CAPP, and computer-aided manufacturing (CAM). As shown in Figure 8.1, computer-aided process planning (CAPP) has been recognized as a key element in fully automated manufacturing systems [1, 8]. Ideally, such approaches can generate a complete process plan from the engineering design data without any human decision making [9]. After a process plan has been made, the operator needs to prepare a part program. Even with

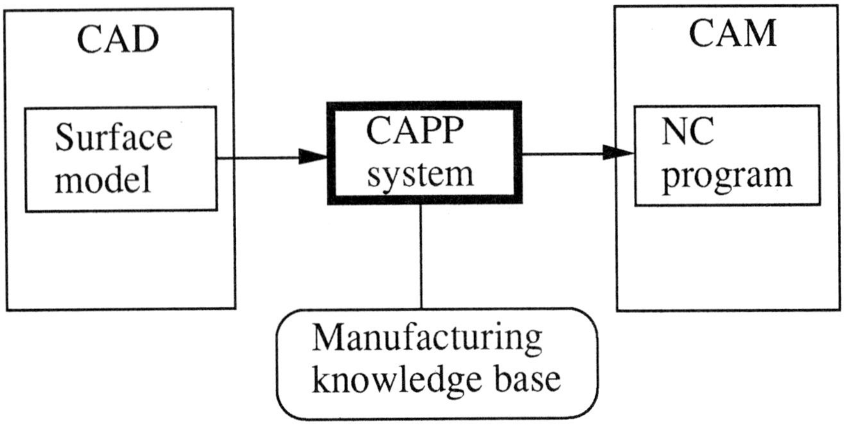

Figure 8.1: Integration of CAD and CAM by CAPP

computer-aided part programming this task can be very tedious and could slow down the entire production system. Therefore, automation of part programming is desirable.

Although there are a number of automatic NC code generation systems, most of these systems deal with only a restricted family of polyhedron features. Very few attempts to automate and integrate these tasks of sculptured surface machining work. To machine a sculptured surface on an NC machine requires not only complex geometric computation but also difficult human decision making. The planning and machining of complex sculptured surfaces are very often the bottlenecks of manufacturing. On a CAD/CAM system, a human operator decides the machining procedure and selects tools to be used; then the system computes the cutter path based on the decisions made. This is often a trial-and-error process. To eliminate the potential errors made by the operator and ease his task, a systematic approach to automate the planning procedures of sculptured surface machining is necessary. The intention of this chapter is to present the techniques that automatically decide the machining procedure, select the best feasible cutters, and generate gouge-free cutter paths for roughing and finishing a sculptured-surface cavity or protrusion. The methodologies can be integrated into an automatic process planner to interface CAD and CAM.

In this chapter, we will first discuss the related NC and part programming of sculptured surface machining. We then discuss the methodologies of automatic process planning and cutter path generation to machine sculptured parts. These methodologies include analytical surface evaluation for collecting machining information, heuristic-based determination of machining strategy, and optimal tool selection. The methodology is able to take a surface defined in a non-uniform rational B-spline (NURB) model and generate an operation plan, as well as the roughing and finishing cutter paths. Both protrusion and cavity surfaces, with or without islands inside, can be handled. Example systems and example parts are shown for demonstration. We also discuss the current trend and future work in sculptured surface machining.

8.2 Sculptured Surface Design and Manufacturing

8.2.1 Sculptured Surface in Product Design

Currently sculptured surface modeling is one of the main fields in computer-aided geometric design and manufacturing. Many systems have been developed for designing sculptured surface [24, 46, 47, 36, 42]. These systems are based on various mathematical expressions proposed by Coons (1967), Bezier (1972), Riesenfeld (1973), B-splines, or recently non-uniform rational B-spline (NURB) [4, 39, 13, 11].

There are several important sculptured surface description models [15, 11]. Bezier surface description is the most famous one [12]. NURB is the most versatile shape-modeling tool, and is gaining more attention in the CAD/CAM world. NURB provides the user more degrees of freedom than other descriptions to describe the product shape. It has a wide range of applications in surface modeling [16]. A cubic NURB surface is defined from the following geometric handlers: control vertices, weights for control vertices, and knot spans along two axes, as shown in Figure 8.2. A general equation used to define a (cubic) NURB surface R(u, v) is shown below [11]:

$$R(u,v) = UN_uH(N_v)^TV^T \tag{8.1}$$

where H is the matrix of control vertices, $U = [1\ u\ u^2\ u^3\ ]$, $V = [1\ v\ v^2\ v^3\ ]$, and N_u and N_v are the matrices related to the knot spans along two axes. More details of the NURB descriptions can be found in [11]. The main advantage of using NURB is that it produces smoother curves and surfaces when constructing composite curves and surfaces from point data. The merit of rational bases (i.e., unequal weights) is in their ability to exactly represent analytic curves and surfaces, especially conic sections and quadratic surfaces [38, 12]. The main drawback of NURB, arising from its versatility, is that it is not easy for the user to understand the geometric meaning of weights and knot spacings [13, 11].

8.2.2 NC and Sculptured Surface Machining

Numerically controlled (NC) machines began basically as machine tools which had servomotors connected to a controller instead of hand-turned cranks to control tool position. At first, the control sequences that were put into these controllers were created entirely by hand. Several computer programs were then developed which allowed the user to enter the tool path data in a shorthand form, which eliminated many of the hand calculations. These shorthand data entry formats evolved into "languages," complete with logical control. Two of the most popular of these were APT and COMPACT II. To manufacture the parts, cutter paths and machining parameters need to be generated and put into a NC part program to drive NC machines.

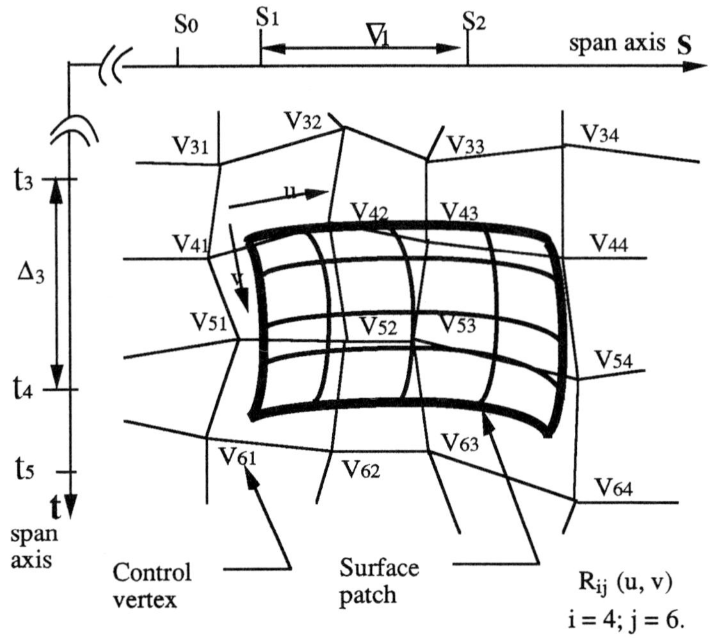

Figure 8.2: Cubic NURB Surface Patch

NC machining can be cost competitive with casting and produces higher quality parts with better material properties. Machining controlled by advanced technology CAM software produces high-precision quality parts without high tooling costs. NC machining is also more flexible than casting. One can easily and quickly produce engineered mechanical components with advanced technology CAM software. Companies that utilize specialized CAM software have a strong competitive advantage. Software designed to address a specific, highly focused application can achieve tremendous savings in time to program a part, evaluate results, and produce a high-quality product. Sculptured surfaces are often machined on CNC milling machining centers. In this chapter, we focus our discussion on the automation of the milling process of surface machining.

Functional free-form surfaces and aesthetic free-form shapes, as in turbine blades and automobile bodies, are commonplace in modern design. Although the development of free-form surface design and manufacturing has been carried out over the past 30 years, there are still some tasks, like the selection of machining procedure, cutter selection and NC programming, where operator interaction is needed. Although there are a number of NC code generation systems which have been built, most of these systems deal with only a restricted family of features [2, 41]. Also very few of the works attempt to automate and integrate these tasks. The major intention of the ongoing research is to develop the methodologies which will automatically decide the machining procedure, select the best possible cutters for machining the sculptured surfaces, and then generate gouge-free cutter paths for roughing and finishing steps. These methodologies will thus be integrated into an automatic process planner to interface CAD and CAM.

8.2.3 Part Programming of Surface Machining

The automatic generation of machine instructions is an important activity that must be automated in a computer integrated manufacturing (CIM) environment. This activity is also vital in today's manufacturing facilities because manual and computer-aided NC programming often proves to be a bottleneck in the manufacturing system. Observing the potential benefits of automated NC programming systems, many researchers have studied the problem and several commercial packages are available. Developing an automated NC system involves two major steps. First a detailed geometric description of the volume to be cleared and the associated machining parameters must be obtained. Once this information has been entered into the computer, a tool path that clears all of the material from the desired area without gouging the machining surfaces must be determined.

Several different methods of input for NC programming have been discussed in the literature. Several early systems, such as Persson's (1978), required the user to manually generate the boundary of the area to be cleared [37]. The computer then read this information from a data file. As interactive graphics techniques developed, researchers began to define the area to be cleared by interactively selecting geometry from a CAD model of the part. This method is by far the most prevalent in the NC programming literature. Some of the systems that utilize this method include Kawabe and colleagues (1983), Ferstenberg and colleagues (1986), Bala and Chang (1991), and Suh and Lee (1990) [22, 17, 2, 45].

Until now, most of the NC systems considered have primarily dealt with 2-1/2-dimensional features. While a large percentage of parts can be produced using 2-1/2-dimensional machining, 3-dimensional features, like sculptured surfaces, are very important, especially for mold and die manufacturers. Many tools, especially four or more axes machines for sculptured surface machining, are nearly impossible to program (direct in their movements) by manually calculating the tool paths. Several commercial NC tool path generation systems developed for 3-D sculptured surface machining are SmartCAM of Point Control, MasterCam of CNC Software, SURFCAM of Surfware, and MAX-5 of NREC. These systems are designed with 3-D geometry capability required for building physical objects as a basic function. These systems need the cutters and cutting planes through human interaction as an input. The CASCAM system developed by Lee and Chang (1991) at Purdue University automatically selects cutters, cutting planes, and produces NC code for 3-dimensional sculptured surface features defined using a non-uniform rational B-spline (NURB) surface definition [26].

8.3 Planning Sculptured Surface Machining

8.3.1 Process Planning and Artificial Intelligence

The idea of using a computer to assist in the process planning task originated in a paper published in 1965 by Niebel [34]. After that, the feasibiility of automated process planning was discussed by Schenk (1966) in his Ph.D. dissertation at Purdue University [44]. As the power of the computer increased during the 1970s and 1980s, research efforts into computerized process planning expanded. Out of this research, two major approaches evolved: variant and generative process planning [6]. The first attempts to introduce computers into the area of process planning centered around improving the productivity of the human process planner. These systems followed the traditional approach for process planning for new parts. This entails identifying and retrieving an existing plan for a similar part and then making the necessary modifications to the plan to accommodate the new part. Some examples of the variant approach are CAPP (1976) and MIPLAN (1980) [32, 19]. While not completely automated, the variant systems offer many benefits over manual process planning. The time to produce a process plan is reduced due to the computer's speed. However, these benefits are not experienced in all manufacturing operations. When dealing with small batches of widely varying parts, the variant approach proves impractical.

The need for a fully automated process planning system sparked research into approaches capable of generating process plans from a description of the part. Since the goal of these systems is to automatically generate process plans from scratch, these systems are called generative process planning systems. Generative process planning systems based on deterministic procedures utilize fixed logic that consists of conditions, actions, and data. This logic is most commonly represented in either a decision tree or a decision table. Systems that use decision trees directly code the manufacturing processes capabilities into a tree structure. Data tables also contain process capabilities but in table form. The table is divided into IF-conditions and THEN-actions.

The development of artificial intelligence techniques gave process planning researchers the power to represent the decision process in a more realistic manner. Recently, expert systems, and knowledge-based expert systems have become very popular methods for implementing the decision logic necessary for generative process planning. The basic idea behind all of these artificial intelligence techniques is to search for a solution through a state space that contains all possible solutions. If the search is directed from initial state to the goal state, a forward-chaining of the rules is executed. If the search is from the goal state to the initial state then the rules are fired using a backward-chaining methodology [1]. The control strategy is used to determine how the state space will be traversed. Examples of common control strategies include depth-first search and breath-first searches. In additional to the search strategy, heuristics may be also used to improve the efficiency of the search.

As more and more advances are made in the field of artificial intelligence, researchers will find ways to adapt them to create more powerful approaches to solving the process planning problem. Regardless of the approach taken, generative process planning systems all develop plans to convert raw stock into a finished part that is described by a part model [8]. While the end goal is the same, the desired level of detail is not. Some of the systems produce very high level process plans with limited details, other systems are more advanced and produce very detailed plans at multiple levels. The Quick Turnaround Cell (QTC) project at Purdue University is an example of the latter type of system [7, 8]. QTC in full is an integrated design-manufacturing-vision inspection system designed for rapid production of one-of-a-kind prismatic parts. In the QTC system, process planning is used to link the design and manufacturing functions together . Through use of a rule-based expert system, the process plan specifies processes, approach and feed directions, machining parameters, tools, operation sequence, and fixturing strategy on a feature-by-feature basis.

Almost all the existing process planning systems are designed for the machining features which are described by polyhedron features, like pocket, hole, slot, and so on. Very little work has been done in generating process plans for sculptured surface machining. Most of the CAM systems deal with the cutter path generation for finishing sculptured surfaces, while the cutter size and cutting planes are given. For sculptured surface machining, part surfaces are described by mathematical equations. The data available from the CAD database are surface control vertices. No direct machining information of sculptured surface is available from the CAD database. To automatically generate a process plan to machine sculptured surfaces directly from a CAD database, the system needs to extract necessary machining information. A constraint-based process planning schema for sculptured surface machining, which was proposed by Lee and Chang (1991) [27], is shown in Figure 8.3. Necessary information and constraints are extracted from the surface description. The geometry and constraint information are used to infer the machining strategy and machining processes. In the following three sections, detailed techniques of machining information evaluation and machining process selection are discussed.

8.3.2 Machining Information Evaluation

In a CAD/CAM system, it is the user's responsibility to determine the cutting planes and the cutters for each cutting plane. The system computes a gouging free cutter path. The decisions are made based on experiences by human expertise. They are arbitrary and nonoptional. The decision is errorprone. The objective is to automate the decision making, and integrate all the planning procedures and cutter path generation for both roughing and finishing sculptured surfaces.

A solution approach called the constraint-based approach has been proposed by Lee and Chang (1991), as shown in Figure 8.3 [27]. The CAD data in the form of control points are the input. A series of hunting planes are used to intersect with the sculptured surfaces to extract the geometric information, as shown in the Figure 8.4. The purpose of machining information evaluation is to provide the geometric and machinability information for the determination of machining procedures and cutting planes.

Initially, a series of hunting planes intersecting with the sculptured surface are used to collect geometric information. On each hunting plane, the intersection contours are constructed by a contour growth method [29]. This geometric information includes size, area of the cross section, and curvatures of the surface. The cutting area on each hunting plane

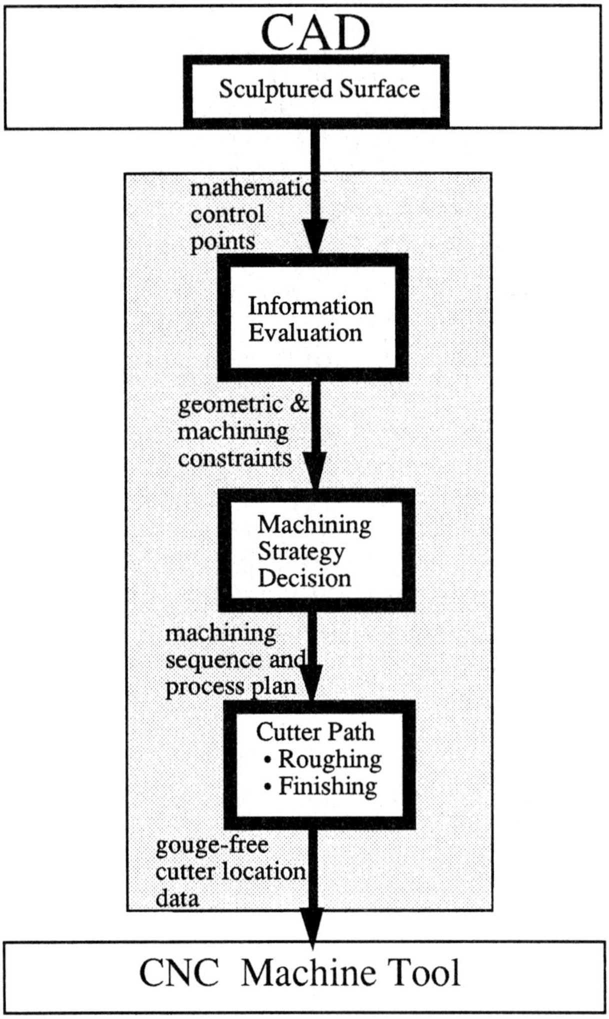

Figure 8.3: Constraint-based Approach to Sculptured Surface Machining

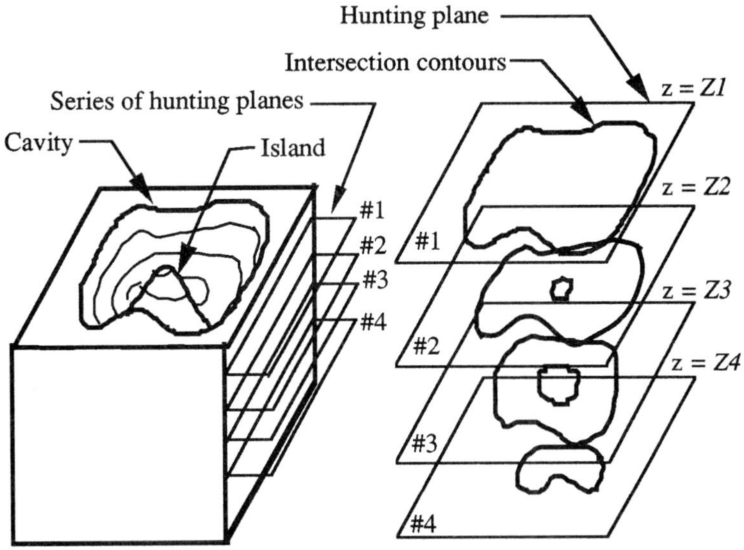

Figure 8.4: Hunting Planes and Intersection Contours

and the total removal volume of the surface are calculated. The general shape of a surface is found by checking the cutting area ratio between cutting planes. Other information, such as size and depth of the surface and the height and cross section of the islands, can also be found by this method. The cutting area and removal volume of arbitrarily shaped contours can thus be calculated [30]. The information extracted from the surface description is used to decide the cutter size for each cutting plane. The information evaluated from the surface description is used to determine the machining processes, select cutters, and optimize cutter selection. The cutter path for roughing and finishing the surface is then generated.

8.3.3 Automatic Cutter Selection and Optimization

Machining process decisions are made based on the size, shape, tolerance, and surface finish of the surfaces. When evaluating the machining cost in sculptured surface machining, we found that the material removal rate (MRR) is a function of the square of the cutter diameter (D) as shown below [28]:

$$MRR_i = f(D^2) \tag{8.2}$$

Since machining time is the inverse of MRR, the larger the cutter, the shorter the machining time. In surface machining, the roughing process has been divided into a series of material removal processes on a series of cutting planes. For each cutting plane, the

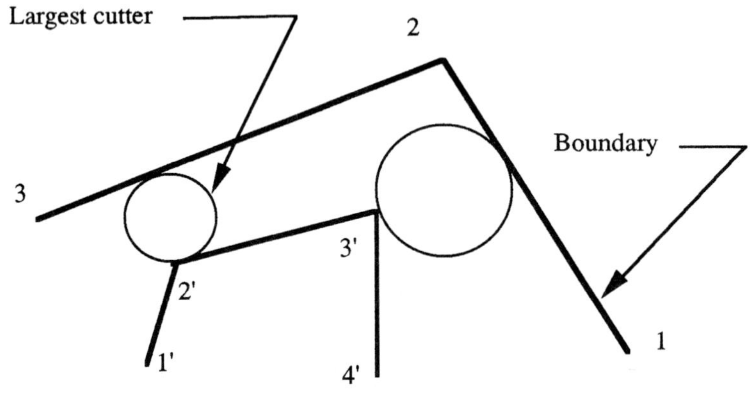

Figure 8.5: Determining the Largest Cutter between Polyons

intersection boundary is constructed and the machining processes on the cutting plane are basically a series of pocketing processes. Both the outer boundary and the islands are considered in cutter selection. The intersection contours on each hunting plane are used to find feasible cutters for such a hunting plane. Bala (1991) proposed an algorithm of circle fitting to polygons to find the biggest cutter that satisfies the given constraints of convex vertices, as shown in Figure 8.5. Details of the technique to find the largest circle between polygons can be found in [2]. A search of the tool database determines that the feasible cutters for a hunting policy for machining is to choose the largest feasible cutter under the geometric constraints of intersection contours. For the roughing process the cutter selected is the largest available tool which can fit into the intersecting boundary. The material left behind must be removed by one finishing pass. Otherwise, another semiroughing process is needed to remove these shoulders left on surfaces. The complete cutter selection procedure for roughing includes the following:

1. Find candidate cutters for each cutting plane.

2. Optimize cutter selection by the merging procedure and reduce unnecessary tool changes.

3. Compute the final cutting planes and cutter size for each cutting plane.

Because there are several cutting planes in roughing, each cutting plane may have different cutter sizes for pocketing. Lee and Chang (1992) proposed an algorithm to optimize cutter selection by a merging procedure for roughing sculptured surfaces [28]. To reduce the tool change time and the number of tools stored in the tool magazine, similarsized cutters are merged. Since the depth of cut (distance between two cutting planes) is affected by the tool size, the new Z-heights of final cutting planes are calculated based on the cutter's maximum depth of cut. As shown in Figure 8.6, cutters are merged based on a checking size-ratio between selected cutters on two adjacent cutting planes [28]. As shown in Figure 8.6c and d, if the checking condition $K_2 < K_1$ is satisfied, the tool selections for adjacent cutting planes are merged to reduce the machining cost. Detailed techniques of optimizing tool selection by merging cutter selection on adjacent cutting planes can be found in [28]. Final cutting planes are adjusted based on the new depths of cut. Figure 8.7 shows an example of deciding final cutting planes after optimizing cutter selection. Figure 8.7a shows the merging of the initial suggested cutters in state (1) to a set of merged cutters in state (2). In Figure 8.7b, the final cutting planes are determined in state (3) based on the merged cutters from state (2).

Cutter selection for semiroughing with ball end-mills is based on the thickness of the shoulders left on the surfaces. Cutter selection for the finishing process is based on the part surface curvature evaluation. Gouging prevention is the most important issue to be considered in finish cutting. The gouging source is the region where curvature is less than the cutter radius. Cutters should be chosen to eliminate the possible gouging source in finishing. An "interference free" cutter is a possible candidate for finishing. The largest cutter with which the given feature can be fully machined is selected [28]. Techniques of finding surface curvature for cutter selection can be found in [11, 12].

8.3.4 Machining Process Selection

To machine sculptured surfaces, there can be three steps: roughing, semi-roughing, and finishing. In order to machine a finished surface starting from raw stock, the following sequence of material removal operations is usually needed :

1. Rough cutting to remove most of the material outside surface on a sequence of cutting planes
2. Semiroughing to remove the shoulders left on the part surface after roughing
3. Finishing to finish the sculptured part surfaces
4. Scraping or grinding, if necessary, to smooth the surface

The robustness of sculptured surface machining depends on the geometric complexity and machining constraints of the part surface. For example, the machining process of a shallow and not-so-curved surface is much simpler than a complex, deep cavity; and the machining of a large, deep cavity is different from the machining of a deep but small cavity. A very curved sculptured surface needs a small cutter and more cutter movements along the surface. The machining procedures can be determined by the complexity and the machinability of the sculptured part surface. Sculptured surfaces can be classified into different types by

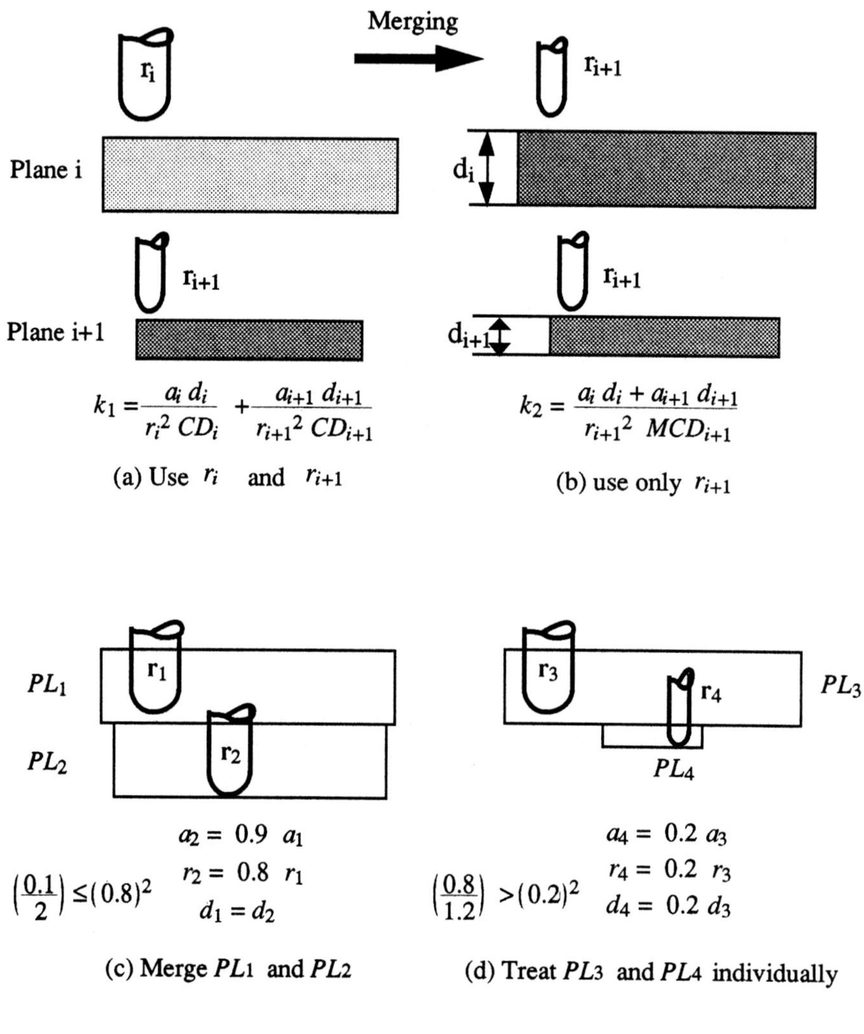

Figure 8.6: Merging Cutters Chosen for Roughing

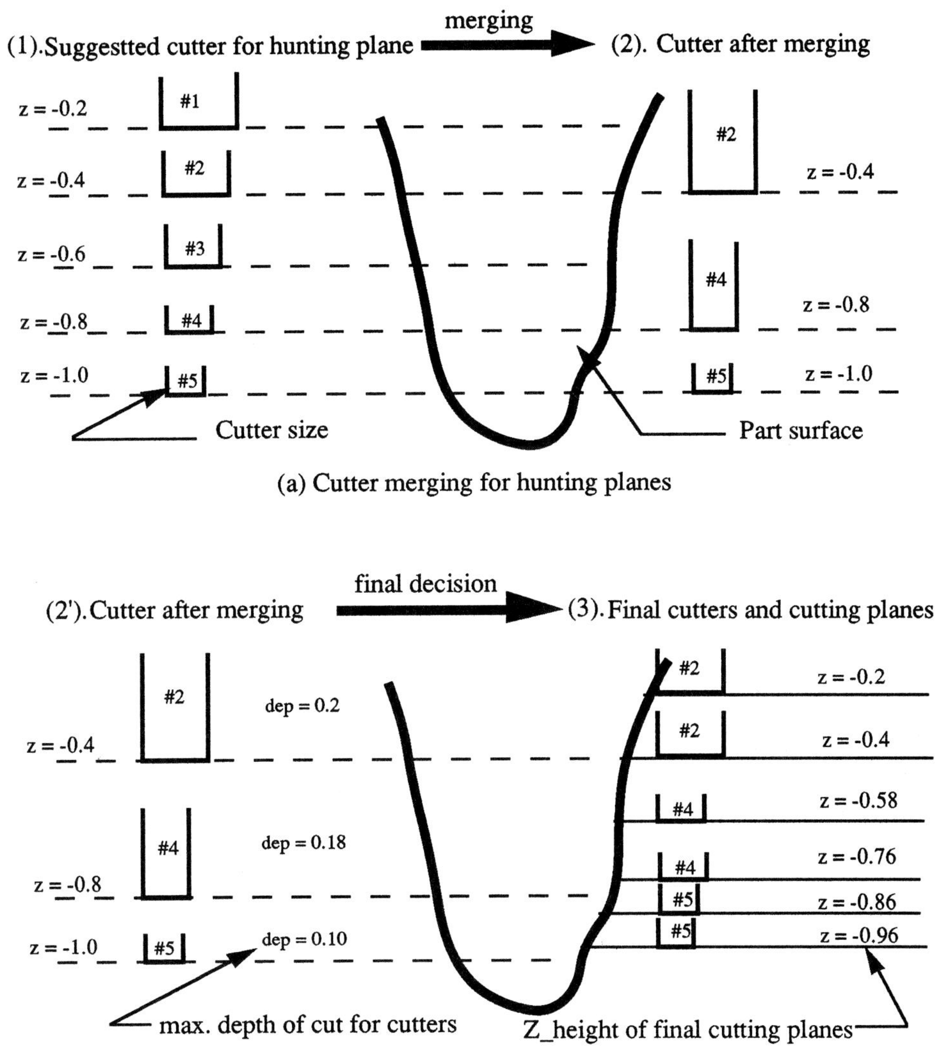

(a) Cutter merging for hunting planes

(b) Decision of final cutting planes

Figure 8.7: Selection of Cutters and Cutting Planes for Roughing considering this machining information and machining cost. The machining procedure is based on the classified type.

For the milling process, machining cost can be calculated by the machining time and the tool life. Tool selection and cutter path generation are decided based on the machining cost. It is assumed that the most economical NC cutter path is the one that requires the minimum machining time. If the radii of cutters for a machining process are r_i $(i = 1, ..., n)$ and the corresponding machining time and tool change time are T_{ri} and T_c, the total machining time T_m can be expressed as

$$T_m = T_{r1} + T_c + T_{r2} + T_c + ... + T_c + T_{rn} \tag{8.3}$$

For a NC machining center that has an automatic tool changer, $T_c << T_{ri}$ and T_{ri} is the major factor for machining cost. However, it is not good machining practice to have many tool changes while machining the part because it increases the machining cost and the number of tools in the tool magazine. Machining time can be evaluated by the material removal rate (MRR) which can be approximated. Chang (1991) pointed out that cutter size is the major factor in MRR [9]. The best policy for machining, therefore, is choosing the highest MRR under the geometric constraints to reduce machining cost. From the machining information evaluated, there are four types of different machining procedures:

- Roughing with large-diameter flat end-mills; semiroughing with large-diameter ball end-mills; finishing with small-diameter ball end-mills.
- Roughing with large-diameter flat end-mills; finishing with ball end-mills.
- Roughing with large ball end-mills; finishing with small ball end-mills.
- Roughing and finishing with the same size ball end-mill.

The classification is made based on the cross-sectional area on each hunting plane, the depth of removal volume, area decreasing rate on the adjacent hunting planes, the depth of hunting plane on which island contours first occur, the number of islands, and the height of the island. The sculptured surface cavities are classified according to the information evaluated from the part design database. An example classification is shown in Figure 8.8. Classification codes can be used to indicate the machining procedures. Figure 8.8 also shows the machining codes of different classification codes. After the determination of classification and machining procedure, cutters can be selected for each machining procedure. The classification of sculptured surfaces in a specific domain can be determined based on the knowledge and experience of sculptured surface machining, and these constants are treated as system parameters.

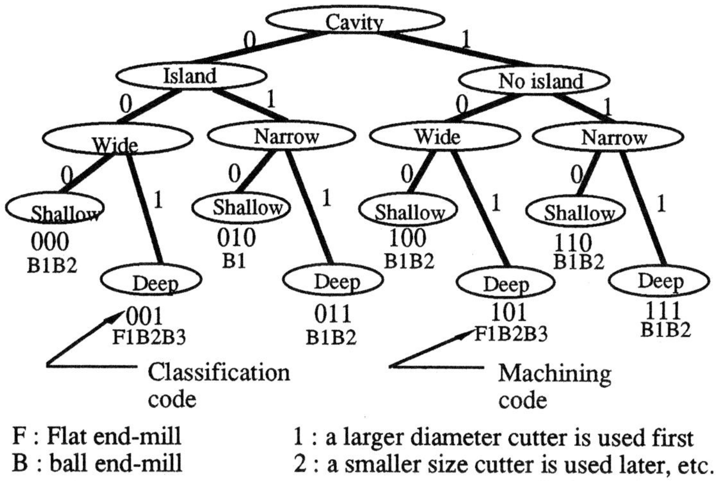

Figure 8.8: Machining Process Classification

8.4 Tool Path Generation for Machining Sculptured Parts

8.4.1 3-D Sculptured Surface Machining

Tool path generation is the last step in process planning. At this stage, most of the process plan has been determined. After cutters and cutting planes have been selected, the tool path for machining can be generated. There are two types of motion in a cutting path: rapid traverse and feed motion. The rapid traverse is used to move the cutter from the end of one feature to the beginning of the next feature. Feed motion does the actual cutting. The collision problem is the main issue that needs to be addressed.

The generated tool path must satisfy the following three conditions: (a) It must be efficient, (b) it must be gouging-free, and (c) it must minimize undercut. To achieve these objectives, the following procedures can be used: roughing with ball end-mills or flat end-mills and generating a gouging-free cutter path for finishing.

From the literature, there are three common approaches to generate cutter location (CL) data for the sculptured surfaces machining starting from a surface model built in a CAD system:

- Bobrow and Loney use the surface model to generate cutter contact (CC) points and use these CC data to decide the CL data for NC machining [5, 33, 25]. The drawbacks of this method are that extensive numerical iterations are involved and it is difficult to remove the undercut volume and interference points.

- The second method is the offset surface method which generates CL data from offset surfaces and uses a surface-surface intersection algorithm to identify interference regions [3, 10, 43].
- The third method is the polyhedron model method which generates CL data from the polyhedron model [14, 35]. This approach prevents the cutter from gouging the polyhedron without a numerical stability problem. It also provides the means to identify interference regions and means for subsequent removal of the undercut volumes using smaller-diameter ball end-mills [10].

Hermann has built a system for cutter path generation [18]. It generates the cutter path by calculating the touching point of the tool and workpiece along the B-spline surfaces. Preiss has developed an expert system for automatic cutter path generation [40]. It uses a set of operations which are actually machining operations to transform the workpiece from raw material to the finished part. Liang built a system called LINC (Logical Integrated NC Program) for NC programming [31]. In his system, the NC programming problem is formulated as a decomposable frame problem. The problem is approached by a sequence of transitions from the initial frame representing the initial part until the final part is obtained. Kanai proposes a hierarchical CAM system based on a cell-constructed-geometric-model [21]. In Kanai's system, the product is constructed by hierarchical modules. The cell-constructed-geometric model is decomposed to several predefined machining face groups which are related to some predefined operation. The cutter path is generated from these machining face groups. Several commercial CAM systems designed for 3-D curved surface machining are SURFCAM of Surfware, MasterCam of CNC Software, and ToolChest of Battelle Memorial. These CAM systems need operators to input the selected cutters and select the cutting region, then generate the cutter path either by the offset surface method or the polyhedron model method. The approach used in the CASCAM system developed by Lee and Chang (1991) at Purdue University involves roughing the feature with an automatically selected large flat or ball end-mill, using a 3-dimensional roughing algorithm or a 2-1/2-dimensional complex pocketing algorithm, depending on the type of selected cutters, on several cutting planes to determine the cutter path [27]. Since roughing often leaves a shoulder on the part surface that cannot be removed by finishing, a semiroughing routine creates a tool path for a large ball end-mill to clean up the roughed surface. The finishing path moves a small ball end mill along the 3-D contours of the feature to generate the proper shape in the workpiece.In the following two sections, we discuss detailed example procedures of generating cutter paths for roughing and finishing using different types of cutters.

8.4.2 Cutter Path Generation for Roughing

Planning cutter path for roughing needs to consider the interference problem. Similar methods of generating a cutter path by the offset surfaces are also found in [22, 23, 24].

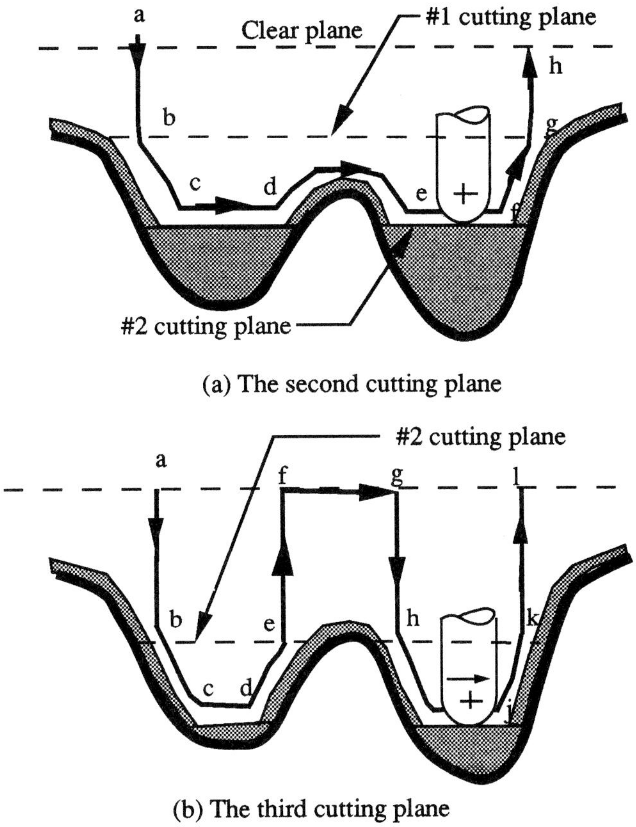

Figure 8.9: Roughing with Ball End-mills

If the part surface is not very wide and the surface is very complex, the sculptured cavity can be rough cut by using large ball end-mills. The procedure of roughing is divided into several machining processes on X-Y planes. The ball end-mill cuts along the sculptured surface until it touches a predefined cutting plane, then the cutter moves along the cutting plane until it touches another part surface, as shown in Figure 8.9. The ball end-mill moves along the sequence of points a-b-c-d-e-f-g-h, and so on, as shown in Figure 8.9a and b. A finishing allowance f is left on the part surface for the finishing process.

This method provides an efficient way of removing most of the material outside the surface because there is not much overlap between cutter paths and most of the sculptured surface features can be machined. The distance between the two cutting planes is the maximum depth of cut of the ball end-mill.

If the surface is wide and a large amount of material needs to be removed, a flat end-mill can be used for roughing. The flat end-mill cutter removes the material on the cutting plane and leaves "shoulders" on the part surface as shown in Figure 8.10a. These shoulders can be removed later in the semiroughing or finishing process. The feasible cutting region, bounded

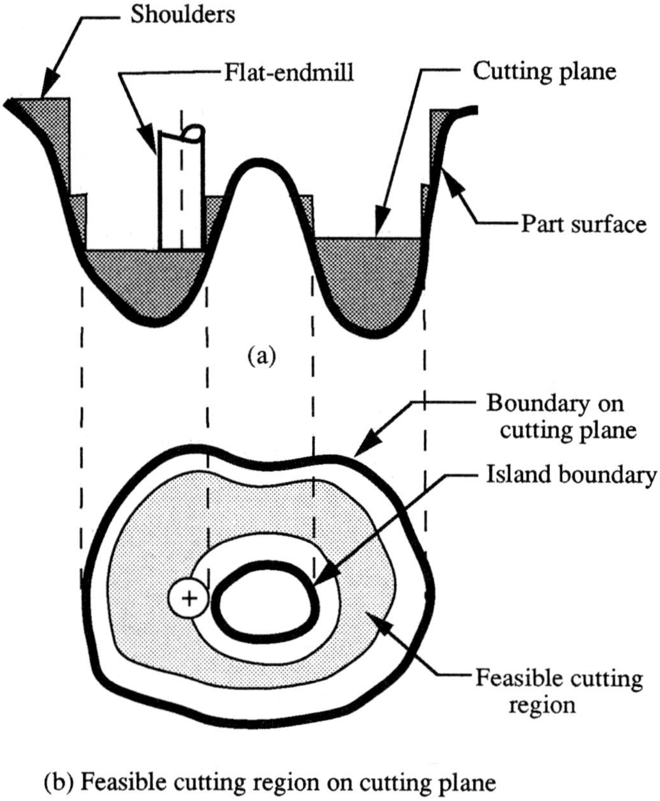

Figure 8.10: Roughing with Flat End-mill

by outer and inner contours on each cutting plane, is used to generate the cutter path for roughing. Figure 8.10b shows the feasible cutting region on a cutting plane.

8.4.3 Gouge-free Cutter Path Generation for Finishing

To control the surface quality of sculptured surface machining, cusps on the surface after machining and tool overcut need to be maintained within surface finishing tolerance. Figure 8.11 shows a cusp on the surface after cutter movements. Details of calculating cusp height can be found in [10, 12]. In sculptured surface machining, cutter interference, or gouging, is the most critical issue. This cutter interference problem can occur when the sculptured surface has high curvature variation, as shown in Figure 8.12. If, at any point, the local radius of curvature of the surface is less than the tool radius, then the locus of the cutter location points generated via these procedures can produce an intersecting cutter path as shown in Figure 8.12a. In this case, the tool gouges into the part surface. There are two ways to avoid cutter interference problems. The first method is to choose an appropriate cutter

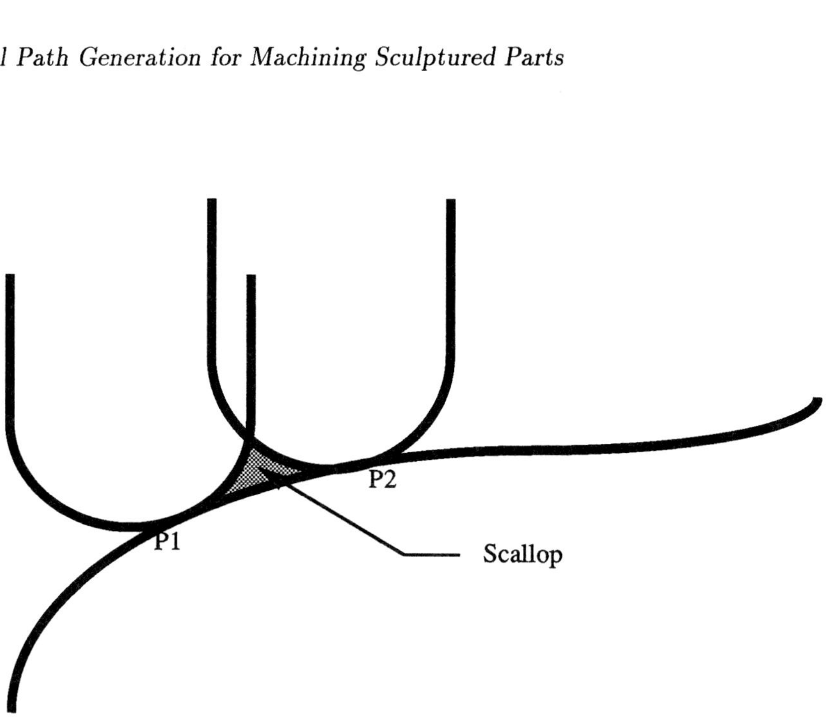

Figure 8.11: Scallop Left between Paths

(a smaller cutter), as shown in Figure 8.12b. The second method is to remove the gouging cutter path by skipping the gouging segment along the cutter path, as shown in Figure 8.12c. The second method will leave undercut material on the surface. Undercut material can be removed by a smaller cutter, so the second method is usually adopted by roughing or semiroughing processes. Choi proposed an interference avoidance method for cutter path generation [10]. Choi's method, combined with a modified scallop height method, is adopted to generate the finishing cutter path. More details can be found in [28].

8.4.4 Verification of Cutter Path

Although most of the cutter path generation is automatically performed by the system software, a part programmer can modify tool paths for a machine tool in a convenient way. This option is available by using NC verification techniques which allow the part programmer to make specific choices if desired, with the advantage of flexibility. Most CAM packages allow for accurate and realistic verification of 2-D-, 2-1/2-D-, and 3-D cutter paths without actually cutting metal. The user can simulate and visually verify the cutter path on a computer screen. The tool path seen on the screen shows the tool actually moving along the part from any perspective. This allows the engineer to check for cutter path correctness and clearance of cutters, components, and fixtures.

Verifying NC codes and editing are an essential parts in the sculptured surface machining,

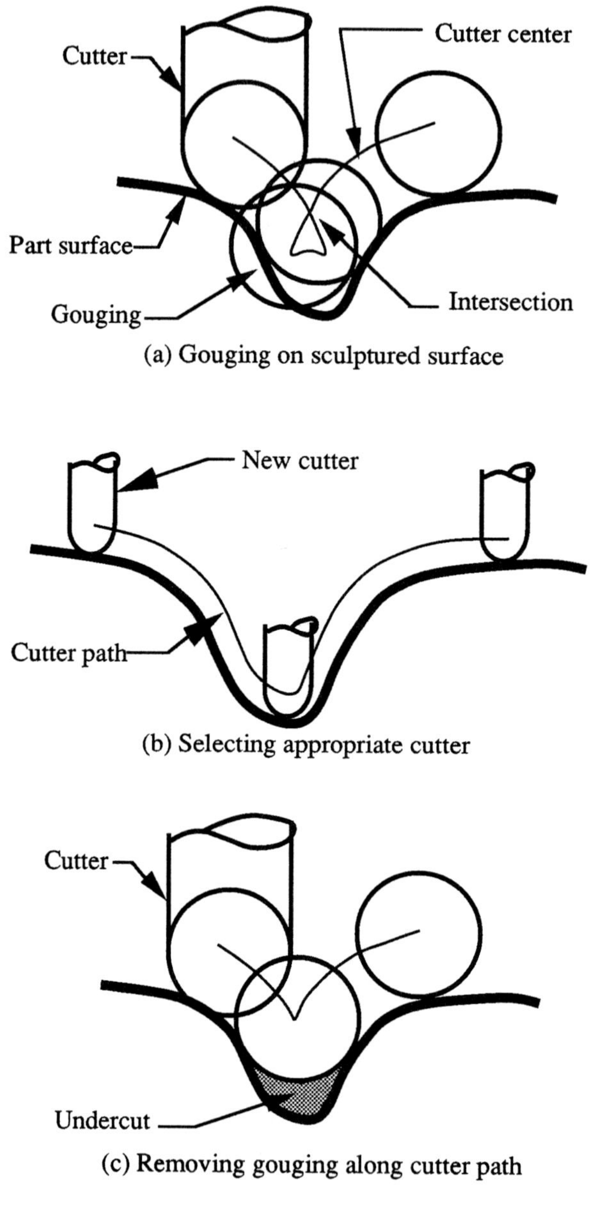

Figure 8.12: Gouging on Sculptured Surface

especially when the NC codes are produced for a complex part surface by a dedicated CAM system [20]. The NC verification and editing functions will make the manufacturing engineer's life easier. Most of the NC verification and editing systems have the following common functions:

- Initial workpiece display
- Cutting process simulation and MRR computation
- Overcut and undercut detection
- Tool paths display and editing
- Correction of feed rates and parameters of machining

The simulation of the cutting process is a computation intensive task and takes a large amount of CPU time. With the advancement of computing speed, NC verification and editing have become more helpful in assisting manufacturing engineers to machine a complex part without error. Several NC verification packages, like the VERICUT of CGTech and NCV module of the QTC system at Purdue University, are available for simulating a cutting process and editing a generated cutter path.

8.5 Examples

8.5.1 The CASCAM System

An example system designed for process planning and machining sculptured surface parts is shown in this section. Lee and Chang developed a system called constraint-based automatic system for sculptured surface cavity machining (CASCAM) for the process planning and cutter path generation of sculptured surface machining at Purdue University [26]. The system uses the control points of a NURB surface as input and generates all the necessary machining information. Currently the CASCAM system is being integrated into the QTC-II system, an integrated design and automatic process planning system developed at Purdue University [7]. Sculptured surfaces can be designed by the QTC-II system, and the NURB control points can be passed to the CASCAM system.

The overall CASCAM system structure is shown in Figure 8.13. Each module of the system corresponds to techniques discussed in the previous sections. The CASCAM system is implemented on a Silicon Graphics Personal IRIS-4D workstation with 20 M bytes of RAM, a networked disk, 0.9 M flops, and 10 Mips speed. The data are passed to the system through a pointer from the QTC-II design module. The results are displayed and saved in a cutter location (CL) data file.

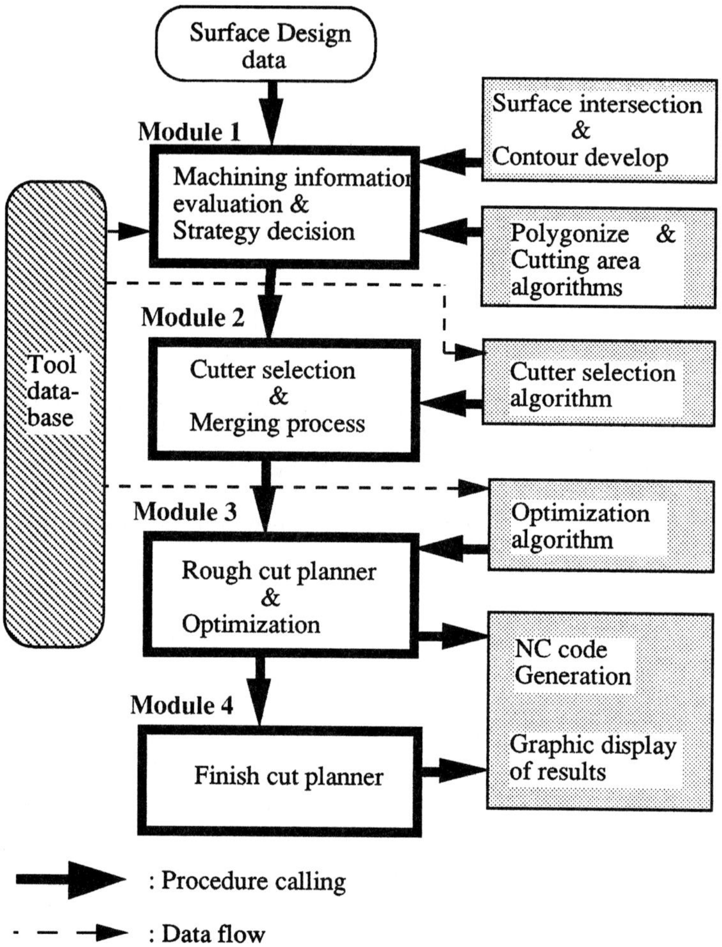

Figure 8.13: Overall System Structure of CASCAM

8.5.2 Example Parts

Figure 8.14a shows an example sculptured surface with an island inside the cavity. The top-right window shows the part surface, and the bottom-right window shows the control points of surface. Figure 8.14b shows the intersection contours on a series of hunting planes. The contours are used for calculating cutting areas and determining the suggested cutter for the cutting plane. Figure 8.15a and b shows the cutter paths generated for roughing a cutting plane and finishing, respectively.

The machining procedure is determined by the machinability evaluations. For this example, the cavity has an island inside and the bottom face of the cavity is complicated. After evaluating and collecting geometric information, the generated procedure for this cavity is to first rough-cut with a large ball end-mill and then finish-cut with a smaller ball end-mill. This procedure is reasonable for a cavity with a large island inside and complicated bottom faces.

The discussed methodology in CASCAM is being extended to handle the protrusion sculptured surfaces by using a virtual boundary approach. A Virtual Boundary is an invisible boundary surrounding the protrusion. It can be violated by cutters during machining. By using a virtual boundary, the protrusion free-form surfaces are treated as islands by the proposed methodology.

Figure 8.16 shows an example part of human face design. The face design has a complex region at the bottom cavity. The cutter path generated for a cutting plane to rough-cut the face design is shown in Figure 8.16. Figure 8.17a and b shows the example parts of the face design and a human body design after being process planned by the CASCAM system and machined on a Cincinnati Milacron T-10 machining center, which is a three-axes horizontal milling machine. The example system is reasonably efficient in solving planning problems for sculptured surface machining within minutes.

8.6 Conclusions

In this chapter we have tried to address the problem of generating cutter paths for machining a generic complex sculptured surface. While most existing systems deal with only the protrusion case or finishing process, the proposed system automatically generates a plan for both roughing and finishing that directly converts a raw stock to a finished sculptured part. The methodologies discussed in this chapter are able to extract geometric information and constraints from mathematical surface description to support process planning for machining. We have tackled the planning problem of complex surface machining which is difficult to solve by existing CAD/CAM systems. We also discussed a method of automatic cutter selection for complex surface machining, which previously needed to be done by human operators and had not been automated in CAD/CAM systems. The presented methods are reasonably general to solve the planning problem of sculptured surface machining, with both roughing and finishing. A detailed discussion of an example CASCAM system, which successfully and efficiently generates operation planning and cutter paths for both roughing and finishing processes, is also presented for demonstrating the feasibility of discussed methods.

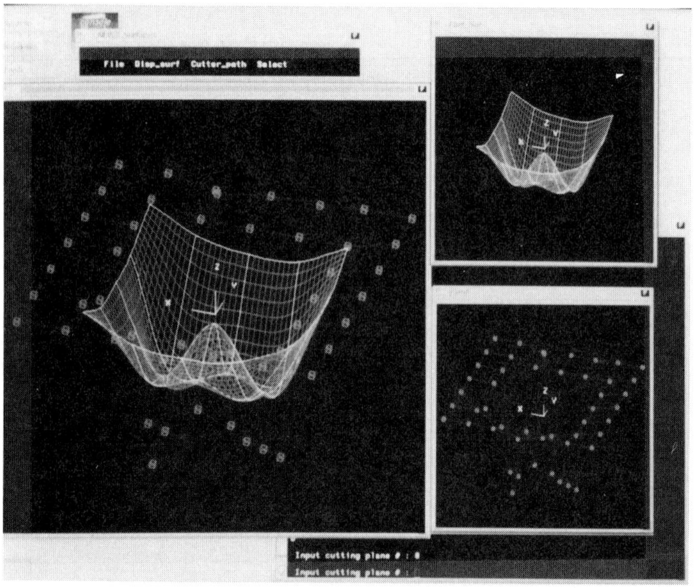

(a) A cavity design on the CASCAM system

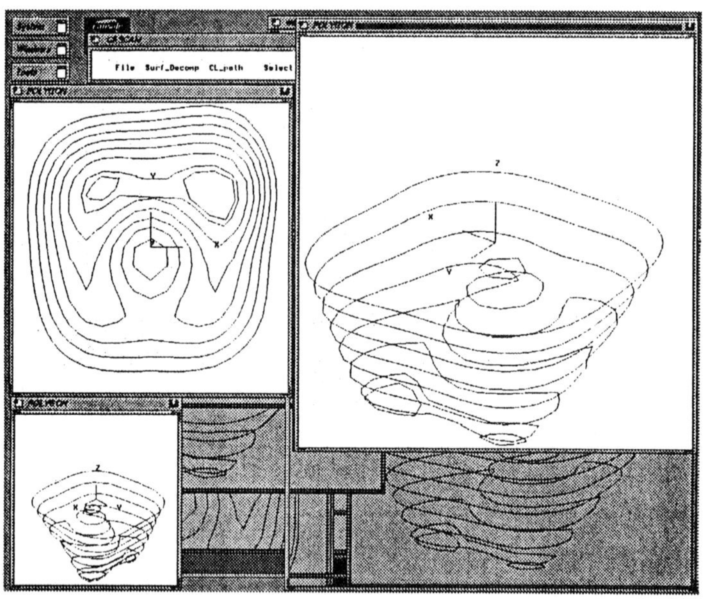

(b) Intersecting contours on a series of hunting planes

Figure 8.14: Example Surface Design and Hunting Planes

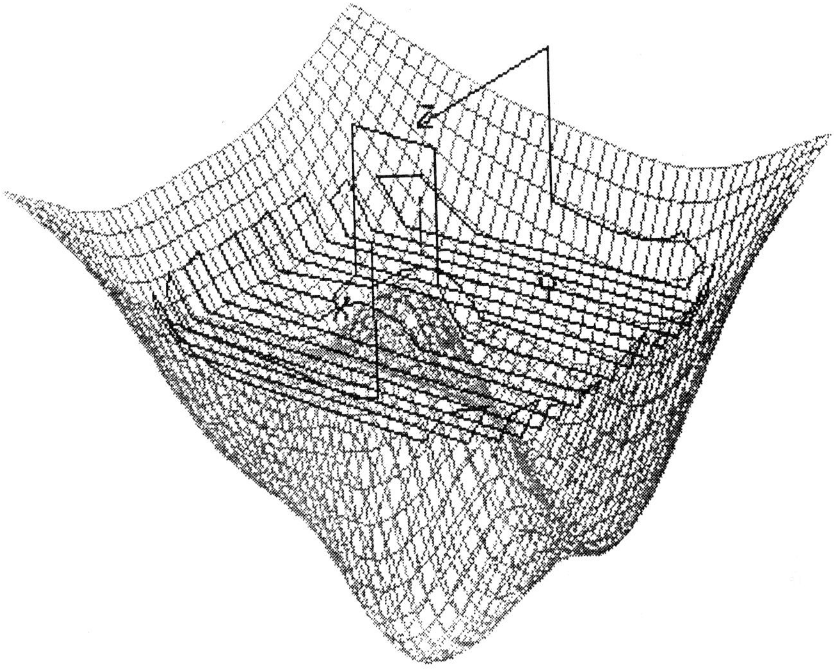

(a) Cutter path generation for roughing a cutting plane

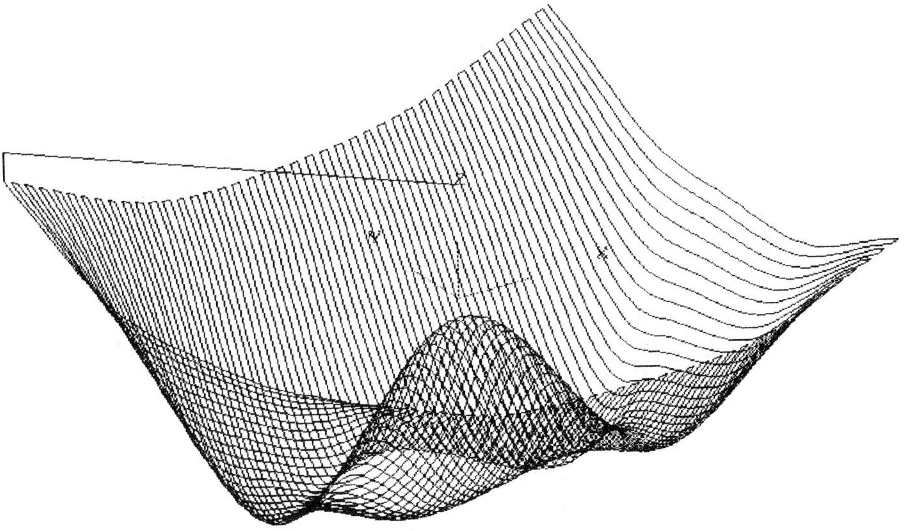

(d) Cutter path generated for finishing

Figure 8.15: Cutter Path Generation for Example Part of Figure 8.14

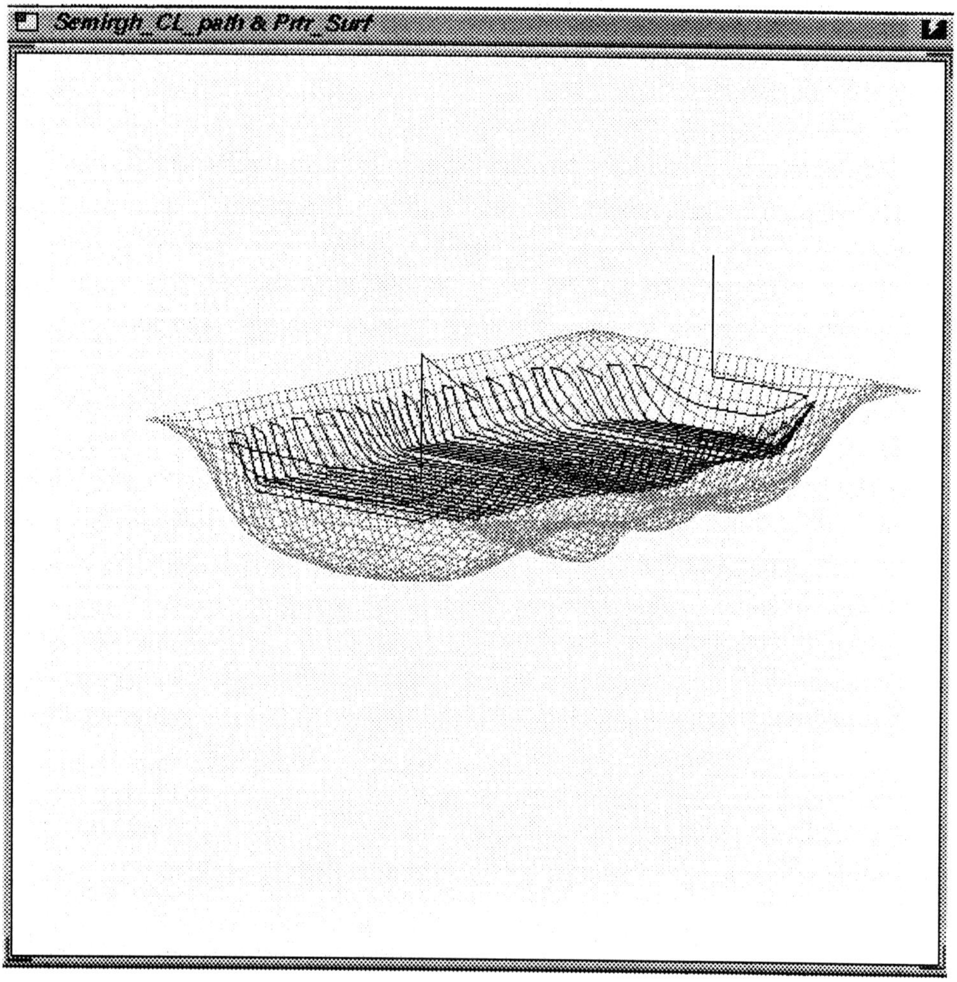

Figure 8.16: Cutter Path Generated for Complex Surface of a Face Design

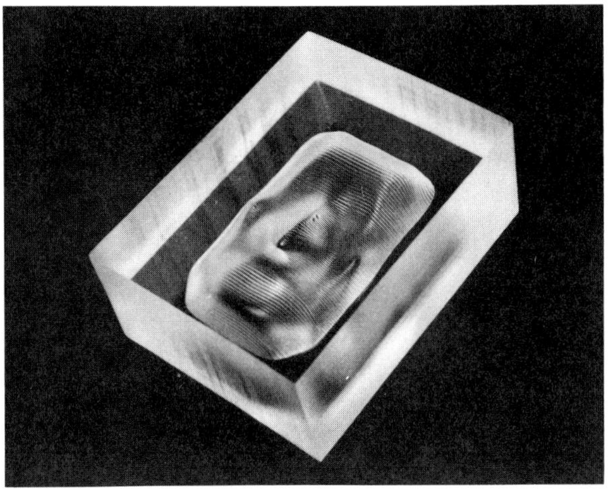

(a) Machined part of the face example of Figure 16

(b) Machined part of a human body design

Figure 8.17: Example Parts after Machining

References

[1] Alting, L., and Zhang, H. (1989). "Computer Aided Process Planning: The State-of-the-Art Survey," *International Journal of Production Research*, Vol, 27, No. 4, pp. 553-585.

[2] Bala, M., and Chang, T.C. (1991). "Automatic Cutter Selection and Optimal Cutter Path Generation for Prismatic Parts," *International Journal of Production Research*, Vol. 29, No. 11, pp. 2163-2176.

[3] Barnhill, R.E., Farin, G., Jordan, M., and Piper, B.R. (1987). "Surface/ Surface Intersection," *Computer Aided Geometry Design*, Vol. 4, pp. 3-16.

[4] Bezier P (1974). "UNISURF System: Principles, Program, Language," *Proceedings of PROLAMAT Conference*, North Holland, Amsterdam.

[5] Bobrow, J.E. (1985). "NC Machine Tool Path Generation from CSG Part Representations," *Computer Aided Design*, Vol. 17, No. 2, pp. 439-444.

[6] Chang, T.C., and Wysk, R.A. (1985). *An Introduction to Automated Process Planning Systems*, Prentice Hall, Englewood Cliffs, NJ.

[7] Chang, T.C., Anderson, D.C., and Mitchell, O.R. (1988). "QTC - An Integrated Design/Manufacturing/Inspection System for Prismatic Parts," *Proceedings of ASME Computers in Engineering Conf.*, Vol. 1.

[8] Chang, T.C. (1990). *Expert Process Planning System*, Addison-Wesley, MA.

[9] Chang, T.C., Wysk, R.A., and Wang, H.P. (1991). *Computer Aided Manufacturing*, Prentice-Hall, Englewood Cliffs, NJ.

[10] Choi, B.K., and Jun, C.S. (1989). "Ball-End Cutter Interference Avoidance in NC Machining of Sculptured Surfaces," *Computer Aided Design*, Vol. 21, No. 6, pp. 371-378.

[11] Choi, B.K., Yoo, W.S., and Lee, C.S. (1990). "Matrix Representation for NURB Curves and Surfaces," *Computer Aided Design*, Vol. 22, No. 4, pp. 235-240.

[12] Ding, Q., and Davis, B.J. (1987). *Surface Engineering Geometry for Computer-Aided Design and Manufacture*, John Wiley, New York.

[13] Dokken, T., Skytt, V., and Ytrehus, A.M. (1989). "The Role of NURB in Geometric Modelling and CAD/CAM," *Proceedings of Advanced Geometric Modelling for Engineering Applications Conference*, Berlin, (November), pp. 95-102.

[14] Duncan, J.P., and Mair, S.G. (1983). *Sculptured Surfaces in Engineering and Medicine*, Cambridge University Press, U.K.

[15] Faux, I.D., and Pratt, M.J. (1978). *Computational Geometry for Design and Manufacture*, Ellis Horwood, Chichester, U.K.

[16] Farin, G. (1989). "Trends in Curve and Surface Design," *Computer Aided Design*, Vol. 21, No. 5, pp. 293-296.

[17] Ferstenberg, R., Wang, K.K., and Muckstadt, J. (1986). "Automatic Generation of Optimized 3-Axis NC Programs Using Boundary Files," *IEEE International Conference on Robotics and Automation* (April), pp. 325-332.

[18] Hermann, G. (1984). "Free-form Shapes: An Integrated CAD/CAM System," *Computers in Industry*, Vol. 5, pp. 205-210.

[19] Houtzeel, A. (1980). "Computer Assisted Process Planning - A First Step towards Integration," *Proceedings AUTOFACT WEST, Society of Manufacturing Engineers*, Anaheim, CA, pp. 801-808.

[20] Jerard, R.B., Hussaini, S.Z., Drysdale, R.L., and Schaudt, B. (1989). "Approximate Methods for Simulation and Verification of Numerically Controlled Machining Programs," *The Visual Computer*, Vol. 5, No. 6.

[21] Kanai, S., Tanaka, F., and Amano, T. (1987). "Hierarchical CAM System Based on Cell-Constructed-Geometric Model," *Bul. Japan Soc. of Precision Engineering*, Vol. 21, No. 2, pp. 125-131.

[22] Kawabe, S., Shimura, S., Miyashita, K., Kimura, F., and Sata, T. (1983). "Programming for Machining Based on Workpiece Models in Computer," *Annals of the CIRP*, Vol. 32, pp. 351-355.

[23] Kishinami, T., Kikuchi, T., Kanai, S., and Saito, K. (1983). "Development of Interactive Mold Cavity CAD/CAM System," *CIRP Annals*, Vol. 32, No. 1, pp. 345-349.

[24] Kuragano, T., Sasaki, N., and Kikuchi, A. (1988). "The FRESDAM System for Designing and Manufacturing Free-form Objects," *USA-Japan Cross Bridge*, Vol. 2, pp. 931-938.

[25] Kuttner, B.C., Majcher, D.S., and Snedecor, P.B. (1985). "Systematic Processing: An Approach to Fully Automatic NC Tool Path Generation," *Proceedings of AutoFact '85*, pp. 18.7-18.22.

[26] Lee, Y.S., and Chang, T.C. (1991). "CASCAM - An Automated System for Sculptured Surface Cavity Machining," *Computers in Industry*, Vol. 16, No. 4, pp. 321-342.

[27] Lee, Y.S., and Chang, T.C. (1991). "Automated Process Planning and Machining for Sculptured Surface Cavities," *ASME, International Computers in Engineering Conference and Exposition*, Santa Clara, CA, (August 18-22), pp. 185-194.

[28] Lee, Y.S., Choi, B.K., and Chang, T.C. (1992). "Cut Distribution and Cutter Selection for Sculptured Surface Cavity Machining," *International Journal of Production Research*, Vol. 30, No. 6, pp. 1447-1470.

[29] Lee, Y.S., and Chang, T.C. (1992). "A Contour Growth Method from Intersections of Hunting Planes to Evaluate Machining Information," to appear in the *Journal of Design and Manufacturing*, Vol. 2, No. 3, September.

[30] Lee, Y.S., and Chang, T.C. (1992). "A Walking-Observer Algorithm to Complex Shape Decomposition for Machining Process Decision," (in review), submitted to the *International Journal of Production Research* (40 pages).

[31] Liang, G.R., and Liu, C.R. (1986). "Automatic NC Programming as a Decomposable Frame Problem," *Integrated and Intelligent Manufacturing*, ASME, New York, pp. 171-179.

[32] Link, C.H. (1976). "CAPP-CAM-I Automated Process Planning System," *Proceedings of the 13th Numerical Control Society Annual Meeting and Technical Conference*, Cincinnati, OH.

[33] Loney, G.C., and Ozsoy, T.M. (1987). "NC Machining of Free Form Surfaces," *Computer Aided Design*, Vol. 19, No. 2, pp. 85-90.

[34] Niebel, B.W. (1965). "Mechanized Process Selection for Planning New Designs," ASME paper, No. 737.

[35] Oetjens, T.J. (1985). "Computer Design of Sculptured Surfaces for CAM," *Proceedings of AutoFact '85*, pp. 14.19-14.36.

[36] Oetjens, T.J. (1987). "Automotive CAD/CAM Die: The Sculptured Surface," *Proceedings of AutoFact '87*, pp. 11.1-11.12.

[37] Persson, H. (1978). "NC Machining of Arbitrarily Shaped Pockets," *Computer-Aided Design*, Vol. 10, No. 3, pp. 169-174.

[38] Piegl, E. (1989). "Key Developments in Computer-Aided Geometric Design," *Computer Aided Design*, Vol. 21, No. 5, pp. 262-273.

[39] Pratt, M.J. (1983). "Three-D Geometric Product Definition for CADCAM: The Present State and Future Trends," *Proceedings of International Mechanical Engineering Conference*, Cambridge, pp. 71-78.

[40] Preiss, K., and Kaplansky, E. (1986). "Automated CNC Milling by Artificial Intelligence Methods," *Proceedings of AutoFact V*, pp. 240-259.

[41] Preiss, K. (1989). "Automated Mill Pocketing Computations," *Proceedings of Advanced Geometric Modeling for Engineering Applications Conference*, pp. 247-260.

[42] Saggu, J.S., and Fielding, J.P. (1989). "An Integrated CADCAM Method for the Design and Construction of Aircraft Wing Components," *Computers in Industry*, Vol. 12, No. 2, pp. 123-130.

[43] Sakuta, T., Kawai, M., and Amano, Y. (1987). "Development of an NC Machining System for Stamping Dies by Offset Surface Method," *Proceedings of AutoFact '87*, pp. 2.13-2.27.

[44] Schenk, D.E. (1966). "Feasibility of Automated Process Planning," Ph.D. Thesis, Purdue University, West Lafayette, IN.

[45] Suh, Y.S. and Lee, K. (1990). "NC Milling Tool Path Generation for Arbitrary Pockets Defined by Sculptured Surfaces," *Computer-Aided Design*, Vol 22, No. 5, pp. 273-284.

[46] Takai, K. (1988). "Free-form Surface Generation System for Camera Design," *USA-Japan Cross Bridge*, Vol. 2, pp. 939-946.

[47] Takeuchi, Y., Morita, K., and Sakamoto, M. (1986). "Development of a Personal CAD/CAM System for Metal Moulds," *Machine Tool Design and Research*, No. 26, pp. 55-60.

Chapter 9

Intelligent Rough Machining of Sculptured Parts — Planning and Programming

HUI LI, ZUOMIN DONG, and GEOFFREY W. VICKERS
Department of Mechanical Engineering, University of Victoria

9.1 Introduction

9.1.1 Background

Sculptured parts are widely used in the aerospace, automobile, shipbuilding and plastic mold manufacturing industries. Increasing applications of sculptured surfaces to the profile of industrial products have been found due to their functional and aesthetic properties.

Sculptured parts are characterized by free-form surfaces and nonanalytical contours. The manufacture of these parts with free-form surface machining is more difficult compared with the production of parts with plane and conic surfaces. Furthermore, planning sculptured part machining with high productivity is a more challenging task.

The planning of sculptured part machining includes machining strategy determination and tool path generation. Significant research has been done in automated tool path generation from CAD models [5, 22, 27, 9]. However, the machining strategy planning often relies on an experienced machinist or the guidelines provided by a machining handbook. These manual approaches frequently lead to inefficient tool paths and require considerable lead time.

Machining sculptured parts is generally completed through two stages: rough machining and finish machining. Finish machining is aimed at generating adequate tool paths for producing the final shape of parts with geometric accuracy and surface finish as the major concerns. Rough machining is used to remove excess stock material, and productivity is most important. When a sculptured part is machined from a prismatic stock a large amount of rough machining, sometimes up to 90 percent of the total machining, is required [18]. This is due to the significant shape difference between the sculptured part and the commonly used parallelepiped stock. Machining time reduction in rough machining can considerably improve the productivity of sculptured part manufacturing and subsequently lower production costs. Unfortunately, most of the previous work on sculptured part design and manufacturing has been focused on the precise description and machining of sculptured surfaces. The productivity improvement of sculptured part manufacturing, considering the shapes of the part and stock, at the rough machining stage has seldom been addressed or systematically studied.

This chapter focuses on the productivity improvement of automated sculptured part rough machining. New methodologies of sculptured part rough machining, based on the quantitative intelligent system approach which combines geometric modeling, artificial intelligence, and numerical optimization, is presented. Starting from the geometric models of a sculptured part and a parallelepiped stock, the approach decomposes a complex 3-D rough machining problem into 2-D contour machining problems of several slicing layers. The efficiency of various tool path patterns for given cutting layer shapes is studied. A method for automatically identifying the optimal tool path pattern for a cutting layer using fuzzy pattern classification and identification is developed. In addition to the optimal tool paths, the approach also optimizes the machining parameters, including feed rate, depth of cut, number of cutting layers and distribution of cutting depths for each cutting layer. The method can automate the optimal CNC tool path generation and significantly reduce the total machining time. Several examples are used to illustrate the approach.

9.1.2 Related Work

Previous work related to optimal sculptured part rough machining can be classified into three major categories: automated NC tool path generation, computer-aided process planning (CAPP) of sculptured part machining, and adaptive control of machining processes.

Research on automated CNC tool path generation was primarily focused on finding the appropriate tool trajectories that can precisely embrace the details of the sculptured surfaces. The quality of machined surface and gouge-free tool paths was the major concern. Productivity of the planned tool path was brought into consideration recently. Guyder proposed an efficient approach for $2\frac{1}{2}$-axis pocketing by automatically reducing the number of cutter retractions [12]. The efficiency of the stair case (or parallel offset) and window frame (or stock offset) tool path patterns for face milling convex polygons was investigated by Wang and colleagues [27]. To develop a general approach which not only automates the CNC programming for sculptured part machining, but also produces the optimal tool paths and machining parameters to achieve the maximum productivity, many fundamental research issues need to be studied.

Research on CAPP of sculptured surfaces machining was concentrated on identifying appropriate machining parameters including feed rate, depth of cut, and cutting speed. Traditionally these machining parameters are determined manually based on experiences and suggested values provided by the machining handbooks. Significant efforts have been made to generate computer databases of production experiences and machinability data

[23, 26, 20]. The selection of machining parameters is automatically conducted by retrieving existing knowledge from these databases and by knowledge reasoning. Artificial intelligence (AI) techniques, particularly the knowledge-based system approach, have been widely used in developing the new generative type CAPP systems [16, 3]. These intelligent systems, functioning as experienced process planners, can automatically generate process plans for sculptured surface machining. Another advantage of this approach is its flexibility in allowing constant updates of the manufacturing knowledge base without modifying the process plan synthesis programs. The drawback of this type of pure knowledge-based systems is the difficulty of comparing all feasible machining plans and generating the optimal plan for a given part based on accurate productivity evaluation.

The adaptive control of the machining process, which seeks the optimal performance of a machining process, has been an interesting research topic in manufacturing engineering. Extensive work has been done in this area [6, 15, 21]. Centner and Idelsohn developed an adaptive control system for the milling process. The system consists of three major parts: numerical control unit, sensors, and adaptive controller. Various input data required by the control programs, including cutter location, feed rate, and spindle speed, are fed into the numerical control unit. The sensors pick up some key feedback signals such as spindle torque, cutting force, spindle vibration, and temperature variations during the sculptured part machining. These signals are sent to the adaptive controller, which in turn estimates the cutting performance in terms of metal removal rate or surface finish. Based upon these assessments, the controller adjusts the feed rate and spindle speed on-line to maintain high cutting performance. The success of this approach relies heavily on the sensors to provide accurate feedback information. However, these sensors are normally expensive to install and maintain. The hostile machining environment frequently makes the reliability of these sensors unpredictable.

In the study of adaptive control of the machining process, various machining process models have been introduced. McGoldrick and Hijazi proposed an analytical cutting process model for predicating the cutting power and cutting torque in the adaptive control of end-milling [15]. A more precise cutting force model, which considered the influence of cutter shape and the intersection geometry of cutter and workpiece, was recently introduced [1]. These machining process models, if combined with the geometric model of sculptured part and stock, can be used to identify the optimal machining parameters for sculptured part rough machining, through machining process simulation. The need for expensive and inconvenient sensors can be eliminated. So far, most of the research has been done under specific machining conditions on special machining tools using dedicated hardware. A generally applicable machining process model and adaptive control facility are not yet available.

9.2 Offset Machining versus Contour-map Machining

Generally speaking, two basic strategies are used in the rough machining of sculptured parts: the offset approach and the contour-map approach.

To machine a sculptured part following the offset strategy, the tool paths are generated on the iteratively formed offset surfaces. The first layer of offset surface is obtained by offsetting the distance of cutting depth from the sculptured surface of a part. The second layer of offset surface is obtained by offsetting the same distance from the formed first layer of offset surface, and so forth. The offsets are done along the 3-D surface normals of the original and offset surfaces. A ball end-mill is normally needed for 3-D offset machining, as illustrated in Figure 9.1a.

To machine a sculptured part following the contour-map strategy, the stock is machined through a number of top-down parallel layers using a flat end-mill, as illustrated in Figure 9.1b. At each cutting layer, tool paths are laid out following a specific 2-D tool path pattern to remove the excess material. A final cutting pass along the surface is usually needed to clear the gouges created during the contour-map machining.

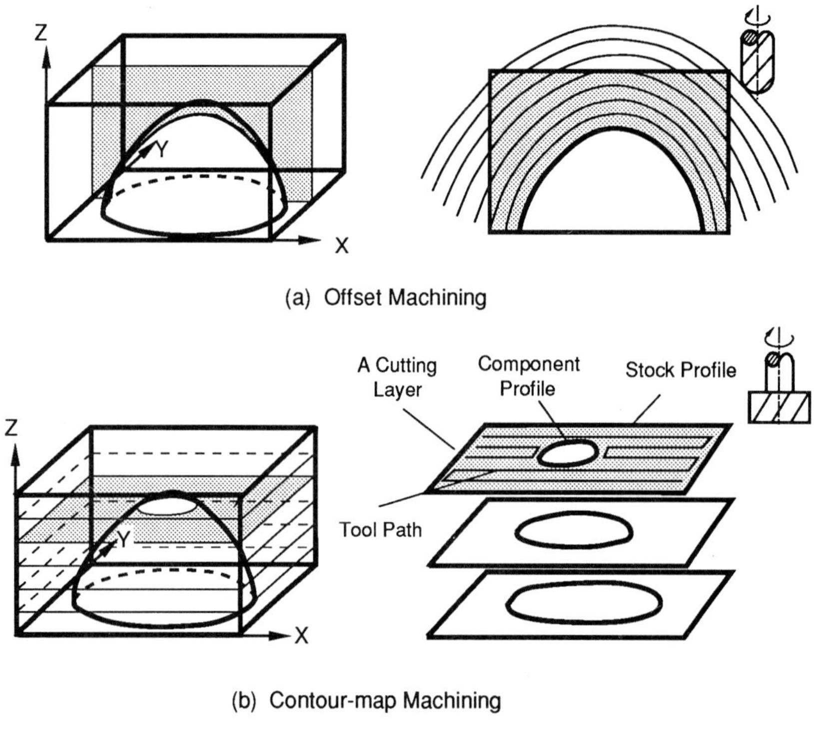

Figure 9.1: Machining Strategy

The choice of machining strategies yields totally different tool paths and heavily influences the productivity of rough machining. In most cases, the 3-D offset approach is less efficient due to the substantial waste cuts, which are introduced when the shapes of the part and the stock are significantly different, as illustrated by Figure 9.1a. But the 3-D offset approach provides a better finished surface, and is more suitable for machining complicated pockets. The contour-map approach, on the other hand, is generally an efficient strategy for rough machining. First, the approach converts a complex 3-D machining problem into a series of simple 2-D machining problems, thus making the planning of productive machining easy to do. Second, a more efficient type of cutter—the flat end-mill—can be used throughout the sculptured part rough machining. Due to these reasons the contour-map strategy is studied in depth. An efficient contour-map approach for sculptured part rough machining is introduced.

9.3 Contour-map Machining and Its Tool Path Patterns

Using the contour-map strategy, a workpiece is machined through a number of parallel cutting layers. At each cutting layer, the cutter follows a specific tool path pattern to remove the excess material. The machining productivity is largely determined by the applied tool path pattern, the starting point of the cutting, and the cutting direction [27].

This section outlines a number of feasible tool path patterns for machining cutting layers with various layer shapes. These shapes cover the single island, no island, and multiple island cases.

9.3.1 Feasible Tool Path Patterns

9.3.1.1 For Cutting Layers with a Single Island

- **Component-offset pattern**—The component-offset tool path pattern, illustrated in Figure 9.2a, is generated along a series of curves offset outward from the component contour. Cutting tool motion starts at the outside and is limited to motion within the stock boundary. Any offset curve paths that go outside the confines of the stock profile are replaced either by noncutting fast feed rate motions (for short distances) or by rapid traverse (and associated retract and plunge) motions along the boundary. This pattern is frequently used in practice due to its simplicity [25]. The disadvantage is that it contains numerous noncutting fast feed rate and rapid traverse motions as well as numerous plunge and retract motions.

- **Stock-offset pattern**—The stock-offset tool path pattern (or window frame pattern [27]), illustrated in Figure 9.2b, is generated by moving the cutting tool along a series of curves offset inwards from the stock contour. Intermediate tool path loops are created in the vicinity of the component as the offset curves and the component shape intersect. Since there is little noncutting motion in the tool path generally, the approach is reasonably efficient for most cutting layer shapes.

- **Stock/component-offset pattern**—The stock/component-offset tool path pattern, illustrated in Figure 9.2c, is generated by a combination of the previous two patterns.

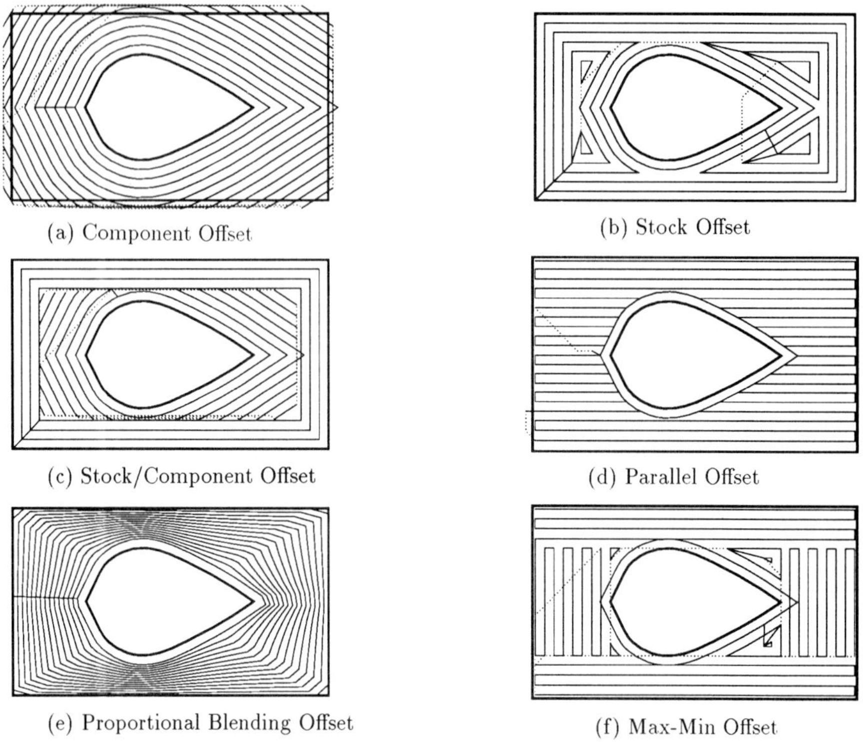

Figure 9.2: Feasible Tool Path Patterns

The stock-offset pattern is initially adopted. At an intermediate point, a switch from the stock-offset pattern to the component-offset pattern is made. The switch point is identified as that which generates the shortest tool path. This point usually occurs when the tool path intersects the first offset curve from the component contour. The pattern is efficient where the stock and component profiles have similar proportions.

- **Parallel-offset pattern**—The parallel-offset tool path pattern (or stair case pattern [27]), illustrated in Figure 9.2d, is generated by moving the cutting tool along a series of offsets parallel to the longest edge of the stock. At the intersection of the parallel offsets with the component contour, the cutter follows the periphery of the component contour until it can continue a parallel offset pattern. The cutter returns to any incomplete areas and continues with a parallel cutting pattern. This pattern is most efficient where the component profile forms a relatively small island.

- **Proportional-blending-offset pattern**—The proportional-blending-offset tool path pattern [24] is illustrated in Figure 9.2e. The tool path is generated by moving the cutting tool along a series of linearly interpolated paths between the stock and component contours. The number of tool path loops is determined by the maximum distance between the stock and component contours and the maximum allowable cross-feed distance (a fraction of end-mill diameter). In this pattern the feed rate is adjusted to maintain a constant maximum material (cutting volume) removal rate. The pattern is efficient for certain intricate pattern shapes.

- **Max-min-offset pattern**—The max-min-offset tool path pattern [24] is illustrated in Figure 9.2f. The tool path is generated by moving the cutting tool along a series of parallel offsets between the maximum stock and component contours and between the minimum stock and component contours. Orthogonal parallel paths with the stock and component contours are also made, and any remaining material left at the corners of the component is cleaned up using the stock-offset pattern. This approach is most efficient for parallelepiped stock and relatively larger center islands.

9.3.1.2 For Cutting Layers with No Island

According to the previous research [27], the most efficient tool path pattern for cutting layers shaped as a convex polygon with no center island is the parallel-offset approach, as shown in Figure 9.3a. For cutting layers with a nonconvex polygon shape and without center island, as shown in Figure 9.3b, the stock-offset approach is generally recommended.

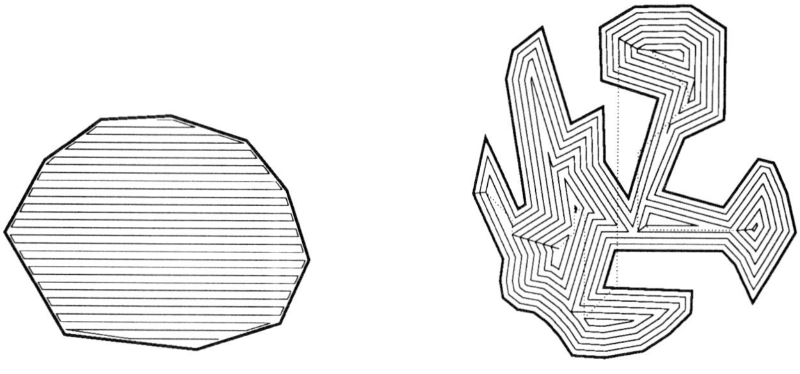

(a) Convex Polygon (b) Non-Convex Polygon

Figure 9.3: Layer Shape without Islands

9.3.1.3 For Cutting Layers with Multiple Islands

In some applications, multiple islands may exist at certain cutting layers, identification of the most efficient tool path pattern for these cases is more complex. In general, the stock-offset and parallel-offset tool path pattern can be applied, as illustrated by Figure 9.4a. Another approach, *convex hull approximation*, is generally applicable for a large variety of cutting layer shapes and leads to efficient rough machining [24]. The approach generates a

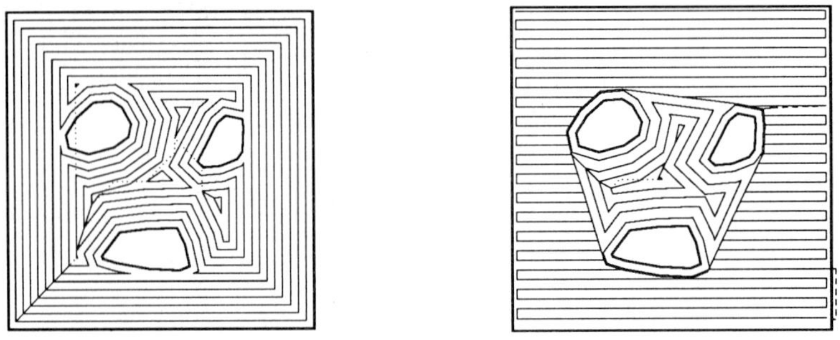

(a) Stock Offset Approach (b) Convex Hull Approach

Figure 9.4: Layer Shapes with Multiple Islands

convex hull that contains all islands as shown in Figure 9.4b and treats the convex hull as a single island to generate an appropriate tool path pattern. The previously discussed rules for cutting layers with a single island can be applied. The parallel-offset tool path pattern is used in in Figure 9.4b, and the stock-offset approach is applied to the area between the convex hull and the island profiles.

9.3.2 Tool Path Pattern Machining Productivity Assessment

For a given cutting layer, application of different tool path patterns will result in different total tool path length and consequently different machining times. To identify the best tool path pattern with maximum productivity, a measure needs to be introduced for the performance or productivity assessment of various tool path patterns.

Total machining time, T, is a commonly used productivity measure. This total machining time is determined by the total tool path length, L, and the adopted feed rate, f. During rough machining, the ideal cross-feed distance is normally given as a fraction of cutter diameter to ensure the residuals left by the last cut can be removed and high productivity can be achieved. For a given cutting layer, the cross-feed distance is adjusted around its ideal value to ensure an integer number of cutting passes. Given the cross-feed distance, the depth of cut, and the maximum material removal rate, which is constrained by the machine tool and the cutter, a nominal feed rate, f_n, is determined. The calculation of total machining time, T ($T = \frac{L}{f_n}$), thus becomes the calculation of total tool path length, L. In rough machining, the cutter motions include cutting, fast move, and rapid traverse (with cutter plunge and retract). Each of these motions has different speeds. Four parameters, λ_{fm}, λ_{rt}, λ_{pl}, and λ_{re} are introduced to compensate for the different rates of tool motion (fast move, rapid traverse, cutter plunge and cutter retract) and to calculate an equivalent tool path length, L_{eq}:

$$L_{eq} = L_{cut} + \lambda_{fm} L_{fm} + \lambda_{rt} L_{rt} + \lambda_{pl} L_{pl} + \lambda_{re} L_{re} \tag{9.1}$$

If a sculptured part and a stock, as well as the depth of cut, cross-feed distance, and maximum material removal rate are given, the equivalent length of tool path for a given tool path pattern at a specific cutting layer can be calculated. The optimal tool path pattern with the minimum equivalent tool path length (or shortest total machining time) can be thus identified based upon the calculations.

9.3.3 An Example of Optimal Tool Path Pattern Identification

The various tool path patterns discussed previously have been tested using various cutting layer shapes. The identification of the optimal tool path pattern and the productivity assessment of various tool path patterns are illustrated using a practical example.

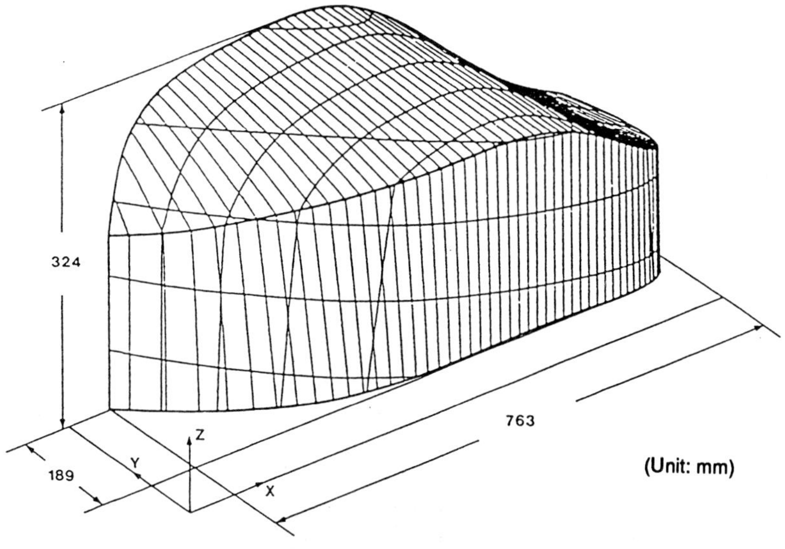

Figure 9.5: Ship Hull Hydrodynamic Model

In this example, the sculptured part to be machined is the ship hull hydrodynamic testing model shown in Figure 9.5. The ship hull model was taken from a lofted table of a hard chine sailing boat [2]. The stock is a prismatic block measuring $850 \times 250 \times 400$ mm. Fifty cutting layers are generated by slicing the part using hypothetical parallel plans. The six tool path patterns are used to generate six different groups of tool paths for each cutting layer. The equivalent tool path lengths of these cutting layers are calculated and plotted in Figure 9.6. The six curves in Figure 9.6 illustrate the performance of the six tool path patterns on various cutting layers. The parallel-offset tool path pattern seems to be the best for the top 17 cutting layers, while the max-min-offset tool path pattern presents shorter total cutting time than the other alternatives for the cutting layers from 18 to 50.

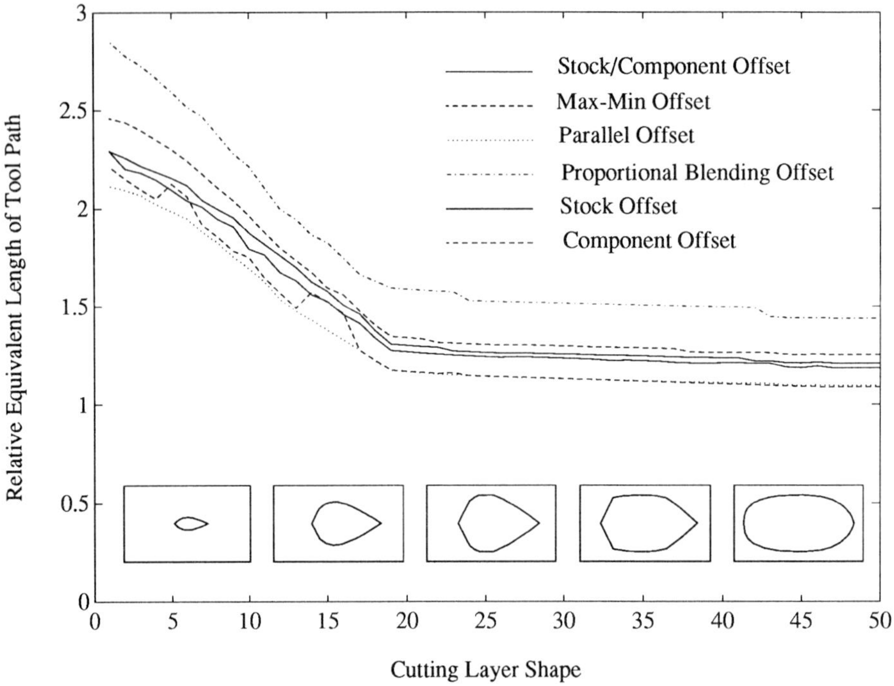

Figure 9.6: Feasibility Study of Tool Path Patterns

9.4 Fuzzy Optimal Tool Path Pattern Identification

To automatically identify the optimal tool path pattern for a given cutting layer, a simple and direct approach is to evaluate and compare the equivalent tool path lengths for all the six known tool path patterns. However, the tool path lengths can only be calculated after the generation of all tool paths. The tool path generation for sculptured part machining requires extensive computation. The exhaustive comparison of all feasible tool path patterns requires all possible tool paths to be generated and their equivalent tool path lengths to be calculated. A significant amount of computation is needed by this approach.

In practice, an experienced machinist applies a straightforward heuristic approach to this problem. After the suggested total number of cutting layers has been picked up from a handbook, these cutting layers are divided into a number of cutting layer groups according to their cross-section shapes. A specific tool path pattern, which is considered to be the optimal approach for a cutting layer group, is assigned to every cutting layers of the groups. This group approach is based on the fact that adjacent cutting layers may have similar shapes and share same optimal tool path pattern.

To bypass the extensive computation required by exhaustive tool path generation and comparison in automated optimal tool path identification, a heuristic-based approach, similar to the above group approach used by an experienced machinist, has been investigated. We have found that different cutting layer shapes and tool path patterns produce a numerous variety of combinations and no clear associations exist among these combinations. It is difficult to derive simple heuristic-based guidelines to identify the optimal tool path pattern and handle the uncertainty and ambiguity of the optimal tool path pattern identification problem. Due to these reasons, the fuzzy cutting layer shape clustering and fuzzy pattern matching techniques are used.

The automated optimal tool path pattern identification is based on the fuzzy pattern clustering and matching principle, and conducted through three steps: (a) cutting layer clustering, (b) tool path pattern analysis for prototype layer shapes with each prototype layer shape as the representative shape for a group of cutting layers, and (c) optimal tool path pattern identification. In the cutting layer clustering, the cutting layers of a sculptured part are classified into a number of cutting layer clusters according to their shapes. The cluster centers of these groups are regarded as the prototype or primitive layer shapes. In the prototype layer shape analysis conducted for each prototype layer shape, the equivalent tool path lengths of all six feasible tool path patterns are calculated and compared to determine the optimal tool path pattern for the prototype layer shape. At the last step, the optimal tool path pattern for each given cutting layer is determined by matching the shape of the cutting layer to the nearest prototype layer shape and assigning the optimal tool path pattern identified for the prototype layer shape.

9.4.1 Fuzzy Set and Fuzzy Pattern Clustering

Before we start to introduce the fuzzy optimal tool path pattern identification method, the fuzzy set and fuzzy pattern clustering method are briefly reviewed.

9.4.1.1 Fuzzy Set

A fuzzy set **A** on set **X** is a set of ordered pairs:

$$\mathbf{A} = \{x, \chi_{\mathbf{A}}(x)\}, \qquad x \in \mathbf{X}.$$

where $\chi_{\mathbf{A}}(x)$ is the membership function. The value of $\chi_{\mathbf{A}}(x)$ for $x \in \mathbf{X}$ (x is called a fuzzy number) represents the degree of belongingness or the grade of membership of x in **A**.

Based upon this definition of fuzzy set, fuzzy mathematics models for physical relations are established. For example, if **S** is a subset of real numbers and $\mathbf{S} \in [0,1]$, a fuzzy set **A** represents the cutting layers with "big island" and x ($x \in \mathbf{S}$) represents the ratio of island area to stock cross-section area, a possible membership function, representing the likelihood of a big island, $\chi_{\mathbf{A}}(x)$, can be given as

$$\chi_{\mathbf{A}}(x) = \left\{ \begin{array}{ll} 0 & 0 \leq x < 0.1 \\ \frac{1}{2} + \frac{1}{2}\sin\frac{\pi}{0.8}(x - 0.5) & 0.1 \leq x < 0.9 \\ 1 & 0.9 \leq x \leq 1 \end{array} \right\}$$

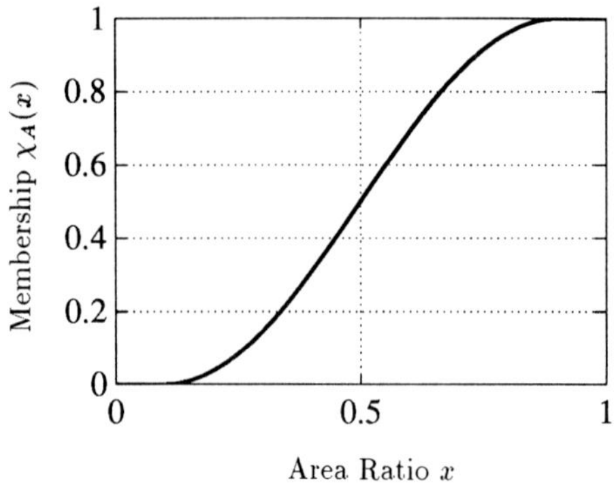

Figure 9.7: Example of Fuzzy Set Membership Function

This membership function is shown in Figure 9.7.

9.4.1.2 Hard Pattern Clustering versus Fuzzy Pattern Clustering

Clustering set $\mathbf{X} = \{x_1, x_2, \ldots, x_n\}$ means to partition $\mathbf{X}$ into a number of subsets or "clusters," $\mathbf{X}_1, \mathbf{X}_2, \ldots, \mathbf{X}_c$, so that the members of each cluster are more similar to one another than to members of other clusters. Suppose $\mathbf{V}_{cn}$ denote the $(c \times n)$ vector space, $\mathbf{M}_c$ denote hard c-partition of $\mathbf{X}$, and $\mathbf{M}_{fc}$ denote fuzzy c-partition of $\mathbf{X}$. Set $\mathbf{M}_c$ and $\mathbf{M}_{fc}$ can be represented in a matrix form [4]

$$\mathbf{M}_c = \{\mathbf{U} \in \mathbf{V}_{cn} \mid u_{ik} \in \{0,1\} \; \forall i,k; \; \sum_{i=1}^{c} u_{ik} = 1 \; \forall k; \; 0 < \sum_{k=1}^{n} u_{ik} < n \; \forall i\} \tag{9.2}$$

$$\mathbf{M}_{fc} = \{\mathbf{U} \in \mathbf{V}_{cn} \mid u_{ik} \in [0,1] \; \forall i,k; \; \sum_{i=1}^{c} u_{ik} = 1 \; \forall k; \; 0 < \sum_{k=1}^{n} u_{ik} < n \; \forall i\} \tag{9.3}$$

where u_{ij} is the degree to which the pattern x_j belongs to class $\mathbf{X}_i$. In hard c-partition, u_{ij} is bounded to 0 or 1, resulting in a finite c-partition for set $\mathbf{X}$. In fuzzy c-partition, u_{ij} can be any value between 0 and 1. This, in principle, produces infinite fuzzy c-partitions for set $\mathbf{X}$. For example, the following two matrices $\mathbf{U}$ and $\mathbf{U_f}$ are typical hard 2-partition and fuzzy 2-partition, respectively, for the set $\mathbf{X} = \{x_1, x_2, x_3\} = \{\text{circle, ellipse, rectangle}\}$ based upon geometrical similarity.

$$\mathbf{U} = \begin{matrix} x_1 & x_2 & x_3 \\ \end{matrix} \begin{bmatrix} 1 & 1 & 0 \\ 0 & 0 & 1 \end{bmatrix}; \qquad \mathbf{U}_f = \begin{matrix} x_1 & x_2 & x_3 \\ \end{matrix} \begin{bmatrix} 0.75 & 0.63 & 0.36 \\ 0.25 & 0.37 & 0.64 \end{bmatrix}$$

9.4.1.3 Fuzzy c-Means Pattern Clustering

Fuzzy c-means pattern clustering is conducted by minimizing dissimilarity. For a given set, $\mathbf{X}$, and a predefined number of subsets to be divided, c, the overall dissimilarity between each data point, x_k, and each fuzzy prototype (the center point of the cluster), v_i, is to be minimized to determine the members of each subset. The objective function of this optimization has the form

$$J_{cm}(\mathbf{U},\mathbf{v}) = \sum_{k=1}^{n}\sum_{i=1}^{c}(u_{ik})^m (d_{ik})^2 \tag{9.4}$$

where $\mathbf{U} \in \mathbf{M}_{fc}$ is a fuzzy c-partition of $\mathbf{X}$; $\mathbf{v} = (v_1, v_2, \ldots, v_c)$, with v_i as the ith cluster center or prototype, $1 \le i \le c$; and

$$d_{ik} = d(x_k, v_i) = \sqrt{\sum_{j=1}^{p}(\chi_j(x_{kj}) - \chi_j(v_{ij}))^2} \tag{9.5}$$

which is the measurement of dissimilarity between the data point x_k and the fuzzy prototype v_i with p being the dimension of feature space and m as the weighting exponent, $m \in [1, \infty)$. The optimal solution $\mathbf{U}^*, \mathbf{v}^*$ is obtained by

$$\min_{\mathbf{U},\mathbf{v}} J_{cm}(\mathbf{U}, \mathbf{v}) \tag{9.6}$$

An algorithm for solving this optimization problem is given by Bezdek [4].

9.4.2 Shape-based Cutting Layer Clustering

To apply the fuzzy clustering and fuzzy pattern matching techniques to automate the optimal tool path pattern identification, a parametric description of the cutting layer shapes is introduced. The study has been focused primarily on the single island cutting layer shapes, which form the majority of the application. The cutting layer is assumed to have a single island part profile with a rectangle stock boundary. This assumed single island shape can be easily extended to more complicated cases with multiple islands. The shape description parameters include:

- Area to dimension ratio: $p_1 = A_i/ab$
- Circumference to dimension ratio: $p_2 = L_i/2(a+b)$
- Length to width ratio: $p_3 = b/a$
- Symmetry along length: $p_4 = u_m/a$
- Symmetry along width: $p_5 = v_m/a$
- Inclination in x direction: $p_6 = x_i/l$
- Inclination in y direction: $p_7 = y_i/w$
- Orientation of the island: $p_8 = \theta$
- Part-to-stock area ratio: $p_9 = A_i/A_s$
- Cutter-to-stock area ratio $p_{10} = d_c^2\pi/4A_s$

where A_i and A_s are the areas of component and stock; L_i, and L_s are the circumferences of the component and stock; parameters a and b are lengths of the long and short edges of the minimum rectangle that encloses the component profile; x_i and y_i specify the center

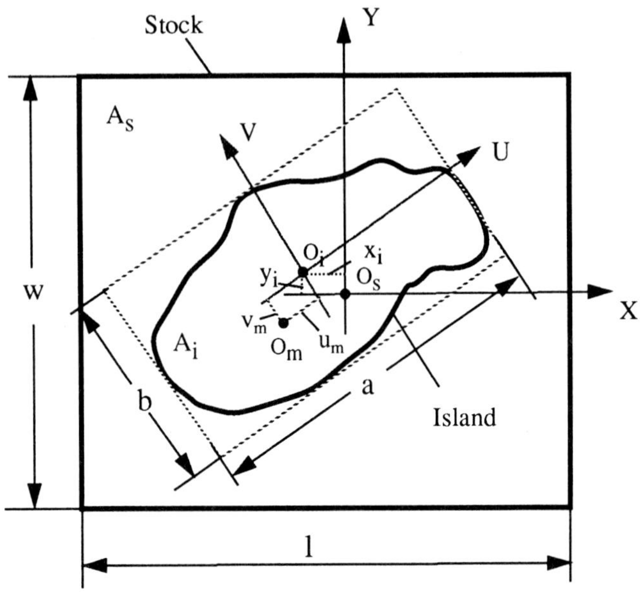

Figure 9.8: Cutting Layer Shape Description

point of the minimum rectangle; u_m and v_m specify the geometric center of the component profile; and d_c is the diameter of the cutter. The relation of these parameters is illustrated in Figure 9.8. Among the ten defined parameters, the first five, $p_1 - p_5$, specify the island shape of the cutting layer, the last five, $p_6 - p_{10}$, specify the relative size of the island, the relative size of the used cutter, and the location and orientation of the island in the cutting layer. First, Bezdek's fuzzy c-means clustering algorithm [4] is used to classify the layer shapes into a given number of prototype layer shapes. The clustering is done by finding a given number of cluster centers or fuzzy prototypes in the parametric space to minimize the overall dissimilarity between each data point and each fuzzy prototype. Second, the number of clusters is determined by cluster validity analysis using the partition coefficient method [4]. The partition coefficient, $F(\mathbf{U}, c) = \sum_{k=1}^{n} \sum_{i=1}^{c} (u_{ik})^2/n$ is a measure of clustering validity. The better the clustering, the larger the value of the partition coefficient will be. The identification of the best fuzzy c-partition is formulated as another optimization problem—the maximization of partition coefficient (or cluster validity) with respect to the number of clusters (or subsets):

$$\max_{c}\{F(\mathbf{U}, c)\} \tag{9.7}$$

where $\mathbf{U}$ is obtained by the fuzzy c-means method.

9.4.3 Prototype Layer Shape Analysis

After all cutting layers are clustered according to their shapes and the center of each shape cluster (or prototype) is identified, the prototype layer shape analysis is conducted. This analysis is used for identifying the optimal tool path pattern to the prototype and finding the application limits of the identified optimal tool path pattern within the shape cluster. Identification of the optimal tool path pattern is conducted by generating the tool paths following the six feasible tool path patterns and by calculating and comparing their tool path lengths. The prototype layer shape is described using the shape description parameters discussed previously. The application limits of the identified optimal tool path pattern is identified by studying the performance of the tool path pattern with the variation of the shape description parameters. The location and orientation parameters, p_6, p_7, and p_8, have been found to have minor influence on the total equivalent tool path lengths. The application limits of the identified optimal tool path patterns to other cutting layers within the shape cluster is thus studied by varying the part-to-stock area and cutter-to-stock area parameters, p_9 and p_{10}. The application limits of the cutting layer shapes sharing the same optimal tool path pattern are calculated based upon the corresponding optimal tool path patterns and are represented by several decision functions. These decision functions are recorded in a database for later use. An example illustrating the analysis results for a typical prototype layer shape is shown in Figure 9.9. In this figure, the applicable domains of parallel offset and max-min offset tool path patterns as the optimal tool path pattern are illustrated. The plot is generated with respect to the two most influential shape parameters, p_9 and p_{10}. The decision function, which represents the application limits, is approximated by a cubic polynomial. Higher-order polynomials can be used for the decision functions to obtain more accurate results. The meaning of point A will be discussed in the following section. In some cases, the application limits of an optimal tool path pattern may not overlap with the shape cluster boundaries. Multiple optimal tool path patterns may exist within a shape cluster, and an optimal tool path pattern may span several shape clusters.

9.4.4 Automated Optimal Tool Path Pattern Identification

The optimal tool path pattern for any given cutting layer is identified through a shape pattern matching process. Matching between the cutting layer to a prototype with the closest layer shape is conducted using the fuzzy pattern matching scheme. This machining is conducted using the weighted relative Euclidian distance in the space formed by the introduced shape description parameters:

$$d_{ik} = d(x_k, v_i) = \sqrt{\frac{1}{p}\sum_{j=1}^{p} w_j(\chi_j(x_{kj}) - \chi_j(v_{ij}))^2} \tag{9.8}$$

where w_1, w_2, ..., w_p are weighting coefficients to include different weights on these fuzzy parameters. The shape description parameters of the cutting layer are then applied to the stored decision functions to access the applicability of the optimal tool path patterns

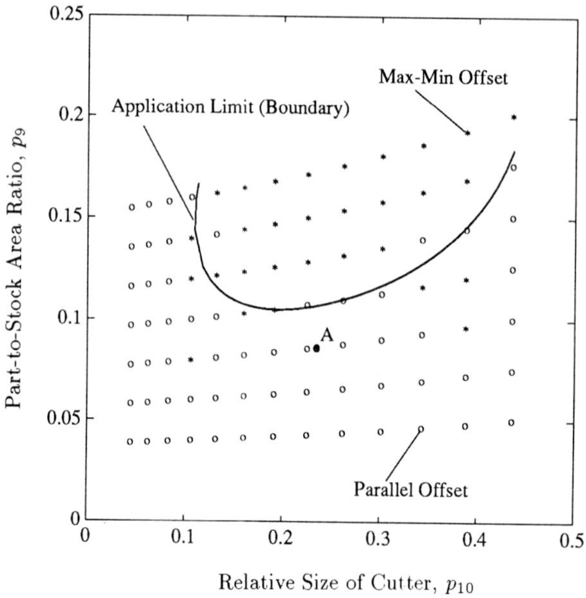

Figure 9.9: Optimal Tool Path Pattern Identification

associated with the prototype. The applicable optimal tool path pattern, recorded in the database, is finally assigned to the cutting layer of consideration.

The cutting layer clustering and optimal tool path pattern identification method focuses on the similarity of cutting layer shapes and several typical cutting layers. The approach works on several cutting layer groups, rather than all cutting layers to avoid the exhaustive tool path generation and tool path length calculation. The method can significantly reduce the required computations in generating optimal tool path patterns.

9.4.5 An Example for Automated Optimal Tool Path Pattern Generation

The procedure for automatically identifying the optimal tool path pattern for a given cutting layer is illustrated using the same ship hull rough machining example discussed previously and shown in Figure 9.5.

Applying the fuzzy clustering approach of Eqs. (9.6) and (9.7), the optimal number of cutting layer shape clusters was found to be three. The three groups of layer shapes and their corresponding prototypes are illustrated in Figure 9.10. The cutting layer shape that is closest to the center of the shape cluster is taken as the prototype and used as the representative of the cutting layer cluster. For instance, layer 9 in cluster II is selected as the prototype cutting layer shape. The tool path pattern analysis is then conducted for

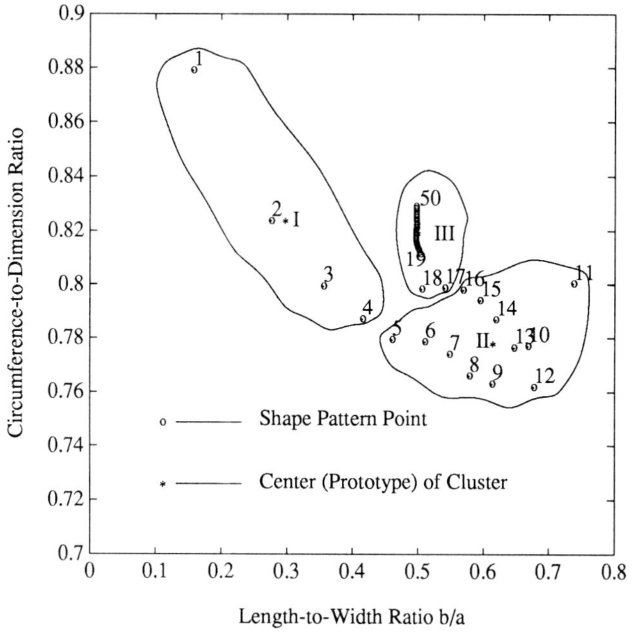

Figure 9.10: Clustering Result

this identified cluster prototype—-layer 9, as shown in Figure 9.11. Two optimal tool path patterns and the application limits of these two tool path patterns are found, as illustrated previously in Figure 9.9.

For a given cutting layer, such as layer 8, the closest prototype layer shape can be matched using Eq. (9.8). The result of this match is layer 9 of group II. Based on the part-to-stock area ratio and cutter-to-stock area ratio, this layer shape is expressed as the point A in Figure 9.9. From the application limits specified by the decision function, the optimal tool path pattern for the given cutting layer is found to be the parallel-offset approach. This result is correct if compared with the direct measurement of equivalent tool path lengths derived from the six feasible tool path patterns.

9.5 Machining Parameter Optimization

Machining parameter optimization is another important part of optimal sculptured part rough machining. Traditional approaches for determining the machining parameters consider only the process-related parameters such as cutting speed, feed rate, and cross-feed distance. In our work, we associate part and stock geometry in the machining parameter optimization to achieve higher productivity. In addition, part geometry-oriented machining parameters, including the total number of cutting layers and the uneven distribution of cutting depths, are included in the optimization. The approach seeks the optimal machining parameters to

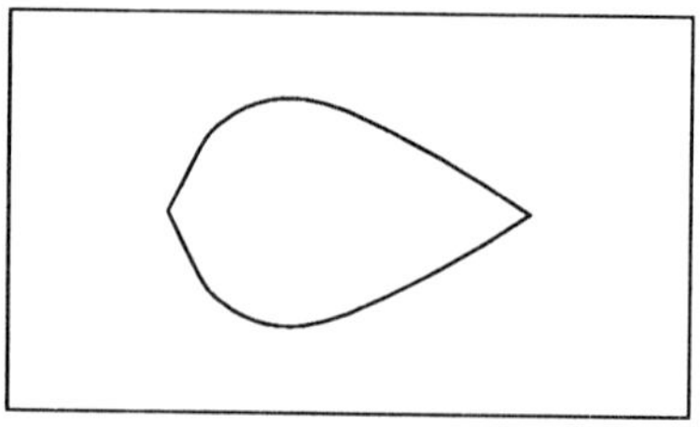

Figure 9.11: A Prototype Layer Shape

maintain the maximum allowable cutting volume per unit time, based upon the material volume derived from the geometric model of the part and stock. The complex and expensive sensors needed in the present adaptive control approach are eliminated.

This new method has revealed significant productivity improvement of sculptured parts rough machining. The method is outlined in this section.

9.5.1 Optimization Formulation

9.5.1.1 Objective Function of the Optimization

The objective of the optimization is to minimize the total machining time with respect to machining parameters. The general formulation of the total machining time, T, found in reference [8] for component machining is given as

$$T = T_c + T_m + T_u + T_d \frac{T_c}{T_l} \tag{9.9}$$

where T_c is the total cutting time, T_m is the tool approaching and retreating time, T_u is the loading and unloading time, T_d is the tool change time, and T_l is the tool life. In sculptured part rough machining, the batch size is usually small (frequently one piece) and the cutting time is usually long so that T is dominated by the first two terms of Eq. (9.9). Therefore the total production time for rough machining of sculptured parts may be approximated by

$$T \approx T_c + T_m \tag{9.10}$$

For a general sculptured part rough machining problem, the total cutting time and noncutting time (rapid traverse, fast move, etc.) are determined by the number of cutting layers N; feed rates applied to each cutting layer $f_1, f_2, \ldots, f_N$; depth of cut of each layer $d_1, d_2, \ldots, d_N$; the geometry of sculptured surface $f(x, y, z)$ and stock $s(x, y, z)$; as well as tool path patterns selected for each cutting layer $P_1, P_2, \ldots, P_N$. In short, the total production time T can be expressed as

$$T = F(N, f_1, f_2, \ldots, f_N, d_1, d_2, \ldots, d_N, f(x, y, z), s(x, y, z), P_1, P_2, \ldots, P_N)$$

This relationship can be expressed in an explicit form as

$$T = \sum_{i=1}^{N} t_i = \sum_{i=1}^{N} t_{ci} + \sum_{i=1}^{N} t_{mi} = \sum_{i=1}^{N} \sum_{j=1}^{n_{ci}} \frac{l_{cij}}{f_i} + \sum_{i=1}^{N} \sum_{k=1}^{n_{mi}} \frac{l_{mik}}{f_m} \tag{9.11}$$

where:

N	—	number of cutting layers
t_i	—	total machining time for the ith layer $(1 \le i \le N)$
t_{ci}	—	cutting time for the ith layer $(1 \le i \le N)$
t_{mi}	—	cutter rapid moving time for the ith layer $(1 \le i \le N)$
f_i	—	feed rate for the ith layer $(1 \le i \le N)$
f_m	—	feed rate of rapid movement
n_{ci}	—	number of cuts for the ith layer $(1 \le i \le N)$
l_{cij}	—	jth cutting length in machining the ith layer $(1 \le i \le N$ and $1 \le j \le n_{ci})$
n_{mi}	—	number of rapid movement for the ith layer $(1 \le i \le N)$
l_{mik}	—	kth moving length in machining the ith layer $(1 \le i \le N$ and $1 \le j \le n_{mi})$
$d_{c,i}$	—	cross feed distance of the ith layer $(1 \le i \le N)$
d_i	—	depth of cut of the ith layer $(1 \le i \le N)$

This formulation of total machining time is used as the objective function of the total machining time minimization. Since the shapes of the sculptured part and the stock are given, feed rate of rapid movement, f_m, is usually set to be a constant and the tool path patterns are determined by part/stock shape and the number of cutting layers N, the minimization is actually conducted with respect to the design variables: $N, f_1, f_2, \ldots, f_N$ and $d_1, d_2, \ldots, d_N$. Among these variables, the number of cutting layers, N and the distribution of cutting depth, $d_1, d_2, \ldots, d_N$, have not been considered in previous work.

9.5.1.2 Constraints of the Optimization

The constraints of the optimization include the physical constraints imposed by the machine tool and the cutting process, as well as the geometric constraints imposed by the consistent conditions among the number of cutting layers, N, cutting depths, d_i $(i = 1, 2, \ldots, N)$, and height of stock H:

Physical Constraints

maximum and minimum feed rate,	$f_{min} < f < f_{max}$
maximum and minimum depth of cut,	$d_{min} < d < d_{max}$
maximum cutting power,	$P_s \le P_{max}$
maximum cutting torque,	$T_s \le T_{max}$
maximum cutting force,	$F \le F_{max}$
maximum material removal rate,	$U \le U_{max}$
meaningful number of cutting layers,	$N_{min} \le N \le N_{max}$

Geometric Constraint $\sum_{i=1}^{N} d_i = H$

The machining tool and cutting process constraints, given by the above formulae, specify the limits on the cutting force, cutting power, and cutting torque. The functional relations of these quantities to the specified design variables need to be given to check these constraints. The empirical cutting power and a cutting torque models reported by McGoldrick and Hijazi [15] are used in the optimization:

$$P_s = 1.4238 n_s^{-0.578} f^{0.234} d^{0.355} d_c^{0.185} \tag{9.12}$$

$$T_s = 1.5371 n_s^{-0.072} f^{0.227} d^{0.324} d_c^{0.169} \tag{9.13}$$

where n_s is the spindle speed of the milling machine, f is feed rate, d is depth of cut, and d_c is cross-feed distance.

These models approximate the nonlinear relationship between the machining parameters and the cutting power or torque. A sensitivity analysis was conducted on these models in the research to find different influences of machining parameters on the cutting power and cutting torque. The analysis agrees with the fact that the unit cutting volume of milling is mainly determined by the feed rate, depth of cut, and cross-feed distance. The influence of cutting velocity to cutting volume (of unit time) is minor when the cutting velocity is high. Thus the spindle speed, n_s (or the cutting speed, v), may be omitted in the optimization.

9.5.2 Solution Method

The optimization formulated in the previous section is a "nested" mixed-discrete-constrained-nonlinear-programming (MDCNP) problem. In this formulation, the number of cutting layers, N, which determines the number of design variables and constraints, is also a design variable of the optimization. Due to this "nested" nature of the problem, the original methods for solving a MDCNP problem, [11, 7] cannot be used directly. In addition, the derivative of the above objective function has severe discontinuities, since the variation in the number of cutting layers and the tool path patterns used for generating the tool paths introduces discontinuous changes in tool path length (or production time). This precludes the use of derivative-based optimization methods.

A new procedure for solving this complex optimization problem is introduced. This procedure consists of two phases. The search for the optima is first conducted with respect to the feed rate and depth of cut for each cutting layer, $R = \{f_1, \ldots, f_N\}^t$ and $D = \{d_1, \ldots, d_N\}^t$. The number of cutting layers, N, is searched afterwards. This can be represented by

$$\min_{w.r.t.N} [\min_{w.r.t.R,D} T(R, D)] \tag{9.14}$$

Due to the discontinuities of the objective function, searches on the optimal feed rates, $\{f_1^*, \ldots, f_N^*\}^T$, and optimal depth of cuts, $\{d_1*, \ldots, d_N*\}^T$, are conducted using the mixed penalty approach [19], and the Powell's conjugate direction method for unconstrained optimization [17]. The search for the optimal number of cutting layers, N^*, at the second phase of the optimization, is conducted using the branch and bound strategy for integer programming [11, 19].

Rough machining of a sculptured part normally requires 10 to 100 cutting layers depending upon the dimension of the workpiece and the acceptable surface accuracy to be achieved. This leads to an optimization problem of 21 to 201 design variables based on Eq. (9.11). Simplification of this large nonlinear programming problem becomes necessary. In our research, the distribution of cutting depths is found to have minor influence on the total machining time. Cutting depths, $\{d_1, d_2, \ldots, d_N\}^T$, are then equally distributed, and excluded from the optimization. Design variables of the optimization are decreased from $2N+1$ to $N+1$. The simplified optimization becomes a very manageable mixed nonlinear optimization problem that is formulated as

$$\min_{w.r.t.N} \left[\min_{w.r.t.f_1, f_2, \ldots, f_N} T(f_1, f_2, \ldots, f_N) \right] \tag{9.15}$$

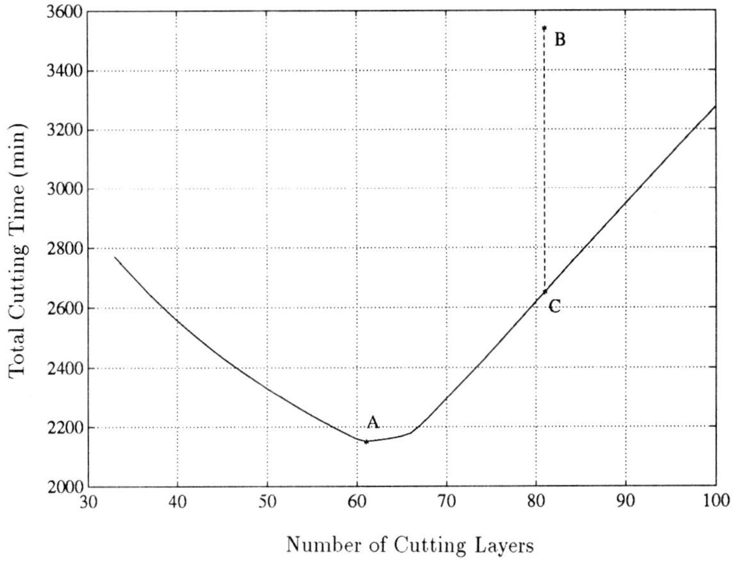

Figure 9.12: Cutting Time versus Number of Cutting Layers

9.5.3 An Example of Machining Parameter Optimization

The optimization formulation has been tested using several practical sculptured part rough machining problems. One of the problems is the ship hull hydrodynamic testing model, used in the previous example. The empirical cutting power and cutting torque models reported in [15] are used in the optimization formulation. These models were obtained on a small milling machine with a maximum power of 450 W, a maximum torque of 7 Nm and a maximum feed rate of 100 mm/min. A 1-inch diameter end-mill is used for the machining, and the stock material is carbon steel.

The results from the total machining time minimization can be illustrated using a plot of total machining time versus the number of total cutting layers, as shown in Figure 9.12. The minimum total machining time is 2,160 minutes (or 36 hours) and 61 equally sized cutting layers, as indicated by point A on the plot. This long total machining time is primarily due to the modest capability of the modeled milling machine. Tests on our Victor 4-Axis CNC machine center require much shorter total machining time. The same rough machining task, if conducted following the data recommended in a machining handbook [14], results in 95 cutting layers. If the feed rates are selected based on the recommended values in reference [14], the total machining time is 3,540 minutes (or 59 hours), as indicated by point B in Figure 9.12. If the feed rates are optimized without considering the optimal number of cutting layers, as with the adaptively controlled machining [15], the actual production time for rough machining is 2,650 minutes (or 44.17 hours), as indicated by point C of Figure 9.12. The optimal rough machining method introduced in this work leads to a 39 percent total machining time reduction over the handbook approach and an 18.5 percent total machining time reduction over the adaptively controlled machining approach.

9.6 Summary

The systematic study on feasible tool path patterns, productivity assessment of different rough machining approaches, and machining parameter optimization provide a solution to the automated optimal sculptured part rough machining. This work also formed a new methodology for approaching an industrial problem through the combined use of geometric modeling, artificial intelligence, and numerical optimization. The research can directly contribute to sculptured part machining, and produce significant economical benefits to the manufacturing industry.

References

[1] Altintas, Y., and Spence, A. (1991). "End Milling Force Algorithms for CAD Systems," *Annals of the CIRP,* Vol. 40.

[2] Bedi, S., and Vickers, G.W. (1988). "Computer-Aided Fairing and Direct Numerically Controlled Machining of Ship Hull Hydrodynamic Testing Models," *Transactions of the CSME,* Vol. 12, No. 1, pp. 43-48.

[3] Berenji, H.R., and Khoshnevis, B. (1986). "Use of Artificial Intelligence in Automated Process Planning," *Computers in Mechanical Engineering,* pp. 47-55

[4] Bezdek, J. C. (1981). *Pattern Recognition with Fuzzy Objective Function Algorithms,* Plenum Press, New York.

[5] Bobrow, J.E. (1985). "NC Machine Tool Path Generation from CSG Part Representations," *Computer-Aided Design,* Vol. 17, No. 2, pp. 69-76.

[6] Centner, R. M., and Idelsohn, J.M. (1964). "Adaptive Controller for a Metal Cutting," *IEEE Transactions on Application and Industry,* Vol. 83, No. 72, pp. 154-161.

[7] Cha, J.Z., and Mayne, R.W. (1989). "Optimization With Discrete Variables via Recursive Quadratic Programming: Part 1 — Concepts and definitions," *Journal of Mechanisms, Transmissions, and Automation in Design,* Vol. 111, pp. 124-129.

[8] Chang, T.C., and Wysk, R.A. (1991). *Computer-Aided Manufacturing,* Prentice Hall, Inc.

[9] Choi, B.K., and Jun, C. S. (1989). "Ball-end Cutter Interference Avoidance in NC Machining of Sculptured Surfaces," *Computer-Aided Design,* Vol. 21, No. 6, pp. 371-378.

[10] Dong, Z., Li, H., and Vickers, G.W. "Optimal Rough Machining of Sculptured Parts on A CNC Milling Machine," to appear in *ASME Journal of Engineering for Industry.*

[11] Gupta, O.K., and Ravindran, A. (1983). "Nonlinear Integer Programming and Discrete Optimization," *Journal of Mechanisms, Transmissions, and Automation in Design,* Vol. 105, pp. 160-164.

[12] Guyder, M.K. (1990). "Automating the Optimization of $2\frac{1}{2}D$ Axis Milling," *Computers in Industry*, Vol. 15, pp. 163-168.

[13] Tipnis, V.A., Filed, M., and Friedman, M.Y. (1975). "Development and Use of Machinability Data for Process Planning Optimization," SME Technical Paper, Series MD75-517, Chicago.

[14] The Machinability Data Center (1989). *Machining Data Handbook*, Metcut Research Associates Inc., 3rd ed.

[15] McGoldrick, P.F., and Hijazi, M.A.M. (1983). "Optimization Strategies for Adaptively Controlled Milling," *Advances in CAD/CAM*, Ellis, T.M.R., and Semenkov, O.I., eds., North-Holland Publishing Company.

[16] Nau, D.S., and Chang, T.C. (1985). "A Knowledge Based Approach to Generative Process Planning," *Proceedings of Symposium of Computer-Aided Intelligent Process Planning*, ASME Winter Meeting, Vol. 1.

[17] Powell, M.J.D., and Ravindran, A. (1964). "An Efficient Method for Finding the Minimum of a Function of Several Variables without Calculation Derivatives," *Computer Journal,* Vol. 7, pp. 155-162

[18] Puttre, M. (1992). "Computer-Aided Manufacturing Sculpting Parts from Stored Patterns," *Mechanical Engineering* (April), pp. 66-70.

[19] Reklaitis, G. V., Ravindran, A., and Ragsdell, K.M. (1983). *Engineering Optimization: Methods and Applications*, John Wiley.

[20] Ruff, K. E. (1979). "Computer Aided Selection of Feeds and Speeds," *Manufacturing Engineering,* pp. 53-55.

[21] Smith, S., and Tlusty, J. (1991). "An Overview of Modeling and Simulation of the Milling Process," *Journal of Engineering for Industry,* Transactions of ASME.

[22] Suh, Y. S., and Lee, K. (1990). "NC Milling Tool Path Generation for Arbitrary Pockets Defined by Sculptured Surfaces," *Computer-Aided Design*, Vol. 22, No. 5, pp. 273-284.

[23] Tipnis, V.A., Filed, M., and Friedman, M.Y. (1975). "Development and Use of Machinability Data for Process Planning Optimization," SME Technical Paper, Series MD75-517, Chicago.

[24] Vickers, G.W., Li, H., and Dong, Z. (1992), "Automated Rough Machining of Curved Surfaces," *Proceedings of CSME Forum, Transport 1992+*, Vol. 2, pp. 593-597.

[25] Vickers, G.W., Ly, M., and Oetter, R.G. (1990). *Numerically Controlled Machine Tools,* Ellis Horwood Limited.

[26] Vogel, S.A. (1979). "Metcut Machinability Process for NC," *Proceedings of the First Annual Conference on Computer Graphics in CAD/CAM Systems*, MIT, Cambridge, MA., pp. 5-15.

[27] Wang, H., Chang, H., Wysk, R.A., and Chandawarkar, A. (1987). "On the Efficiency of NC Tool Path Planning for Face Milling Operation," *Journal of Engineering for Industry*, Transactions of the ASME, Vol. 109, pp. 370-376.

Chapter 10

Artmap Neural Networks for Closed-loop Welding Process Control

H.-H. TOM HUANG, H.-P. BEN WANG, and GARY W. FISCHER
Department of Industrial Engineering, University of Iowa

10.1 Introduction

Welding in a manufacturing context is much more complex than the familiar concept of the fusion process for joining materials. This is illustrated in Figure 10.1 which is intended to emphasize the elements requiring consideration in most applications. Generally the planning and design begin with specification of the weld joint geometry and environment and final required characteristics from which selection of a suitable power source can be made. However, consideration of the interaction between a particular power source and the joint requirements can or may lead to joint redesign or even to the selection of a different power source. Once a particular welding process has been selected there may be questions about the behavior (weldability) of the materials. This requires analysis of the heat transfer to and within the workpiece and determination of the resulting metallurgical structure, residual stresses, and deformation. Moreover, in order to make good welds consistently, consideration must be given to monitoring and controlling machine performance continuously.

The gas metal arc welding process (GMAW, formerly called metal inert-gas or MIG welding) is extensively used in the metalworking industries to weld a variety of ferrous and nonferrous metals. The process uses a continuous consumable wire electrode. The molten weld puddle is completely covered with a shield gas, which is also fed through the welding gun.

The GMAW process has been widely used as a metal joining technique because of two facts [15]:

- Its potential for increasing the productivity and quality of welding
- Its use by most arc welding robots

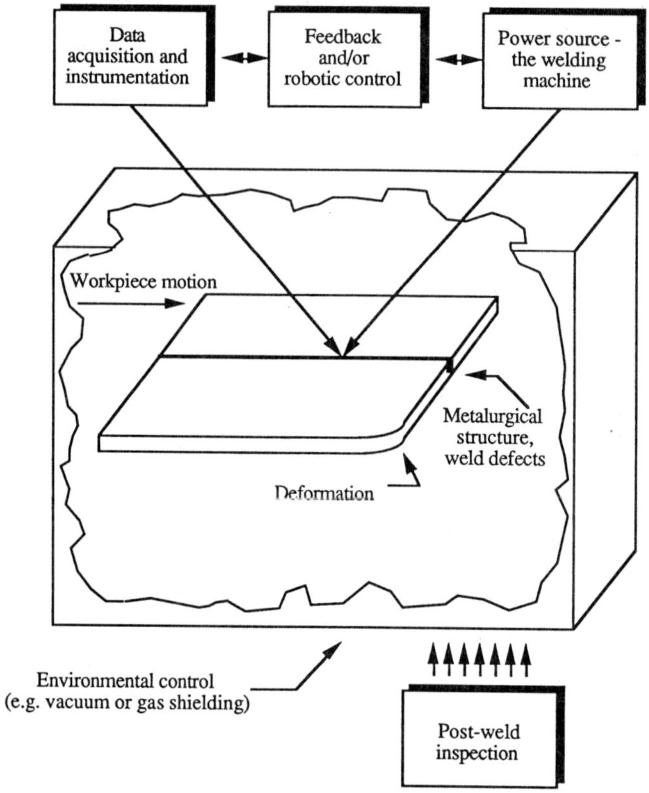

Figure 10.1: Welding Process in Manufacturing

Significant advancements in several technologies have made high-speed gas metal arc welding a reality. However, to produce quality welds using high-rate welding equipment, careful control of the welding process variables on a real-time basis is essential.

Such control in manual welding is derived from the skill of the person performing the welding. The basic welding parameters are established by the settings on the welding power supply, but fine-tuning of the process is done by the person (welder). After the weld material, wire size, gas type, gas mixture, gas flowrate, wire feedrate (welding current), and welding voltage are preset, the welder maintains a consistent welding process by adjusting the gun angle, welding direction, and stand-off distance, as well as the travel speed of the welding gun. The welder uses his or her senses of sight, touch, and sound to detect when adjustments should be made to the process variables.

High-performance automated welding offers the potential for excellent consistency, increased levels of productivity, and reduced cost. For example, it can be shown that with an increase in gun travel speed from 11 ipm to 30 ipm for a weldment with 100 inches of weld, the production rate can increase by over 170 percent and manufacturing cost per piece can be reduced by as much as 35 percent. To achieve a high level of welding performance and quality at high gun travel speeds, the weld path must be well defined and the appropriate welding process variables must be correctly adjusted along the weld path. The multivariate environment suggests the need for an intelligent control system that can evaluate the process and determine the "best" adjustments and do so in real time.

The full potential for an intelligent control system to evaluate and control the welding process goes beyond the shop floor. Such technology makes it possible to integrate the weldment design and process planning activities more completely.

To establish a philosophy of how this proposed research approaches the welding process control, the following passage is quoted from the Summary of the 1987 Conference on "Design and Analysis of Integrated Manufacturing Systems" sponsored by the National Academy of Engineering [19]: "the complexity of discrete manufacturing is such that one cannot effectively design and control a system without understanding the totality of it."

Also quoted is the concluding remark of an article by Professor W.H. Giedt of the University of Delaware [9]:

"Though experimental studies of weld temperatures, residual stresses and distortions have been made, none appears to describe the weld process and the material properties in sufficient detail to adequately define the finite element thermal and mechanical models. It is clear that further assessments of both models can be made only when numerical results are correlated with experimental data generated from test welds designed specifically for this purpose."

Hale and Hardt [10] proposed a control model for the arc welding process to control the outputs of bead width, height, and penetration. They found that the control authority available for decoupled two-output control is very limited. Several other numerical models have been devised to model the weld bead geometry [17]. However, these models usually require more calculations; therefore, they are not suited for real-time control purposes.

Arata and colleagues [2] did a series of studies to investigate the characteristics of the welding arc sound. They concluded that the welding arc sound is a good information signal to express the welding condition. Yu [21] conducted a set of experiments to study the feasibility of using the welding arc sound as a means of feedback to control the welding process. The results clearly indicated the excellent repeatability of the arc sound signal and the characteristic response for various process conditions—the potential for using sound as the basis of feedback control. In order to be able to measure the weld condition and control the welding process in real time, a special pattern classification technique is required to classify welding arc sound in order to assess the condition of the weld.

In the past few years, neural networks have been recognized and accepted as a powerful tool for correlating data without making strong assumptions about the underlying distribution of the problem. One of the most widely used neural network configurations is a multi layer, feed-forward network structure, which is normally trained through an optimization technique such as back propagation. Several studies have been conducted to apply the back propagation neural network approach for real-time arc welding process control [1]. In this investigation, a modification to a relatively new neural network architecture, called ARTMAP, is employed.

10.2 Neural Networks

An artificial neural network is a simulation of the biological neural network of the human brain. It has been studied in the hope of achieving humanlike performance in solving problems with complex, incomplete, or seemingly unrelated data that cannot be solved by conventional programming techniques.

The power of artificial neural networks lies in their computational speed and ability to provide a greater degree of robustness or fault-tolerance [16]. Because artificial neural networks are formed and operate in a parallel fashion, they have been found to be able to perform complex computations simultaneously and rapidly. In addition, because a network consists of a number of neurons (processing elements), when even a few neurons or their interconnections are damaged, the network can still maintain its regular performance [12, 16].

A neural network consists of a large number of simple processing units which communicate in parallel through weighted connections. These processing units are characterized by a state of activation, which is a function of the inputs to the units. Each unit consists of the following two functions: integration and activation. As the name implies, the integration function integrates all weighted inputs to a node. Mathematically, an integration function is shown as follows:

$$I_{pj} = \sum_i O_{pi} W_{ji} \tag{10.1}$$

where I_{pj} is the integrated input of neuron j corresponding to input pattern p; O_{pi} is the output from neuron i; and W_{ji} is the connection weight between neurons i and j. An activation function, defining the node response to the integrated input, usually takes the following form:

$$O_{pj} = \frac{1}{1 + e^{-(I_{pj} - \alpha)}} \tag{10.2}$$

where α is the bias of neuron j.

10.3 Methodology

10.3.1 Architecture

Figure 10.2 shows the tandem neural network structure adopted in this investigation for closed-loop welding process control. The first neural network, signal classification network (SCN), is used to classify the welding arc sound signal in order to assess the quality of the weld. The second neural network, feedback control network (FCN), provides the feedback control signal to adjust and control the welding process.

The closed-loop control system starts with setting the controllable variables of the welding system. The sound signal corresponding to this setting is then collected. The collected sound signal is fed into the SCN. The output of this network indicates the present quality of the weld which is determined by four measures: penetration, spatter, weld bead width, and weld bead height. After comparing the present weld quality with the baseline weld quality, a quality difference is obtained and fed into the FCN. The FCN then in turn provides the

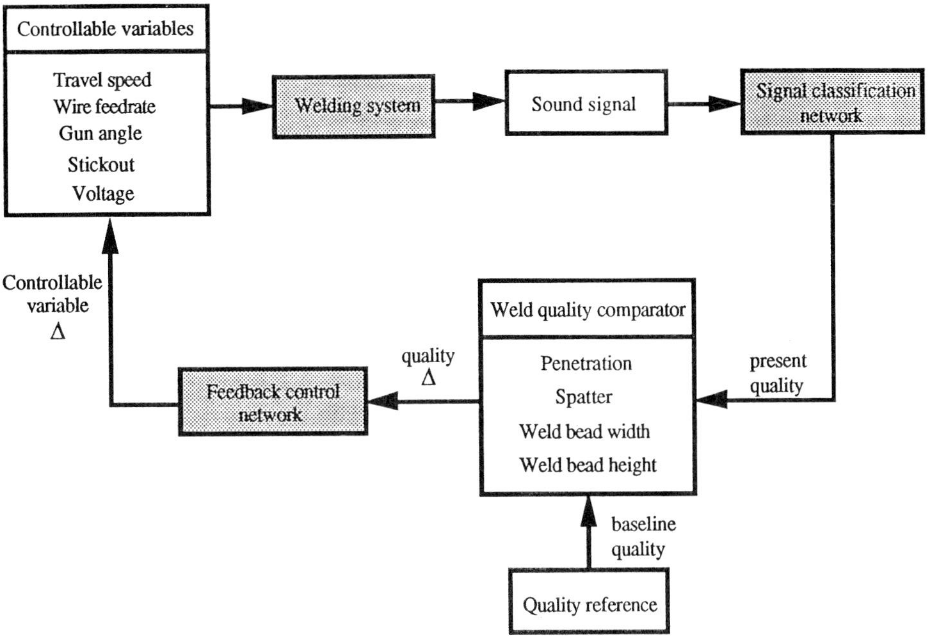

Figure 10.2: Tandem Neural Network Structure for Closed-loop Arc Welding Process Control

feedback control signal to adjust the controllable variables and forms a closed-loop control system.

10.3.2 Modified ARTMAP Neural Networks

An ARTMAP neural network which is developed by Carpenter [6] is an extension of the ART (Adaptive Resonance Theory) network which autonomously learns to classify arbitrarily ordered vectors into recognition categories based on predictive success. Like back propagation, ARTMAP is a supervised learning network. However, ARTMAP has several superior properties to back propagation (BP), such as fast learning and incremental learning. Therefore, ARTMAP neural networks are adopted for this application.

An ARTMAP network consists of two ART modules and a map field which controls the mapping between two ART recognition categories (details can be found in Carpenter [6]). As mentioned earlier, ARTMAP is capable of learning the association between two different input patterns. However, in this application, only one input pattern (i.e., sound signal or quality discrepancy) is used for the network, therefore, a modified ARTMAP architecture is adopted in order to perform the supervised learning. Figure 10.3 shows the modified ARTMAP architecture in which the second ART module is replaced by a target output

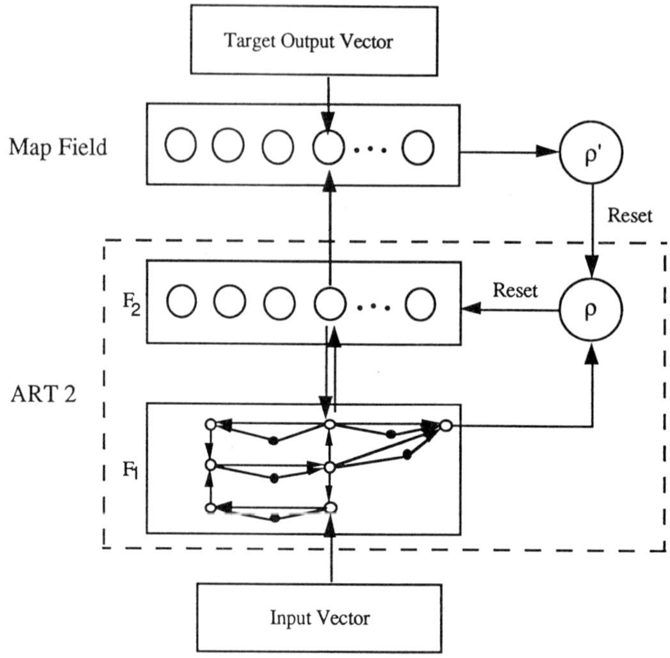

Figure 10.3: Modified ARTMAP Architecture

vector provided by the user.

Two classes of ART networks have been developed by Carpenter and Grossberg" ART 1 is capable of processing arbitrary sequences of binary input patterns and ART 2 is capable of handling either binary or analog input patterns [4, 5]. Because the energy level in a sound signal and the value of quality discrepancy are analog (continuous), an ART 2 network is used as the underlying ART module.

The F_1 layer of ART 2 network (see Figure 10.4) includes three processing levels which enable the network to separate signal from noise and, in addition, to enhance the contrast of activation signals. Each level performs two kinds of computation: integration of intrafield and interfield inputs to that level and normalization of the integrated activation signal. The filled circles are the gain control systems which normalize the integrated signals. For example, at the lower level of the F_1 layer, vector w_i is the integration of an intrafield input vector I_i and the interfield feedback signal au_i, that is,

$$w_i = I_i + au_i \tag{10.3}$$

where i is the ith node at the F_1 layer and a is a constant. Once the vector w_i is obtained, it can be normalized to yield x_i from the following equation:

$$x_i = \frac{w_i}{e + ||w||} \tag{10.4}$$

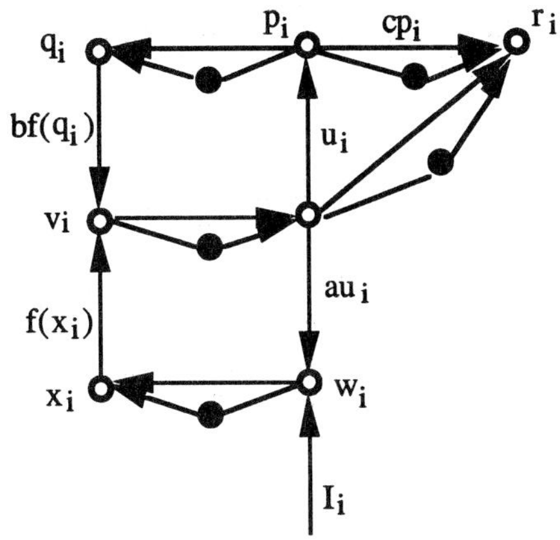

Figure 10.4: F_1 Layer of ART 2 Network

where e is a constant close to zero, and $||w||$ denotes the L_2 norm of a vector w.

The rest of the activities in the F_1 layer can be calculated from the following equations:

$$v_i = f(x_i) + bf(q_i) \tag{10.5}$$

$$u_i = \frac{v_i}{e + ||v||} \tag{10.6}$$

$$q_i = \frac{p_i}{e + ||p||} \tag{10.7}$$

$$p_i = u_i + \sum_j g(j)T_{ji} \tag{10.8}$$

where b is a constant, $g(j)$ is the activation of the jth F_2 node, and T_{ji} is the top-down weight between the jth F_2 node and the ith F_1 node. The linear signal function f in Eq. (10.5) is

$$f(x) = \begin{cases} 0 & \text{if } 0 \le x < \theta \\ x & \text{if } x \ge \theta \end{cases} \tag{10.9}$$

where q is a threshold value.

Once F_2 nodes receive an input signal from F_1, the matching score for F_2 nodes is then computed from the following:

$$\mu_j = \sum_i p_i B_{ij} \tag{10.10}$$

where B_{ij} are the bottom-up weights. Then, an F_2 node with the largest matching score is activated. The activation of F_2 is given below:

$$g(j) = \begin{cases} d & \text{if the jth } F_2 \text{ node is active} \\ 0 & \text{otherwise} \end{cases} \tag{10.11}$$

where d is a constant between 0 and 1.

At this point, the activation in the F_2 layer is propagated back to the F_1 layer. A vigilance test is then carried out to determine whether the top-down signal is matched with the input pattern. The vigilance test is given as follows:

$$r_i = \frac{u_i + cp_i}{e + ||u|| + ||cp||} \tag{10.12}$$

$$\frac{\rho}{e + ||r||} > 1? \tag{10.13}$$

where c is a constant and $0 < r < 1$. If the match fails to pass the vigilance test, a reset is sent to F_2 which forces F_2 to deactivate the selected F_2 node and to search for the next best match. Otherwise, the bottom-up and top-down weights are adapted from the following equations:

$$B_{ij^*}(t+1) = d[p_i - B_{ij^*}(t)] \tag{10.14}$$

$$T_{j^*i}(t+1) = d[p_i - T_{j*i}(t)] \tag{10.15}$$

where j^* is the selected F_2 node.

Once the ART module is presented with an input vector, it selects an F_2 node which passes the vigilance test. Then, the F_2 activations are propagated to the map field through the weighted connections between F_2 and the map field. The signals received from F_2 are calculated from the following equation:

$$X = w_{j*} \tag{10.16}$$

At the map field, a second vigilance test is performed to determine the level of match between the predictive output from F_2 (X) and the target output pattern (Y). A mismatch between X and Y will trigger a map field reset to the underlying ART module. This occurs whenever the following condition holds:

$$\frac{||X||}{||Y||} < \rho' \tag{10.17}$$

where ρ' denotes the associative memory vigilance parameter. If the Map Field reset occurs, the vigilance of the underlying ART module is raised to prevent the system from making repeated errors. At this point, a reset is sent to F_2 and forces F_2 to find another best match. This process will continue until the second vigilance test succeeds. Once the second vigilance test is passed, the top-down and bottom-up weights between F_1 and F_2 are adapted according to Eqs. (10.14) and (10.15). The weights between F_2 and the map field are updated from the following equation:

$$w_{j^*k} = Y_k \tag{10.18}$$

10.4 Experimental Apparatus

A "semi-automatic" welding facility is used in this study [21]. The welding arc sound is measured by a condenser microphone. Through an amplifier, the sound signal is amplified and transferred to a four-channel tape recorder. The recorded signal is transferred to an oscilloscope for visual display of the signal. Meanwhile, the signal is also transferred to a Type 2131 digital frequency analyzer where the signal is transformed into a digital representation of its magnitude. The amplitude of the signal is calibrated with dB in 42 channels having central frequencies from 1.6 Hz to 20 kHz. Finally, a microcomputer is used to store all the digital signals.

Tables 10.3 and 10.4 (see appendix) illustrate the process variable settings and respective weld quality measurements in weld bead width, weld bead height, penetration, and spatter formation. For each experiment, a sound spectrum is recorded.

10.5 Computational Experiences and Discussion

In this section, we describe the network training and testing procedures and discuss the computational results. Figure 10.5 illustrates the framework of the tandem ARTMAP networks for welding process control.

For each of the four quality indicators and each of the five process variables, there is a modified ARTMAP neural network. For the sake of simplicity, "Penetration" is used as a quality criterion for signal classification and "voltage" as a process variable for "feedback control."

10.5.1 Signal Classification (Penetration) Network

For the signal classification (penetration) network, as mentioned earlier, the F_1 layer consists of 42 nodes, each for a sound frequency channel. The entire frequency spectrum ranges from 1.6 Hz to 20 kHz. As far as the output layer is concerned, ARTMAP is different from back propagation networks in that, for each test, only one output node will be turned on in an ARTMAP network. As such, the F_2 layer is designed to consist of 14 nodes, each covering a small range of weld penetration (see Table 10.1). For example, if node 9 is turned on after a test, we know that the predicted penetration reading is in the bin of 0.155″ to 0.160″. The penetration resolution, depending on the quality requirement of an application, can be adjusted by changing the number of output nodes.

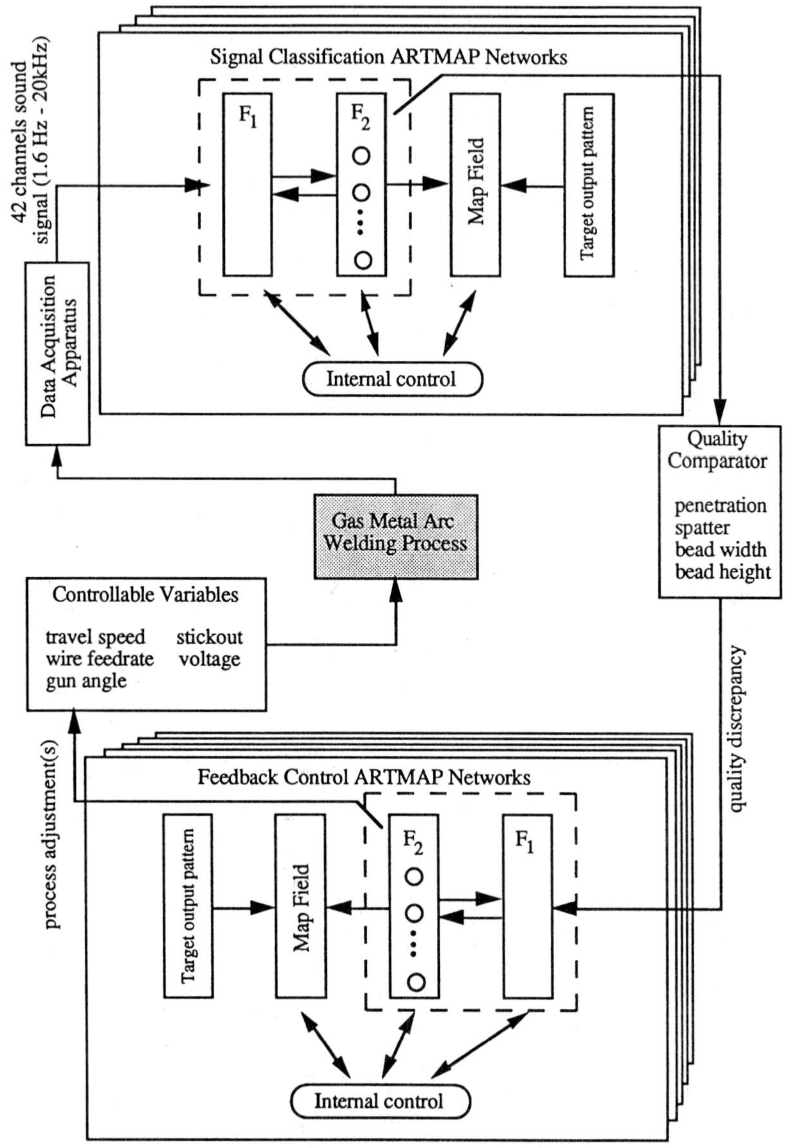

Figure 10.5: Framework of Tandem ARTMAP Networks

Table 10.1: Output Layer of Penetration Prediction Network

Node no.	Penetration range (in.)
1	less than 0.100
2	0.100 - 0.110
3	0.110 - 0.120
4	0.120 - 0.130
5	0.130 - 0.140
6	0.140 - 0.145
7	0.145 - 0.150
8	0.150 - 0.155
9	0.155 - 0.160
10	0.160 - 0.165
11	0.165 - 0.170
12	0.170 - 0.175
13	0.175 - 0.180
14	more than 0.180

10.5.2 Feedback Control (voltage) Network

A quality discrepancy may have a positive value (e.g., $+0.005''$ in penetration comparison meaning deeper than desired penetration), or a negative value (e.g., $-0.002''$ in width, meaning too small a weld bead). Because an ART 2 network is not able to properly recognize negative values, a normalization process has to be applied in order to make sure that the ART 2 network can perform correctly by getting proper inputs. This normalization process takes two steps. First, each quality discrepancy is associated with two nodes. The ARTMAP system then decides whether a quality discrepancy element is positive or negative. If an element has a positive value, then the negative part will be assigned to zero, and vice versa. Second, it needs to scale the elements by dividing each element by the maximum element value. Thus, for the feedback control (voltage) network, the F_1 layer consists of eight nodes (two nodes, a positive and a negative, for each one of those four quality criteria). The F_2 layer is designed to have nine nodes; each presents the amount of adjustment (see Table 10.2).

10.5.3 Impact of the Network Vigilance Value

Figures 10.6, 10.7, and 10.8 are plots of the testing results of the penetration prediction network when the vigilance is set to 0.7, 0.8 and 0.9, respectively. Each figure consists of two sets of readings: target and predicted penetration values. Note that among those 20 data sets, 2, 3, 4, 7, 8, 9, 10, 11, 13, 14, and 15 are used in both training and testing sessions,

Table 10.2: Output Layer of Voltage Control Network

Node no.	Voltage adjustment (volt)
1	0
2	2
3	4
4	6
5	8
6	-2
7	-4
8	-6
9	-8

whereas 1, 5, 6, 12, 16, 17, 18, 19, and 20 are used only in testing.

Figure 10.6 indicates that the network is able to make perfect prediction for those patterns (2, 3, 4, 7, 8, 9, 10, 11, 13, 14, and 15) that the network has seen. Out of those nine unseen patterns, 1, 5, 12, and 16 are correctly predicted, whereas 6, 17, 18, 19, and 20 witness some error, the largest being 0.025″ for patterns 17, 18, and 19. Among those incorrectly classified patterns (6, 17, 18, 19, and 20), 17, 19, and 20 are the only patterns in their respective penetration range bins. In other words, the network has not seen any example of that particular classification.

The vigilance was then increased from 0.7 to 0.8. Figure 10.7 shows the results. It indicates that the network is again able to make perfect prediction for those patterns (2, 3, 4, 7, 8, 9, 10, 11, 13, 14, and 15) that the network has seen. Out of those nine unseen patterns, 1, 5, 12, 16, and 18 are correctly predicted, whereas 6, 17, 19, and 20 are classified to incorrect bins. Based on the correct number of classifications, the performance of the network is slightly improved by increasing the vigilance. However, the largest penetration error remains 0.025″ for patterns 17 and 19. Again, among those incorrectly classified patterns (6, 17, 19, and 20), 17, 19, and 20 are the only patterns in their respective penetration range bins.

The vigilance was then further increased from 0.8 to 0.9, and the test results are shown in Figure 10.8. As can be seen, it indicates that the network, similar to the previous two batches of testing is, able to make perfect prediction for those patterns that the network has seen. For the nine unseen patterns, 1, 5, 12, 16, and 18 are correctly predicted, but 6, 17, 19, and 20 are classified incorrectly. As far as the correct number of classifications is concerned, the performance of the network is the same as the previous one when the vigilance was 0.8. The largest penetration error remains 0.025″ for only one pattern (17). Again, among those incorrectly classified patterns (6, 17, 19, and 20), 17, 19, and 20 are the only patterns in their respective penetration range bins.

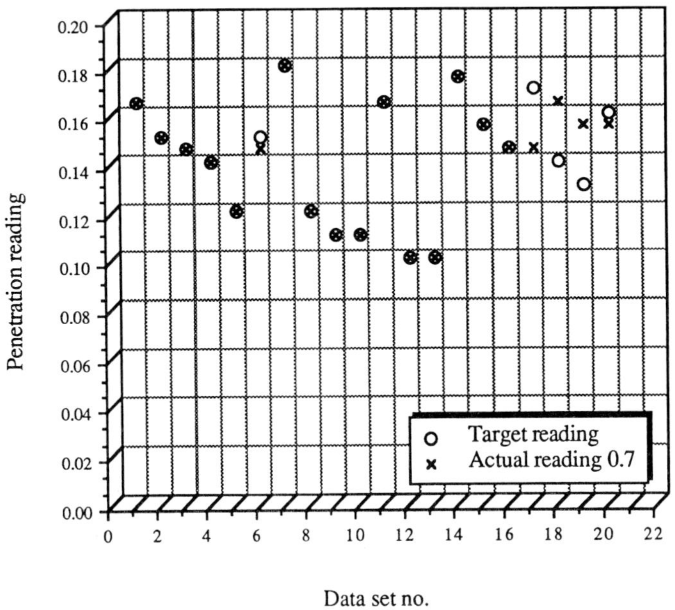

Figure 10.6: Penetration Prediction (ARTMAP, vigilance = 0.7)

10.5.4 Comparisons with Back Propagation, Simulated Annealing, and Tabu Search

In order to benchmark the performance of the ARTMAP network, a multilayered, feed-forward network was developed and trained using the back propagation (BP), simulated annealing (SA), and tabu search (TS) network learning paradigms, respectively [13]. The details of them can be found in [14], and are not described in this paper. As it was done in the previous training and testing sessions, 2, 3, 4, 7, 8, 9, 10, 11, 13, 14, and 15 are used in both training and testing sessions, whereas 1, 5, 6, 12, 16, 17, 18, 19, and 20 are used only for testing. The results are shown in Figures 10.9 and Figure 10.11.

For BP, note that not all the data patterns used in the training session have been correctly classified; 8, 13, 14, and 15 are correct, whereas 2, 3, 4, 7, 9, 10, and 11 are wrong. Among the unseen patterns, only 1 and 6 are precisely placed in the correct bins. Others have errors ranging from 0.005″ to 0.05″.

The simulated annealing (SA) network was trained and tested using the same sets of data patterns. The results are shown in Figure 10.10. Even for those data patterns that have been used in training, the network is still not able to make close predictions. As we can see, the prediction accuracy appears not as good as that of the BP network. However, according to a statistical analysis reported in [13], as far as prediction accuracy is concerned, there is no statistical difference among BP, SA, and TS.

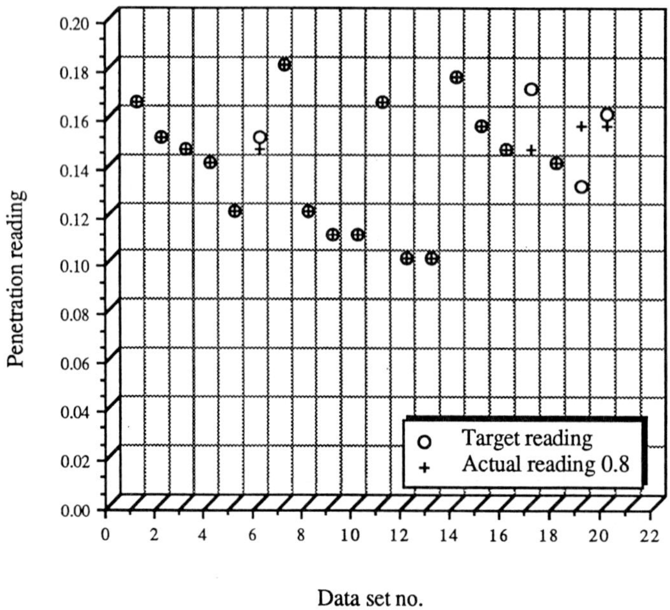

Figure 10.7: Penetration Prediction (ARTMAP, vigilance = 0.8)

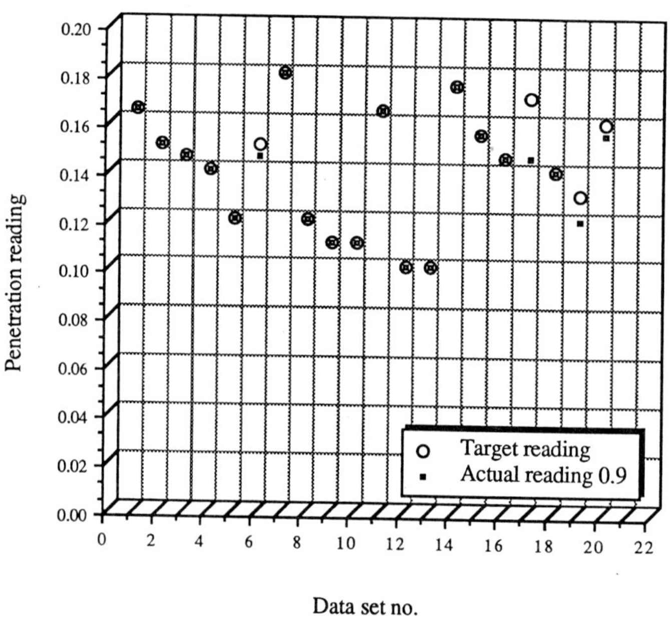

Figure 10.8: Penetration Prediction (ARTMAP, vigilance = 0.9)

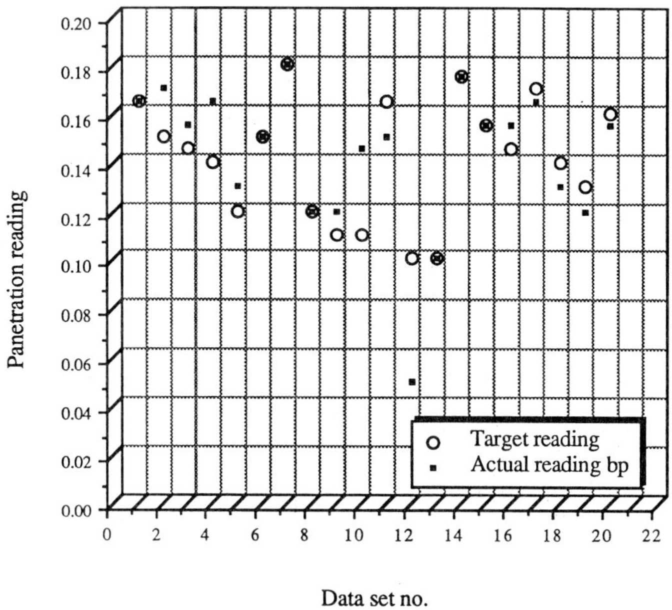

Figure 10.9: Penetration Prediction (back propagation –)

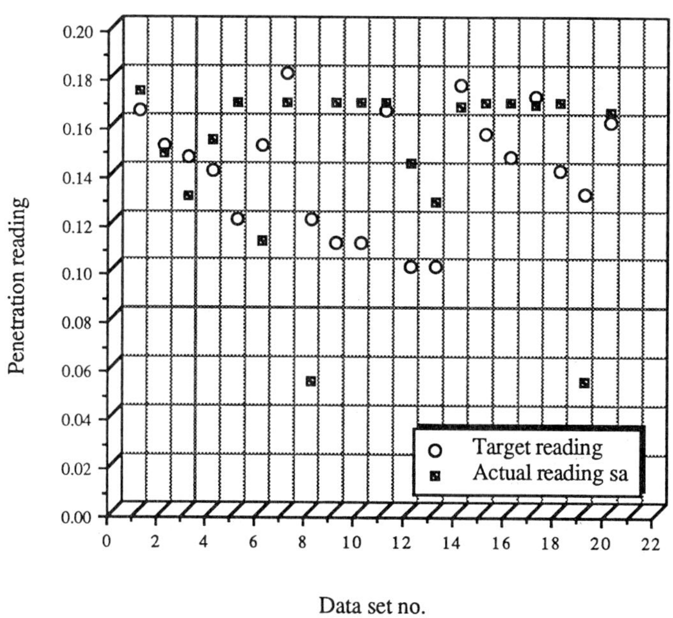

Figure 10.10: Penetration Prediction (simulated annealing –)

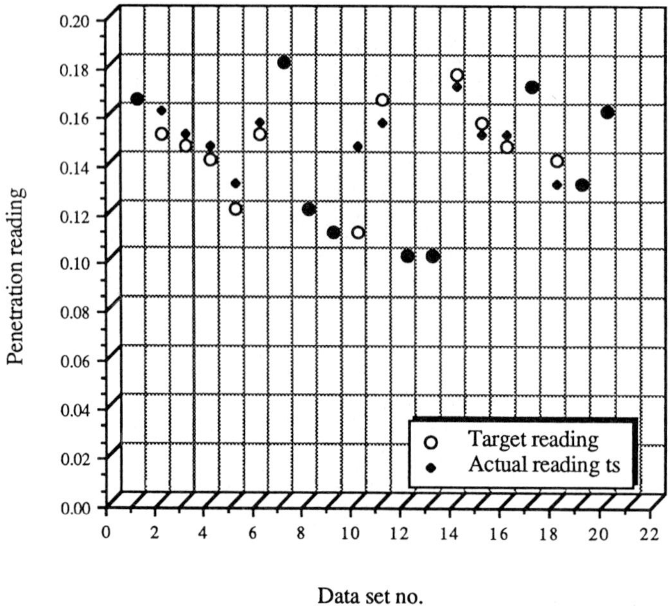

Figure 10.11: Penetration Prediction (tabu search –)

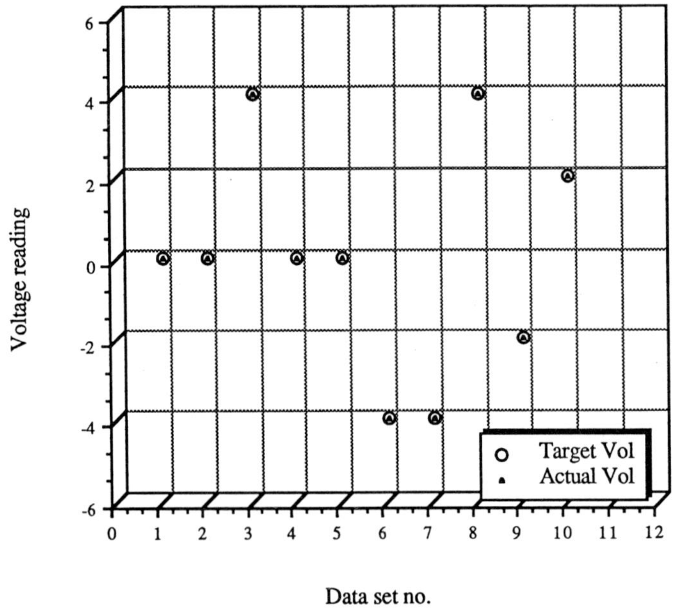

Figure 10.12: Voltage Prediction (ARTMAP, vigilance = 0.9)

For the tabu search (TS) network, the data patterns used in the training session are not all correctly identified. However, the prediction errors are smaller than those of the BP network. As for those unseen data patterns, the predictions are more accurate as well.

Figure 10.12 illustrates a testing session of the feedback control (voltage) network, when the vigilance was set to 0.9. As can be seen, the network is able to make perfect predictions, even for every unseen data pattern. The network shows comparable performance to the signal classification networks. For the sake of simplicity, the comparisons with BP networks are not discussed.

10.6 Concluding Remarks

Several conclusions came from this investigation:

- Training efficiency. A typical training session for a modified ARTMAP network takes about 10 to 12 seconds per pattern. However, for a BP network, it generally takes more than 10 hours to complete a session, involving 120 patterns. Both are trained on an HP/Apollo 9000/425T workstation. As far as training times are concerned, the modified ARTMAP is clearly better than BP. According to a statistical test conducted by [13], BP, SA, and TS exhibited about the same learning speed. Therefore, it is fair to say that ARTMAP is more efficient than BP, SA, and TS.

- Training effectiveness. It can be seen from Figures 10.6 through 10.11 that the ARTMAP network is able to make a perfect prediction for each and every one of the patterns that the network has seen. The BP, SA, and TS networks, however, have some problem with those patterns which are used in training. For those testing data patterns not used in training, ARTMAP still outperforms all three other learning paradigms.

- Based on the shown efficiency and effectiveness, the ARTMAP network should be able to meet the requirements for real-time welding process control.

Currently, the training and testing of all the SCNs and FCNs are done sequentially. Future study will be focused on the application of the parallel programming technique to speed up the network performance. By using this technique, all of the SCNs or FCNs can be trained and tested simultaneously.

10.7 Acknowledgements

This research was supported in part by the National Science Foundation under Contract DDM-9113770 and the U.S. Army FCIM Program. Experimental data are taken from [21].

References

[1] Andersen, K., Cook, G.E., Karsai, G., and Ramaswamy, K. (1990). "Artificial Neural Networks Applied to Arc Welding Process Modeling and Control," *IEEE Transactions on Industry Applications*, Vol. 26, No. 5, pp. 824-830.

[2] Arata, Y., Inoue, K., and Toh, T. (1979). "Investigation on Welding Arc Sound (Report I)– Effect of Welding Method and Welding Condition of Welding Arc Sound," Research Report, Institute of Osaka University, Osaka, Japan.

[3] Bates, B.E., and Hardt, D.E. (1989). "A Realtime Calibrated Thermal Model for Closed loop Weld Bead Geometry Control," *Journal of Dynamic Systems, Measurement, and Control* (March), pp. 25-33.

[4] Carpenter, G.A., and Grossberg, S., (1987). "A Massively Parallel Architecture for a Self-Organizing Neural Pattern Recognition Machine," *Computer Vision, Graphics, and Image Processing*, Vol. 37, pp. 54-115.

[5] Carpenter, G.A., and Grossberg, S., (1987). "ART 2: Self-Organization of Stable Category Recognition Codes for Analog Input Patterns," *Applied Optics*, Vol. 26, No. 23, pp. 4919-4930.

[6] Carpenter, G.A., Grossberg, S., and Reynolds J. (1991). "ARTMAP: Supervised Real-time Learning and Classification of Nonstationary Data by a Self-Organizing Neural Network," *Neural Networks*, Vol. 4, pp. 569-588.

[7] Eraslan, A.H., Zacharia, T., and Aidun, D.K. (1986). "WELDER: A Computer Code for Simulating Fast-Transient, Three-Dimensional, Three-Phase, Flow, Temperature and Material Composition Conditions during Welding," Research Report, Dept. of MIE, Clarkson University, Potsdam, NY.

[8] Fischer, G.W., Yetukuri, N.V., Yu, M.-L., and Prescott, W.C. (1991). "A Knowledge-Based System Concept for Process Planning and Programming of Industrial Robots," *Proceedings of the 1991 International DADS Users' Conference*, Iowa City, IA, October, pp. B1-B16.

[9] Giedt, W.H. (1987). "Welding An Interdisciplinary Science and Technology," *Interdisciplinary Issues in Materials Processing and Manufacturing*, Samanta et al., eds., 1987 ASME Winter Meeting (December 13-18), pp. 403-418.

[10] Hale, M.B. and Hardt, D.E. "Multivariable Geometry Control of Welding. Part I. Process Modeling," *Dynamic Systems and Control*, Vol. 22, ASME, New York, pp. 1-9.

[11] Hanright, J. (1986). "Robotic Arc Welding under Adaptive Control— A Survey of Current Practice," *Welding Journal* (November), pp. 19-23.

[12] Hopfield, J. J. (1982). "Neural Networks and Physical Systems with Emergent Collective Computational Abilities," *Proceedings of the National Academy of Sciences*, Vol. 74, pp. 2554-2558.

[13] Huang, H.H., and Wang, H.P. (1991). "Tandem Neural Networks for Welding Process Control," *Working Paper* No. 91-32, Department of Industrial Engineering, University of Iowa.

[14] Huang, H.H., Zhang, C., Lee, S. and Wang, H.P. (1991). "Implementation and Comparison of Neural Network Learning Paradigms: Back Propagation, Simulated Annealing and Tabu Search," *Proceedings of the Artificial Neural Networks in Engineering (ANNIE '91) Conference*, November 10-13, St. Louis, MI., ASME Press, New York, pp. 95-100.

[15] Jablonowski, J. (1983). "Robots That Weld," *American Machinist* (April), pp. 113-128.

[16] Lippmann, R.P. (1987). "An Introduction to Computing with Neural Nets," *IEEE ASSP Magazine* (April), pp. 4-22.

[17] Paley, Z., and Hibbert, P.D. (1975). "Computation of Temperature in Actual Weld Designs," *Welding Journal*, Vol. 54, pp. 385s-392s.

[18] Rumelhart, D.E., and McClelland, J.L. eds. (1986). *Parallel Distributed Processing*, Vol. 1, MIT Press, Cambridge, MA.

[19] *The Bridge.* (1987). "Design and Analysis of Integrated manufacturing Systems–Status, Issues, and Opportunities," Suppl. to *The Bridge*, Vol. 17, No. 2, pp. 1-8.

[20] Verbeck, W.J.P.A. (1984). "Arc Welding Process Control by Preview Sensor," *Industrial Robot*, Vol. 11, No. 2, pp. 86-88.

[21] Yu, M.-L. (1990). *An Investigation of Process Variable Relationships for High Performance Gas Metal Arc Welding (GMAW)*, Master's Thesis, Department of Industrial Engineering, University of Iowa.

Appendix

Welding Process Parameters and Weld Quality Measurements

Table 10.3: Settings for Testing Arc Welding Sound [21]

Experiment No	Travel Speed (in/min)	Wire Feedrate (ipm)	Gun Angle (degree)	Stickout (in)	Voltage (volt)
1	11.0	370	10	0.3750	23
2	13.2	370	10	0.3750	23
3	15.6	370	10	0.3750	23
4	21.8	370	10	0.3750	23
5	11.0	185	10	0.3750	23
6	11.0	278	10	0.3750	23
7	11.0	463	10	0.3750	23
8	11.0	740	10	0.3750	23
9	11.0	370	0	0.3750	23
10	11.0	370	20	0.3750	23
11	11.0	370	10	0.3750	23
12	11.0	370	10	0.3750	19
13	11.0	370	10	0.3750	21
14	11.0	370	10	0.3750	25
15	11.0	370	10	0.3750	27
16	11.0	370	10	0.3125	27
17	11.0	370	10	0.3125	23
18	11.0	370	10	0.4375	23
19	11.0	370	10	0.5000	23
20	11.0	370	10	0.3750	23

Table 10.4: Measurements of Welding Specimens (Arc Sound Tests) [21]

Exp. No.	WI (in)	W II (in)	P.I (in)	P II (in)	H.I (in)	H II (in)	A.W (in)	A.P (in)	A.H (in)	S.
1	0.302	0.303	0.165	0.166	0.117	0.118	0.303	0.166	0.118	0
2	0.291	0.290	0.154	0.154	0.104	0.107	0.291	0.154	0.106	0
3	0.276	0.275	0.147	0.148	0.103	0.106	0.276	0.148	0.105	0
4	0.240	0.241	0.144	0.142	0.087	0.087	0.241	0.143	0.087	0
5	0.217	0.218	0.129	0.129	0.056	0.055	0.218	0.129	0.056	0
6	0.287	0.288	0.153	0.152	0.073	0.074	0.288	0.153	0.074	0
7	0.366	0.369	0.181	0.182	0.150	0.151	0.368	0.182	0.151	0
8	0.315	0.356	0.136	0.117	0.246	0.227	0.336	0.127	0.237	1
9	0.311	0.312	0.110	0.112	0.108	0.109	0.312	0.111	0.109	0
10	0.291	0.290	0.109	0.116	0.126	0.120	0.291	0.113	0.123	0
11	0.301	0.302	0.167	0.165	0.117	0.117	0.302	0.166	0.117	0
12	0.189	0.223	0.045	0.054	0.163	0.144	0.206	0.050	0.154	1
13	0.223	0.208	0.080	0.095	0.136	0.121	0.216	0.088	0.129	1
14	0.358	0.360	0.181	0.182	0.101	0.093	0.359	0.183	0.097	0
15	0.397	0.399	0.156	0.159	0.078	0.074	0.398	0.158	0.076	0
16	0.401	0.394	0.145	0.146	0.096	0.091	0.398	0.146	0.094	0
17	0.292	0.293	0.176	0.170	0.101	0.106	0.293	0.173	0.104	0
18	0.277	0.274	0.144	0.142	0.135	0.137	0.276	0.143	0.136	0
19	0.200	0.203	0.130	0.137	0.146	7.137	0.202	0.134	0.142	0
20	0.305	0.304	0.166	0.164	0.118	0.118	0.305	0.165	0.118	0

Exp.No. : Experimental Number
W.I : Width I — W.II : Width II
P.I : Penetration I — P.II : Penetration II
H.I : Height I — H.II : Height II
A.W : Average of Width — A.P : Average of Penetration
A.H : Average of Height — S. : Spatter

Part IV

Motion Planning for Intelligent Robotic and Manufacturing Systems

Part IV consists of three chapters organized in tandem. Chapter 11 discusses the on-line environment perception and modeling from the 3-D range data of a machine vision system. Chapter 12 focuses on motion planning using nonlinear programming for a given geometric environment. Chapter 13 introduces the on-line collision identification and distance calculation methods using artificial neural networks to speed up the nonlinear programming-based motion planning. A combination of the techniques presented in these three chapters forms an innovative approach for the on-line motion planning of intelligent robotic and manufacturing systems. Specifically, the contents of these three chapters include the following.

Chapter 11, Object and Workspace Modeling Based on a 3-D Vision System, by Helai Yao, Ron Podhorodeski, and Zuomin Dong, introduces an approach for generating geometric models of existing objects and workspaces from 3-D range images of object surfaces. The previous approaches on geometric modeling, machine vision, and active 3-D range data acquisition are reviewed. A method for generating geometric models from 3-D range data, based on the cross-section geometric representation and the device frame data processing, is presented. Advantages of the approach in reducing computation time for model generation and supporting sensor fusion and continuous observation of objects and workspace are illustrated with examples. The method can be used in environment perception and modeling for intelligent robotic and manufacturing systems.

Chapter 12, Motion Planning and Geometric Design, by Chung-Yuh Liu and Roger W. Mayne, considers the problem of robot path planning and illustrates the integration of path planning with geometric design via nonlinear programming. Collision avoidance is considered directly by using distance-to-contact information in inequality constraints along with an explicit definition of clearance requirements. Path planning optimization is discussed with both motion criteria and geometric design criteria. Distance-to-contact and interference calculations are described in detail, as well as gradient computations which are critical to the application of efficient nonlinear programming methods. Symbolic computation has been applied to obtain general six degree-of-freedom transformations for the robot links and to obtain analytical derivative expressions. Example problems in path planning and geometric design are presented for a simple 3-D robot with three links and four joints.

Chapter 13, Collision Identification Between Convex Objects Using Neural Networks, by Jing Yuan, Yury Stepanenko, and Zuomin Dong, introduces a new parallel approach to collision identification and distance calculation in motion planning using a modified Hamming net. The method supports parallel processing for object-to-obstacle distance calculation, interference detection, and optimal collision removal direction identification. The method can be used for evaluating geometric constraints in the optimization-based collision-free trajectory planning approach. It can considerably decrease required computation time and has the potential for being applied to on-line motion planning of intelligent robotic and manufacturing systems. Application of the method to motion planning is illustrated using several examples.

Chapter 11

Object and Workspace Modeling Based on a 3-D Vision System

HELAI YAO, RON PODHORODESKI, and ZUOMIN DONG
Department of Mechanical Engineering, University of Victoria

11.1 Introduction

A description of the geometric characteristics of a working environment is a prerequisite for planning the motion of robots, CNC machine tools, coordinate measuring machines, and other manufacturing facilities. Intelligent robotic and manufacturing systems further require the geometric modeling of objects and workspace to be conducted in conjunction with the machine vision system and to be performed on-line to support on-line motion planning and decision making.

Geometric models for the working environment, including objects and workspace, may come from a CAD model built by a designer or from the geometric data acquired by a machine vision system. Traditionally, geometric modeling is mainly concerned with geometric description. Many geometric modeling techniques have been developed and used in computer-aided design and manufacturing (CAD/CAM), computer graphics, graphical animation, and other related applications over the last three decades. A geometric modeling system is supported by a geometric representation, algorithms for geometric manipulation, and a database for geometric entities. Such a system provides applications with object models that convey various geometric information.

11.1.1 Information Embedded in a Geometric Model

A geometric model describes objects by their shape, dimension, area, volume, position, orientation, and topology. In general, various geometric models for objects can be classified into three major categories: wireframe model, surface model, and solid model.

A wireframe model has the simplest model form, composed of object edges that are surface intersections and/or surface outlines. Wireframe may include straight lines, analytical curves, and free parametric curves. Object descriptions provided by a wireframe model are primarily limited to edge-related features. To overcome the ambiguity introduced by such a model human intelligence is often needed.

In contrast, a surface model is able to intrinsically provide more detailed characteristic data, such as surface shape, curvature and surface points. A surface model can be built using either regular analytic surfaces of a plane, a cylinder or a sphere, or sculptured surfaces defined by spline functions or surface meshes. Both wireframe and surface models supply few of the topological relationships among their component elements: points; boundaries and surfaces.

A solid model, as a more sophisticated geometric model, provides a complete description of an object with sufficient geometric and topological information. The dimensions, areas, volumes, surface shape descriptions, surface neighborhood relationships, and relative object positions and orientations can be obtained from the database of a solid model. The inner and outer spaces of an object can also be detected.

Different types of geometric models have different capabilities of representing geometric information. A geometric model with a simple form is usually at a low level in terms of the geometric information it can supply, while a sophisticated geometric representation at high level can provide much more.

11.1.2 Traditional Approach to Geometric Model Generation

Construction of a geometric model requires initial data input. Sophisticated modeling approaches such as solid modeling require more detailed information about the object to be modeled. These initial data include the geometry and topology information of the object.

Traditionally, geometric models for various objects are created using a CAD/CAM system. The modeling approach has four primary components: (1) symbol structures which represent solid objects; (2) processes which use such representations for answering geometric questions about the objects; (3) input facilities for creating and editing object representations and evoking processes; and (4) output facilities and representations of results [19].

The creation of a geometric model is usually done interactively by users of the geometric modeling system through graphics input and output devices such as a mouse, a 2-D digitizer, a trackball, a joystick, a lightpen, a keyboard, a graphics terminal, or a plotter. Since the users of the graphics systems well understand the geometry and topology of objects that they want to model, they can effectively use the input devices to provide the object's geometry and topology to the geometric modeling system.

This modeling procedure can be illustrated using the example in Figure 11.1. When a solid model for the flange is built, the shape and location of various surfaces representing the part, surface adjacency relationships indicating object topology, and surface orientation identifying the interior and exterior of the object are specified by a designer. This traditional approach to geometric model generation is covered by many CAD and computer graphics textbooks (e.g., [8, 14, 25]).

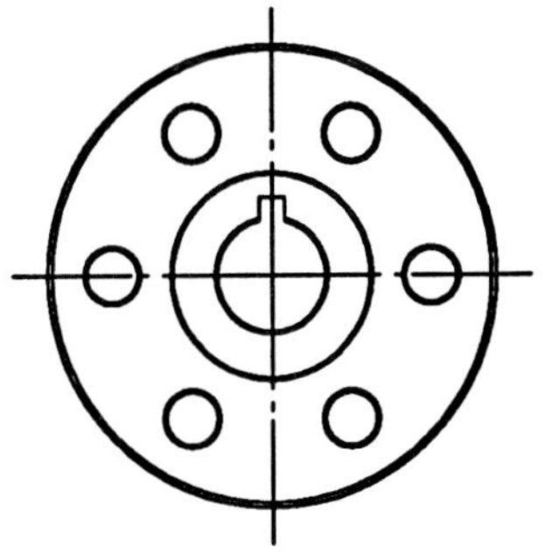

Figure 11.1: Modeling with Known Object Geometry and Topology

11.1.3 The Need for More Flexible Modeling Approaches

Although the traditional geometric modeling approaches have been used broadly in design and manufacturing, they cannot satisfy some key requirements of geometric model generation imposed by intelligent robotic and manufacturing systems. These requirements include: (1) modeling objects and workspace with known geometries, but in unstructured environments and (2) modeling existing objects whose geometries are not known.

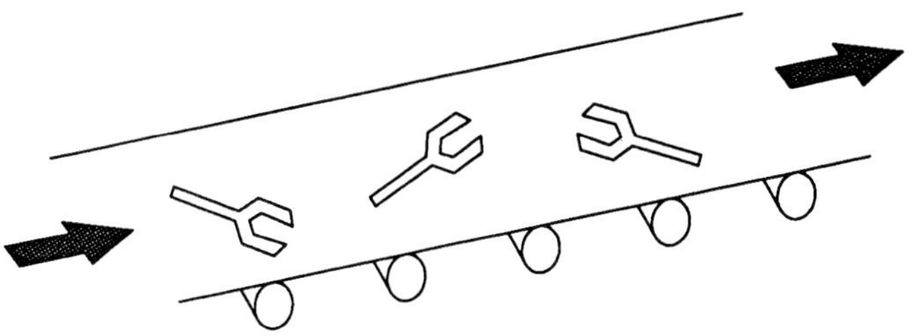

Figure 11.2: Objects with Known Geometry in an Unstructured Environment

Figure 11.2 shows an example for case one. A cast part has been designed and a geometric model for the ideal geometry of the part has been created. In manufacturing, the shape and dimensions of produced parts need to be measured automatically and be compared with design dimensions given in the ideal model. However, no direct match between the real cast part and its ideal model exists due to the different orientation and/or scale between these two. On-line creation of a geometric model based upon the real geometry of the produced part is needed for inspecting the dimensions of the manufactured part. Without human interaction the traditional modeling approaches are not able to construct this type of real geometric model. Other similar situations may include robots picking up parts from a bin and automatic vehicles moving in unstructured surroundings.

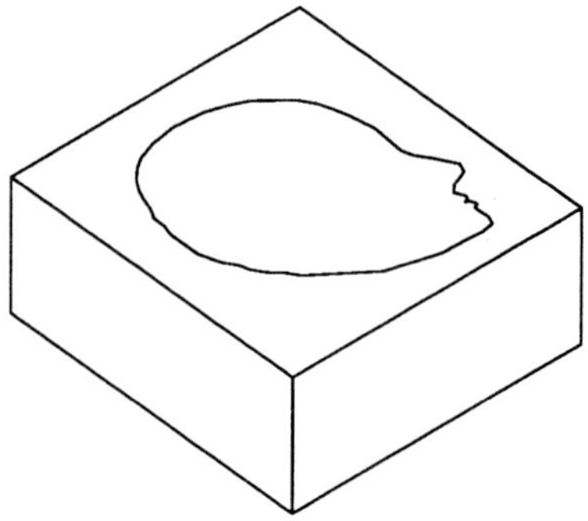

Figure 11.3: An Object with Unknown Geometry

An example of the second case is illustrated in Figure 11.3 where an injection mold of a face sculpture is desired. Since the shape is irregular and the exact geometry and topology of the object are unknown, the initial object description cannot be explicitly input into a CAD/CAM system with traditional approaches. Without a geometric representation, planning for machining and generation of CNC codes cannot be conducted. The solution to this problem is to model the object based on object surface measurements. Although coordinate measuring machines (CMM) are used for measuring object profiles in industry, applications of CMM are limited by its mechanical contact working principle and the degrees of freedom of the machine and its touch probe.

Apparently a noncontact surface measurement-based, on-line geometric modeling approach is needed. To measure object surfaces for such a geometric modeling system, vision sensors have been applied and relevant modeling algorithms have been developed. To understand how a vision sensor works with geometric modeling, a brief review of the basic functions of a machine vision system will be helpful. The remainder of this section serves this need. For further details on the subject a reader can refer to [2, 4, 6, 9, 11, 18, 22].

11.1.4 A Brief Review of Machine Vision

Machine vision is an industrial practice of computer vision that is concerned with automating and integrating a wide range of processes and representations used for vision perception. Its goal is to allow a machine to understand its environment and perform required tasks using perceived information. The major tasks for machine vision are image processing and scene analysis.

Image processing takes an image as its input in order to generate another image with noise suppressed and useful image regions enhanced. In Figure 11.4, several application examples of image processing are shown. The original image (top left) looks very gloomy because of poor illumination and has noise. Applying histogram equalization to this image improves image visibility (top right). Reducing noise is one of major purposes for image processing. In this example, high frequency noise is suppressed by a low-frequency-pass filter (bottom left). There are many types of more complex image noises that affect information extraction

Figure 11.4: Top Left: Original Image; Top Right: Histogram Equalized Image; Bottom Left: Image after Using Median Filter; and Bottom Right: Image after Using Laplacian Filter

from images. Therefore, noise suppression is very important and is sometimes difficult. Another purpose of image processing is to enhance and extract useful information, such as edges and image pixels representing the same scene surface. Usually, a significant change in brightness in an image corresponds to an object's surface boundaries. This attribute is used for enhancing and extracting information. Using a Laplacian filter, which is a differential operator, enhances image region border (bottom right).

Scene analysis provides applications with structured description of a scene. This description is often concerned with a three-dimensional (3-D) scene. In industry applications, geometric descriptions in both low and high level are needed. Scene analysis approaches obtain information from a description in relatively low level, and generate a more sophisticated description at high level. For instance, the low-level wireframe description shown in Figure 11.5 can first be obtained from the images given in Figure 11.4. From this description, middle-level surface representation can be obtained by analyzing which edges enclose a surface. From the wireframe representation, a high-level solid model might also be built depending upon the completeness of the wireframe model.

Three-dimensional scene analysis relies on 3-D vision sensors which are able to generate initial descriptions for a scene. A 3-D vision sensor belongs to either the active or the passive type [18]. A passive sensor detects reflections of illumination provided by the environment lights, and an active sensor receives reflected light that is controlled by a special light source such as a laser. Passive sensors are usually appropriate for military use or applications where active light sources are not allowed for security or environmental constraints. However, active sensors can be used in many industrial applications. Since it is very difficult and not robust

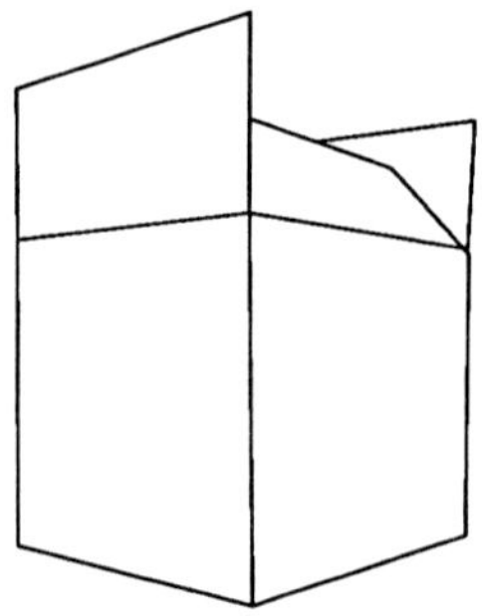

Figure 11.5: Wireframe-based Geometric Model is Built from the above Image

to extract 3-D information using a passive sensor, the research described in this chapter has considered the use of active sensors. More details on active sensors will be found in the next section.

11.2 Active 3-D Surface Data Acquisition

Since an active sensor usually measures the distance (range) between the sensor and object surface, it is referred to as a range sensor, a range finder, a range-imaging sensor, or a laser range scanner. In the following discussion, the term range sensor will be used. A range sensor is capable of generating a range image of a scene with dedicated hardware and software. A range sensor generates a range image that is a 2-D array. Each element of the image array records a number to represent a range (a coordinate value). This element is indicated by two subscripts to represent the range measurement direction. Commonly used range sensors are the triangulation-based and time-of-flight- (or phase-shift-) based sensors.

11.2.1 A Triangulation-based Range Sensor

The triangulation principle is straightforward, and is frequently used for measurement. The principle can be mathematically represented by the law of sines as

$$\frac{sin(\theta_{laser})}{S} = \frac{sin(\phi)}{T} = \frac{sin(180^\circ - \theta_{laser} - \phi)}{B} \tag{11.1}$$

The triangulation-based range sensor shown in Figure 11.6 employs a HeNe laser source with an optical straightline generation lens, a charged coupled device (CCD) camera, and a nodding mirror. The laser source, object surface point, and the camera form a triangle. By applying the law of sines the distance between an object surface point and the camera is obtained.

The laser beam is extended into a line-structured light (not shown in the Figure 11.6) by going through the straightline generation lens. This line-structured laser light is projected by the nodding mirror which rotates to perform horizontal scanning of the laser light. Then

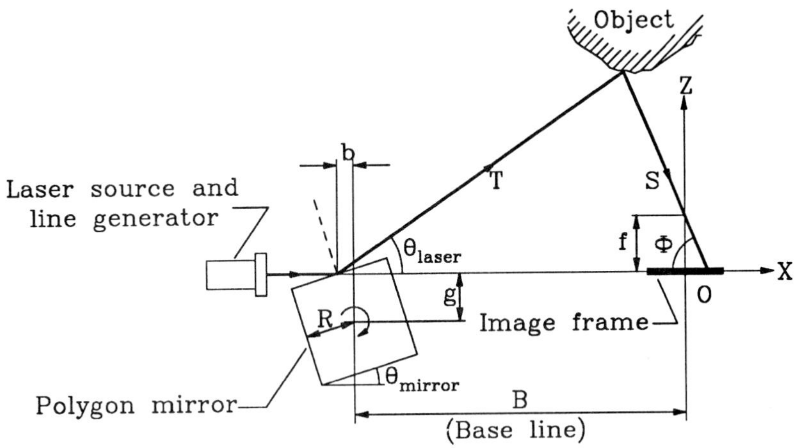

Figure 11.6: Geometry of a Triangulation-based Range Sensor

the reflected laser light, after hitting the object surface, is caught by the CCD camera. An image of an object with the laser light is shown in Figure 11.7.

The laser light is extracted from the image by subtraction from an image of the same scene without the laser light and by taking thresholds. The resulting laser light image is scanned row by row to find each pixel corresponding to the laser light. The image frame's row and column numbers of each found pixel are used for the range calculation.

The camera is considered as a pinhole camera whose geometry is simple. The geometric position, in terms of the image plane, of a pixel corresponding to laser light is obtained by the transform

$$x_{plane} = (x_{center} - x_{frame}) \cdot aspect\_ratio_x \tag{11.2}$$
$$y_{plane} = (y_{center} - y_{frame}) \cdot aspect\_ratio_y \tag{11.3}$$

where, x_{center} and y_{center} represent the center of the image frame, and $aspect\_ratio_x$ and $aspect\_ratio_y$ are the interval distances between two pixels in the horizontal and vertical directions, respectively.

The angle ϕ is found as

$$\phi = tan^{-1}(\frac{f}{x_{plane}}) \tag{11.4}$$

The coordinates of a surface point with respect to the frame $X - Y - Z$ are obtained from

$$X = (1 - \frac{S}{\sqrt{x_{plane}^2 + f^2}}) \cdot x_{plane} \tag{11.5}$$

$$Y = (1 - \frac{S}{\sqrt{x_{plane}^2 + f^2}}) \cdot y_{plane} \tag{11.6}$$

$$Z = \frac{S}{\sqrt{x_{plane}^2 + f^2}} f \tag{11.7}$$

where, according to the law of sines, S is found as

$$S = \frac{(B + b + x_{plane}) \cdot sin(\theta_{laser})}{sin(\theta_{laser} + \phi)} \tag{11.8}$$

To a triangulation-based range sensor, "missing data" are a significant drawback which is caused by separation of the laser source and the camera. Part of the laser light may be occluded from the viewing direction of the camera by other objects in the scene. In Figure 11.7 discontinuous laser light can be seen, since the scene hides part of the light. This disadvantage exists in almost all sensors which are based on the structured light and/or triangulation measurement.

Figure 11.7: Image of a Scene with the Laser Light

11.2.2 A Time-of-Flight-based Range Sensor

The structure of a typical time-of-flight range sensor, the ERIM (Environmental Research Institute of Michigan) range finder [21], is illustrated in Figure 11.8. A laser beam is generated by a laser diode and is projected onto a polygon mirror through a collimator and an expansion telescope. The polygon mirror performs horizontal scanning by reflecting the beam onto the nodding mirror. The nodding mirror controls the vertical scanning of a scene. The reflected beam from the object surfaces is received by a receiver through the two mirrors and a similar optical path.

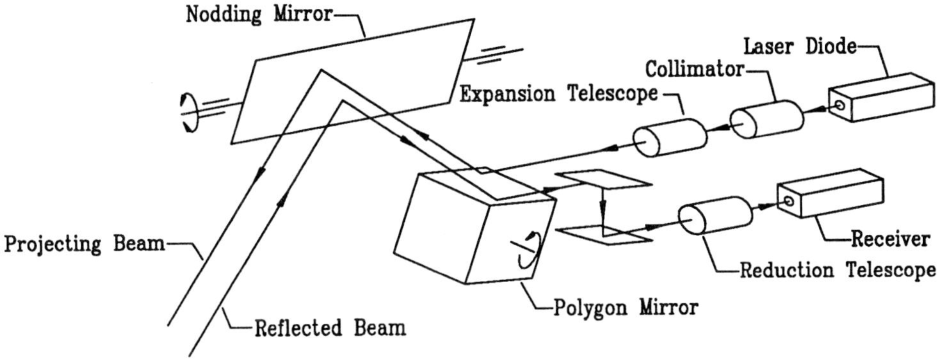

Figure 11.8: Structure of a Time-of-flight-based Range Sensor

The ERIM range sensor uses active image acquisition with the laser scanning. The range data are acquired at equal angular intervals of the nodding and polygon mirrors. The laser beam is sent out and reflected back to the range sensor, and the time of flight is measured to determine the range of the projected point.

The optical paths of the range sensor can be divided into exterior and interior portions, defined by the extrinsic and intrinsic parameters of the time-of-flight sensor. The exterior working space of the sensor is formed by rotations of the polygon and nodding mirrors. The interior optical paths can be ignored to consider the sensor as a point, if the sensor is not located very close to the object surface [23].

This introduces a spherical coordinate system, as illustrated in Figure 11.9, where, α, β, and r are extrinsic parameters of the sensor. The parameter r is the distance between the projected point on the scene and the sensor. The coordinates of the projected point P on the object, specified in this spherical coordinate system, can be transferred back into a Cartesian coordinate system aligned with the range sensor by

$$x(\alpha,\beta,r) = r\sin(\beta)\cos(\alpha) \tag{11.9}$$

$$y(\alpha,r) = r\sin(\alpha) \tag{11.10}$$

$$z(\alpha,\beta,r) = r\cos(\beta)\cos(\alpha) \tag{11.11}$$

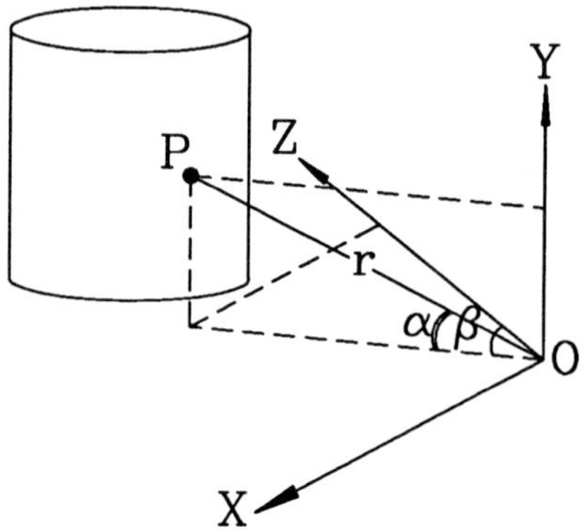

Figure 11.9: Cartesian and Spherical Coordinate Systems of the Range Sensor

11.2.3 Pseudo Range-image Generation

In research on geometric modeling with range data, a tool for generating pseudo range data is very helpful. The generated pseudo range data allow: (1) algorithm testing with images of objects of various shapes in a variety of desired positions and orientations, and (2) the control of the affect of image noise for the purpose of research.

The pseudo range-data generation system includes a solid modeling subsystem, a range-data generation subsystem, and a display subsystem as shown in Figure 11.10. The solid modeling subsystem, which has the basic functions of geometric modeling and interfaces to a CAD system, provides geometric models of desired objects and workspace [24]. Primitive solids of various types are represented by B-rep using the winged-edge data structure [8]. More complex objects can be created using a CAD system in polyhedra form and incorporated into the system through metafiles. The logical operations of OR on the primitives are conducted by a modified z-buffer algorithm. This approach greatly simplifies the generation and manipulation of a solid model.

The modified z-buffer algorithm, called r-buffer, is applied to generate the depth map of surface points, that is, range data. It simulates a scanner composed of two scanning mirrors that perform horizontal and vertical scanning. Therefore the r-buffer works in the spherical coordinate system. The r-buffer approach consists of a matrix, the elements of which are indexed by the polygon mirror angle (column) and the nodding mirror angle (row). Each element of the matrix corresponds to a "light beam" with fixed orientation. In range-data generation, the beam scans through the surfaces of all pertinent objects. When the beam strikes a surface, the range of the struck point is calculated. If this range is less than the previously calculated range (of another surface) stored in this matrix element slot, the newly calculated range replaces the old one in the slot. Eventually, each element of the matrix contains the minimum range, r, along the light beam between surfaces of the pertinent objects and the range sensor. The range data of an example scene generated by the simulation system are illustrated in Figure 11.11. The scene is about a composite object composed of a polyhedron and an ellipse-shaped cone. The display subsystem uses a floating horizons algorithm or intensive image display to illustrate the range-data values.

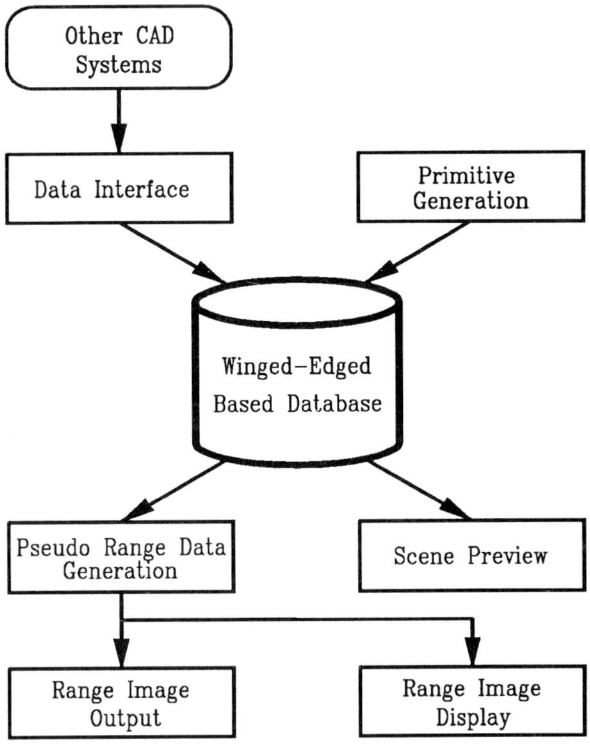

Figure 11.10: Structure of the Pseudo Range-data Generation System

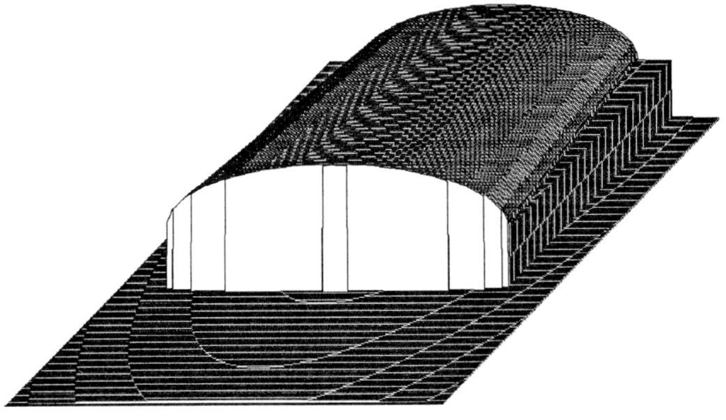

Figure 11.11: Pseudo Range-data Values Illustrated in a Floating Horizons Representation

11.3 Object and Workspace Modeling from Range Data

To obtain 3-D geometric models for existing objects and workspace, the initial description of a scene is required. In many situations, on-line creation of geometric models may be required. Active sensors are able to directly provide the required initial description by providing discrete range data representing object surface points. A key point is how to convert these discrete surface points into geometric models. This requires a transformation from a 3-D image representation of objects to a relatively concise geometric representation of the objects.

Due to the limited viewing field of a sensor and due to the possibility of object occlusion, acquiring range images for objects requires viewing from multiple range sensor locations to allow a complete geometry description to be obtained. Therefore, the modeling system needs to perform multiple-view fusion in its modeling processing. Flexible selection of viewing locations is quite important in order to handle objects and workspace with complex surface shapes. Guaranteeing this flexibility increases the difficulty of modeling from range images.

11.3.1 Review of Previous Methods

Previous work on geometric model generation from active range sensor image data utilizes geometric representations that require no sophisticated topologic relationships for modeling and facilitating the conversion from discrete 3-D range data to geometric models. Previous research has used mesh surface based representation, cross-section-based representation, and segmentation associated (surface or feature) representation. Most of these approaches manipulate range data in a Cartesian frame.

A mesh surface-based representation can have either regular mesh node distribution or irregular distribution. Faugeras and colleagues used a triangular facet-oriented polyhedral approximation to model sculptured surfaces [7]. The range data were represented and processed in a Cartesian coordinate system. Henderson used a graph—a spatial mesh to organize discrete surface points in 3-D space and obtain a surface representation through range data segmentation based on this graph [10]. To generate a workspace model for mobile robots, Asada developed a height map that was a mesh representation with regularly distributed nodes [1]. Despite the differences of their methods, these researchers all used multiple views to cover the entire object. Each of these views was usually taken using an auxiliary view selection device on a fixed trace. Examples of auxiliary devices include turntables, linear moving platforms, or high marks.

A cross-section-based representation is frequently used to generate a 3-D description in the volumetric rendering of computer tomography (CT) and/or magnetic resonance (MR). Since the CT/MR scanners conduct well-structured scanning, the acquired data are in ideal slice form. The 3-D volume can be formed by stacking all slices and interpolating between two neighboring slices [16]. A similar work on approximating 3-D surfaces using serial microscopic sections was reported in the research of biomedical engineering [13]. Theodoracatos used a laser plane-of-light projector for obtaining range surface contours and for building the model of 3-D objects [20]. In these applications the form of the acquired image data is consistent with the cross-section representation. Different from the geometric representation using polyhedrons, the cross-section-based representation simplifies a complex 3-D description problem by handling the task in 2-D space.

Geometric modeling with image data is also found in the area of close range photogrammetry [5]. In the work of [5] one cross section or profile was captured by an angular scanning of the laser beam. With the step motion of the measuring system a tunnel surface was described by a series of profiles.

To process the raw range images for obtaining surface models, various segmentation algorithms are used. These algorithms divide an image into several local regions. All surface points in each of these local regions belong to a certain surface function. A surface-based representation such as a bicubic spline surface can thus be used to fit the regional image. Processed region by region, the object surface, initially recorded by the range images, is transformed into the surface-based representation. In-depth discussion of this type of technique can be found in [15] and [17].

11.3.2 Device Frame of Active Range Sensors

Both time-of-flight and triangulation-based range sensors work primarily by a central projection mechanism. In effect, this generates a device frame. The coordinates of the device frame can be two scanning angles and a range. Range data for each acquired surface point is a distance value between the sensor and the surface point and is associated with the direction pointing from the sensor to the projected surface point in the device frame. The direction can be defined by two angular scanning components, horizontal and vertical. When scanning, the angle changes equally in each scanning direction, and range data are stored into a rectangular array, called a depth map or a range image. Each element of this 2-D range image array is addressed by the horizontal and vertical scanning angles. In general, the device frame can be considered to be basically a spherical frame due to the central projection mechanism [12, 21, 23]. Hence, each surface point is expressed by its range value and two scanning angles in the object range image array. In the device frame array the data are dense and regularly distributed. The global coordinates of surface points can be calculated by transforming the corresponding range data from the device coordinate frame to the global coordinate frame.

In present practice, range image processing and/or range-data-based modeling are usually conducted in a Cartesian coordinate frame. Surface points are transformed from the device frame to the Cartesian frame, and are indicated by their x, y and z coordinates. Because such a transformation is nonlinear, the transformed surface points with x, y and z coordinates can no longer be uniformly represented in a 2-D array. Although some alternative data structures (such as a sparse 2-D array) can be used to represent these discrete points, computation costs will increase due to the complexity of handling a sophisticated data structure. A sparse 2-D array or a specially designed data structure also makes many present image processing methods either inapplicable or very costly to use. The coordinate transformation of the massive amount of range data also contributes to high computation costs.

The remainder of this section discusses a representation allowing the characteristics of the data in the device frame to be exploited. Further details on basic representations can be obtained from [3, 10].

11.3.3 Selection of Representation

Different representations handle geometric data on different levels. To describe objects and workspace based on discrete range data of surface points and to provide applications with relatively concise geometric models of objects and workspace, an appropriate geometric representation must be considered and applied in the modeling system. The criteria for selecting such a representation primarily include (1) compatibility with discrete range data; (2) suitability for modeling in the device frame; (3) capacity for facilitating multiple-view fusion and model updating; and (4) efficiency of converting range data into the representation.

With the above requirements various representations, classified into several types with respect to their levels to represent objects, can be evaluated.

Point based— such as 3-D voxel array representation (a digitized space). Objects are depicted by a set of discrete surface pixels in low level. One voxel is adjacent to its neighbors in the digitized space; otherwise the description is not complete. It is difficult to adjust the resolution of the digitized space by changing the distance between the range sensor and surface to be scanned. This kind of adjustment is important to cope with surface details carried in the range images. Point-based representations also require a large amount of memory.

Line based— such as wireframe representation. Wireframes are intersections of adjacent surfaces. Surface properties are normally not described in this representation. The major disadvantage of a line-based representation is that range data do not explicitly provide intersection lines of surfaces and end points of each line.

Surface based— such as mesh, sweep, generalized cylinder (GC), and cross-section contour representations. Surface-based representation is harmonious with surface range data.

Mesh represents a surface by surface points and spatial grids. With the other representations, object shapes are primarily represented by stacking a series of planer contours at intervals. These representations are a combination of discrete and continuous modeling. Mesh can be used by linking the surface points with a graph to form a polyhedron of triangular- or rectangular-shaped boundary elements. Object modeling based on this representation requires relatively extensive computations. Cross-section contours can be formed from object surfaces described by surface points with moderate computations. The representation is compatible with range data. Difficulties of handling shape details, data condensation, and fusion are reduced due to the use of the 2-D space of cross-section planes as the working space. Sweep and

GC are similar to the cross-section form except GC contains fewer geometric details and sweep is inadequate for representing irregular shapes.

Volume based— such as octree representation. Not only the surface but also the volume of an object is modeled by this representation in a discrete form. The major problem with using this type of representation with range data is that the involved recursive spatial subdivision requires complete data acquisition before modeling. However, this requirement cannot in general be satisfied.

Boundary based— such as B-rep representation. A boundary-based representation of an object includes surface facets and surface borders. Linkages between borders and facets are required. Models in this representation are concise and in high level. However, this representation is not compatible with range data due to the required linkages.

Object based— such as CSG representation. This type of representation describes an object using geometric primitives. The geometric and topologic properties of the object are incorporated into the representation. The topologic relations at all levels of the model are explicitly required. These requirements make an object-based representation unsuited for direct use with range data.

Based on this comparison, the surface-based representations are most appropriate for representing range data. Of surface-based representations the cross section is superior to others in terms of the computational efficiency in multiple-view fusion and the ability to describe shape details and complex surfaces. Therefore, a cross-section-based approach is used in this work to serve as the representation for building geometric models from range data.

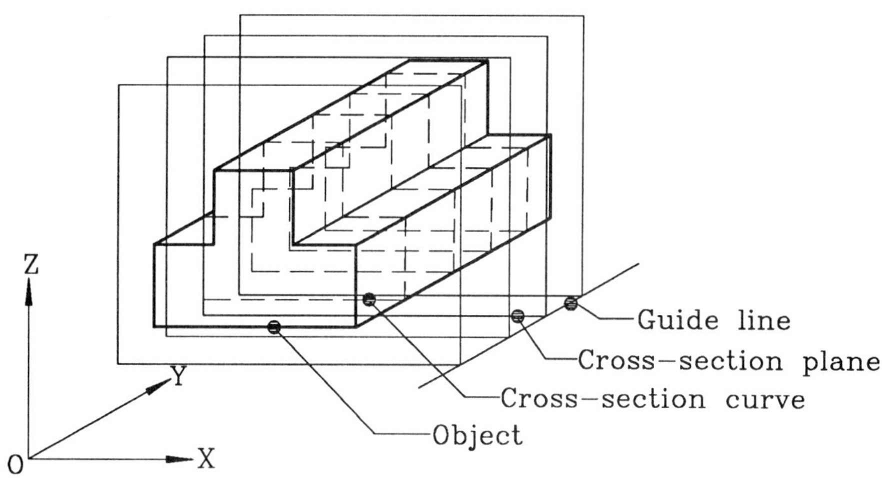

Figure 11.12: Cross-section-based Representation

Cross-section representation uses 2-D surface contours lying on a series of planes to represent object shapes. These contours are called cross-section curves, and the planes are called cross-section planes. Cross-section planes are arranged along a guide curve. The planes are arranged to be always perpendicular to a guide curve. If the guide curve is a straight line, the planes form a group of parallel planes.

A modeling frame, which is a global 3-D Cartesian coordinate system, is introduced for using the representation. The guide curve and the cross-section planes are specified in this Cartesian modeling frame.

A 2-D cross-section curve on a cross-section plane is the intersection curve of the plane and the surfaces of the 3-D objects to be modeled. If the modeled object is bounded, the corresponding cross-section curves will be closed when complete. The cross-section curves for an unbounded surface, for example, the workspace surroundings, will be open. The surfaces of modeled objects can be approximated by interpolating the cross-section curves. Since the outer and inner sides of object boundary are identified by the curves, the volume of the modeled objects can be computed. When the object surface changes greatly, surface details can be preserved by decreasing the interval of the cross-section planes.

A simple object represented by cross sections is illustrated in Figure 11.12 where the cross-section curves, the cross-section planes, and the guide curve are shown.

11.3.4 Modeling in Device Frame with Cross-section-based Representation

The generation of cross-section curves can be conducted in the Cartesian modeling frame. However, the range data are nonuniformly distributed in a Cartesian frame, and the processing will be complex. Therefore, in this work cross-section curve generation in the device frame is introduced to preserve the uniform distribution of the surface points (raw range data) in a 2-D array. This approach requires the cross-section planes to be transformed into the device frame in order to find the intersections (cross-section curves) of these planes and the object surfaces represented by the range data.

Three coordinate systems are involved in this modeling approach: the global Cartesian modeling frame; a device Cartesian frame attached to the range sensor; and the device frame of the range sensor, which is in spherical form. Mapping between the global Cartesian modeling frame and the device Cartesian frame can be conducted by a homogeneous transformation, that is,

$$\{\,X,\ Y,\ Z,\ 1\,\} = \{\,x,\ y,\ z,\ 1\,\}\begin{bmatrix} \left[\,\mathrm{R}\,\right]^{T}_{3\times 3} & \begin{matrix}0\\0\\0\end{matrix} \\ \{\,\mathrm{t}\,\}^{T}_{1\times 3} & 1 \end{bmatrix} \tag{11.12}$$

where $[R]$ is a 3×3 rotation matrix describing the orientation of the range sensor with respect to the global modeling frame, $\{t\}$ is a vector describing the location of the range sensor in the global modeling frame, and $\{X, Y, Z\}$ and $\{x, y, z\}$ represent the coordinates of a point in the global Cartesian modeling frame and the device Cartesian frame, respectively.

The coordinates of the device frame are the two scanning angles, θ and ϕ, and the range, r. Transformation from the device Cartesian frame to the device frame has the analytical form given by Eqs. (11.9), (11.10), and (11.11).

The analytical expression of a cross-section plane in the global modeling frame can be represented by

$$Ax + By + Cz + D = 0 \tag{11.13}$$

Transformation to the device Cartesian frame yields

$$ax + by + cz + d = 0 \tag{11.14}$$

where $\{a, b, c\} = \{A, B, C\}\,[R]$ and $d = D - \{t\}^T\{a, b, c\}^T$.

Substituting for x, y and z from Eqs. (11.9) to (11.11) yields

$$ar\cos\theta\cos\phi + br\cos\theta\sin\phi + cr\sin\theta + d = 0 \tag{11.15}$$

Due to the nonlinear form of Eq (11.15), it is difficult to use it directly for finding the plane-to-object intersections and for conducting range image interpolation. In particular, searching the solution all over the image would be a very time consuming task. In this work, a quick estimation method for finding plane-to-object intersections using discrete values is introduced. This method is explained below.

A cross-section plane transformed to the device frame forms a range image in a 2-D array. This image will be named the plane image. The plane image can be generated with identical size and the same subscripts of the object range image using the two scanning angles as indices. As illustrated in Figure 11.13a, the plane and object can be assumed to simultaneously exist in 3-D space. Since the existence of a cross-section plane is hypothetical, the range images of the object and the cross-section plane can be generated separately as illustrated in Figure 11.13b and c. It should be noticed that the plane range image is not produced by the sensor with physical perceiving, but by the modeling system.

With these two identical indexed range images, it is easy to estimate the object surface to cross section plane intersection, that is, the cross-section curve. A simple solution would be scanning the two range images pixel by pixel and connecting these pixels where the discrete object image and plane image share a same value, as shown in Figure 11.13c.

In practice, the object-to-plane intersection may occur at a subpixel location. When the range of the plane changes from larger (or smaller) than the range of the surface to smaller (or larger) at two neighboring pixels, an intersection exists somewhere between these two pixels. The intersection can be generated by interpolating the two continuous surface patches formed by the range values of the two images at the four neighboring pixels, as shown in Figure 11.14d.

The generated curve section is then transformed to the global Cartesian modeling frame to update the present model. Through successive viewing and updating multiple-view fusion

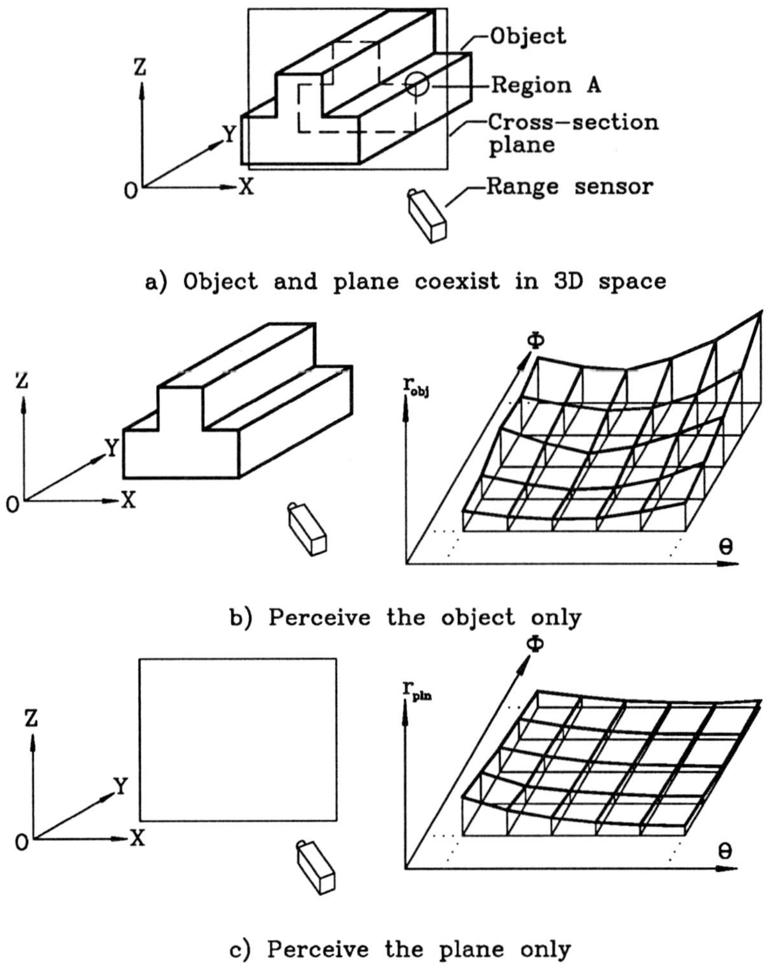

Figure 11.13: Generation of the Object and Cross-section Plane Range Images

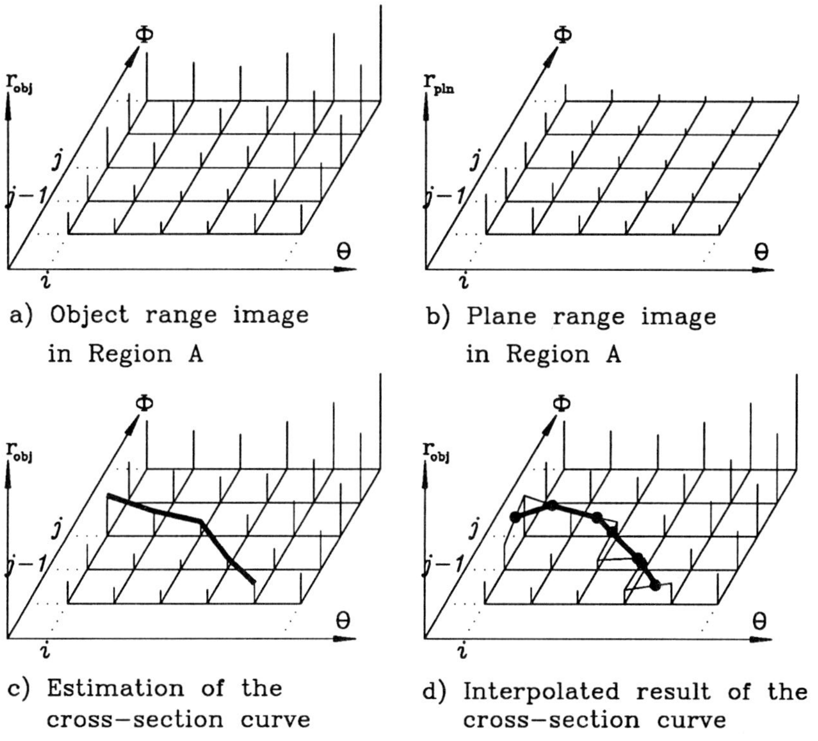

Figure 11.14: Estimation and Interpolated Solution of the Cross-section Curve

and a complete geometric description can be achieved. Successive viewing and updating also allow basing the next view to be selected on interpretation of the updated model (intelligent multiple-view selection).

11.3.5 Advantage of Range-data Processing in Device Frame

In previous research, range-data processing and object modeling using a cross-section representation are conducted in a Cartesian coordinate frame. The computational cost of these approaches is quite high due to the need for searching data with indirect addressing and for massive data transformation, as discussed previously. The device-frame-based range-data processing method introduced in this work directly uses the device frame for range-data processing. Significant savings in computational effort can be found after comparing the data processing procedures of these two approaches. This saving is primarily due to (a) only the solved intersection curves (several control points), rather than all of the raw range data, need to be transformed between different coordinate frames; (b) the "transform" of the involved cross-section planes from the global Cartesian modeling frame to the device frame can be conducted using a simple analytical formula; and (c) the range data are preserved in their well-distributed 2-D array form. This well-distributed form can allow efficient existing image processing techniques to be applied.

11.3.6 Complete Description for Objects and Workspace

A complete geometric model for 3-D objects and workspace using the proposed cross-section representation can be built by generating all cross-section curves from multiple-view range data.

After a range image is acquired, a partial surface model that consists of a series of incomplete cross-section curves is constructed. This partial model is added to and integrated with the present object model, which was generated previously by the earlier processed partial models. The multiple-view fusion is implicitly accomplished. The partial model merging activity can eliminate any redundant surface data caused by overlaps of different range images and identify surface discontinuities caused by gaps between range images. The model updating procedure can be expressed by the OR Boolean operation

$$S_{obj_n} = S_{obj_{n-1}} \bigcup s_n \qquad n = 1, 2, \ldots$$

where S_{obj_n} is the updated object surface model, $S_{obj_{n-1}}$ is the previous object surface model, and s_n is a partial surface model. Initially, when $n = 1$, $S_{obj_{n-1}} = \emptyset$.

Overlaps between partial models (partial redundancy) exist when

$$S_{obj_{n-1}} \bigcap s_n \neq \emptyset$$

Gaps exist between between partial models or s_n is perfectly adjacent with $S_{obj_{n-1}}$ when

$$S_{obj_{n-1}} \bigcap s_n = \emptyset$$

When the current surface measure is totally redundant to the previous ones, we have

$$S_{obj_{n-1}} \bigcap s_n = s_n$$

Specifically, updating the object model is performed on the involved cross-section planes in the global Cartesian modeling frame. On each cross-section plane any overlaps between the current curve section and the previously generated curve section(s) are tested. If an overlap of curve section is found, the redundant part is eliminated, and the newly generated curve section is connected to the curve section(s) generated previously. If the newly generated curve section is separated from the previously generated curve sections in the current object model, the new curve should be stored separately. For a complete object model, gaps between separated curve sections should all have been filled. Once the partial modeling for an acquired range image and the model updating using the newly created model are completed, the next range image can be taken and processed. By this updating approach, as long as the observation for object or workspace is complete, the geometric descriptions will be complete too.

11.4 An Example

Illustrated in Figure 11.15 is an object that will be modeled using the cross-section representation. The object consists of a truncated cone merged with a rectangular parallelepiped. Range data from three sensor locations will be used to construct the cross-section model.

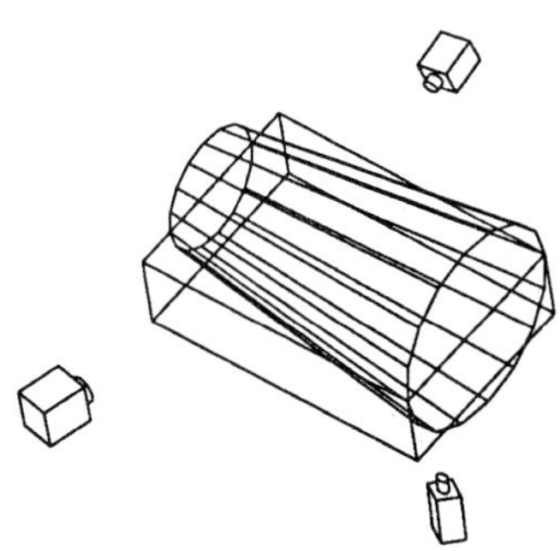

Figure 11.15: Three Different Sensor Locations

Figure 11.16 shows an intensity visualization of the range image for the first sensor location. Based on this range data the partial cross-section curves illustrated in Figure 11.17 were found. Based on data from a second viewing location (range intensity visualization in Figure 11.18) the model was updated yielding the curves illustrated in Figure 11.19. A final sensor location yielded sufficient range data (intensity visualization in Figure 11.20) to complete the cross-section model. The completed cross-section model based on all three views is shown in Figure 11.21.

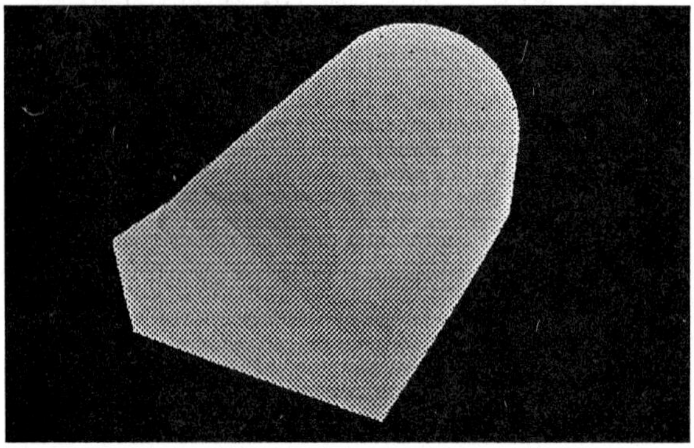

Figure 11.16: Range Image (Intensity Visualization) from a First Sensor Location

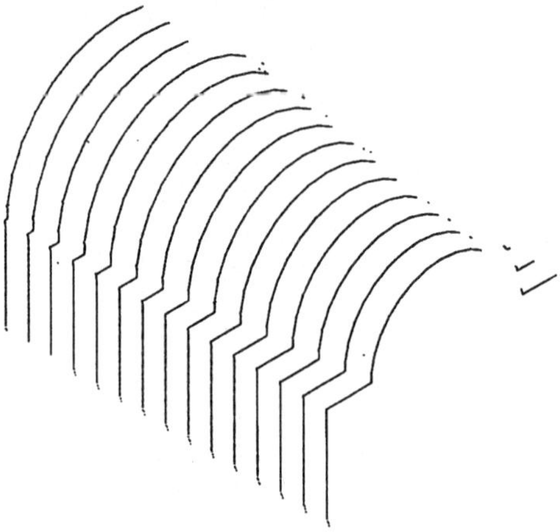

Figure 11.17: Partial Model Based on the Range Image from a First Sensor Location

Figure 11.18: Range Image (Intensity Visualization) from a Second Sensor Location

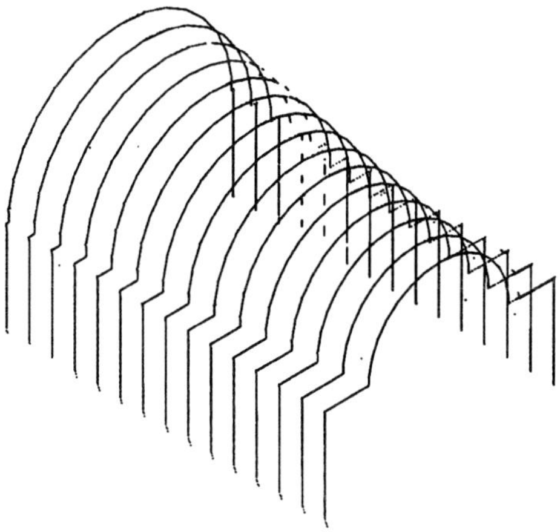

Figure 11.19: Updated Model Based on a Second Sensor Location

Figure 11.20: Range Image (Intensity Visualization) from a Third Sensor Location

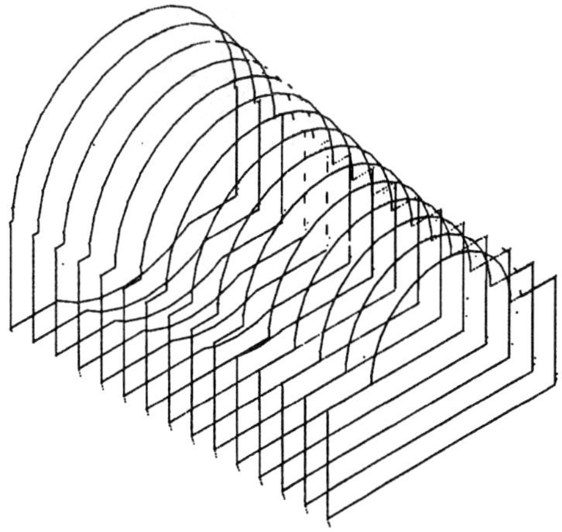

Figure 11.21: Complete Model Based on the Three Sensor Locations

11.5 Summary

Geometric data input is important for applications of intelligent robotic and manufacturing systems. In many cases, these applications depend on the understanding of the work environment. This dependence has strongly activated research and development of 3-D machine vision and geometric modeling based on vision data. This chapter has reviewed the basic principles of active range sensors and previous approaches of forming geometric representations based on range information. Furthermore, recent research on a new approach for generating a representation based on 3-D range information has been discussed.

The discussed research focused on a geometric modeling approach for representing multiple-view range data. Acquiring object range images requires viewing from various sensor locations (multiple views) to allow a complete geometric description. A flexible multiple-viewing approach requires a modeling technique that is able to integrate partial models together. Effective on-line application of a modeling approach for such range data requires the capability of simultaneous performance of acquisition and modeling.

A cross-section-based representation has been used for supporting implementation of the introduced modeling approach. This representation uses a series of object contours to describe geometric shapes. Each contour curve is built from partial model curve sections by model updating in a global Cartesian modeling frame. The partial model curve sections can be formed efficiently in the device frame. Low computational cost of range-data processing in the device frame has been found possible due to fast searching for adjacent surface points and transforming generated curves rather than raw data to the Cartesian modeling frame.

The research introduces a new geometric modeling approach for representing range data. The approach allows partial modeling and flexible multiple-view selection. The concept of modeling after the acquisition of each view allows decreased memory requirements. The approach should provide subsequent researchers with a platform for the consideration of intelligent view acquisition schemes. In addition, the new device-frame-based range-data processing method increases data processing speed by avoiding computationally expensive range-data processing in a Cartesian frame.

References

[1] Asada, M. (1988). "Building a 3D World Model for a Mobile Robot from Sensory Data," *Proceedings of the IEEE International Conf. on Robotics and Automations*, Vol. 2, pp. 918-923.

[2] Ballard, D.H., and Brown, C.M. (1982). *Computer Vision*, Prentice-Hall, Inc.

[3] Besl, P., and Jain, R. C. (1985). "Three-Dimensional Object Recognition," *Computing Surveys*, Vol. 17, No. 1, pp. 73-145.

[4] Besl, P. (1989). "Active Optical Range Imaging Sensors," *Advances in Machine Vision*, J. Sanz, ed., Springer-Verlag, pp. 1-63.

[5] Clarke, T. (1990). "The Use of Optical Triangulation for High Speed Acquisition of Cross Sections or Profiles of Structures," *Photogrammetric Record*, Vol. 13, No. 76, pp. 523-532.

[6] Fan, T.J. (1989). *Describing and Recognizing 3D Objects Using Surface Properties*, Springer-Verlag, London.

[7] Faugeras, D., Hebert, M., Mussi, P. and Boissonnat, J. (1984). "Polyhedral Approximation of 3D Objects without Holes," *Comput. Vision, Graphics, and Image Process.*, Vol. 25, pp. 169-183.

[8] Foley, J. D., Dam, A. V., Feiner, S.K. and Hughes, J.F. (1990). *Computer Graphics Principles and Practice*, Addison-Wesley, Reading, MA.

[9] Gonzalez, R.C., and Wintz, P. (1987). *Digital Image Processing*, Addison-Wesley.

[10] Henderson, H. (1983). "Efficient 3-D Object Representations for Industrial Vision Systems," *IEEE Trans. Pattern Anal. Machine Intell.*, Vol. PAMI-5, No. 6, pp. 609-618.

[11] Horn, B. K. P. (1986). *Robot Vision*, MIT Press and McGraw-Hill, New York.

[12] Kaisto, I., Kostamovaara, J., Moring, I., and Myllyla, R. (1990). "Laser Rangefinding Techniques in the Sensing of 3D Objects," *SPIE Sensing and Reconstruction of Three-Dimensional Objects and Scenes*, Vol. 1260, pp. 122-133.

[13] Mercer, R., McCauley, G., and Anjilvel, S. (1990). "Approximation of Surfaces in Quantitative 3D Reconstructions," *IEEE Trans. Biomed. Eng.*, Vol. 37, No. 12, pp. 1136-1146.

[14] Mortenson, Michael E. (1985). *Geometric Modeling*, John Wiley.

[15] Neik, S.M., and Jain, R.C. (1988). "Spline-Based Surface Fitting on Range Images for CAD Applications," *Proceedings CVPR '88: The Computer Society Conf. on Computer Vision and Pattern Recognition*, pp. 249-256.

[16] Ney, D.R., Fishman, E.K., Magid, D., and Drebin, R.A. (1990). "Volumetric Rendering of Computed Tomography Data: Principles And Techniques," *IEEE Computer Graphics & Applications* (March), pp. 24-32.

[17] Orr, M.J.L. (1991). "World Modeling from Range Images," *International Conf. on Control '91*, pp. 807-812.

[18] Poussart, D., and Laurendeau, D. (1989). "3-D Sensing for Industrial Machine Vision," *Advances in Machine Vision*, J. Sanz, ed., Springer-Verlag, pp. 122-159.

[19] Requicha, A.A.G. (1980). "Representations for Rigid Solids: Theory, Methods, and Systems," *Computing Surveys*, Vol. 12, No. 4, pp. 437-464.

[20] Theodoracatos, V., and Katti, R. (1991). "An Automated and Interactive Approach for Fitting B-Spline Surfaces through 3D Planer Visual Data," *Proceedings of the 1991 ASME Advances in Design Automation*, Vol. 2, ASME, New York, pp. 23-31.

[21] Veatch, P. A., and Davis, L. S. (1990). "Efficient Algorithms for Obstacle Detection Using Range Data," *Computer Vision, Graphics, and Image Processing*, Vol. 50, pp. 50-74.

[22] Wang, Y.F., and Aggarwal, J.K. (1988). "An Overview of Geometric Modeling Using Active Sensing," *IEEE Control Systems Magazine*, Vol. 8, No. 3, pp. 5-13.

[23] Yao, H., Podhorodeski, R., and Dong, Z. (1991). "Modeling and Simulation of a Range-Finding Device," *IEEE Pacific Rim Conference on Communications, Computers and Signal Processing*, IEEE, New York, pp. 166-169.

[24] Yao, H., Dong, Z., and Podhorodeski, R. (1991). "Simulation of Range Finding Devices by Geometric Modeling," *Proceedings of the 1991 ASME Computers in Engineering Conference and Exposition*, pp. 327-332.

[25] Zeid, I. (1991). *CAD/CAM Theory and Practice*, McGraw-Hill, Hightstown NJ.

Chapter 12

Motion Planning and Geometric Design

CHUNG-YUH LIU
Department of Mechanical Engineering, University of Michigan - Dearborn

ROGER W. MAYNE
Department of Mechanical and Aerospace Engineering, SUNY at Buffalo

12.1 Introduction

This chapter approaches the problem of robot path planning via nonlinear programming and also illustrates the possible integration of motion planning with geometric design. The motion planning problem is formulated as a problem in design optimization, and emphasis is placed on recursive quadratic programming (RQP) as a particularly effective solution strategy. Collision avoidance is considered directly by using distance-to-contact information in inequality constraints along with the definition of specific clearance requirements. Distance-to-contact and interference calculations are developed in detail as well as gradient computations which are critical to applying RQP and other efficient nonlinear programming methods. A kinematic objective function is used in this work which is based on a summation of weighted joint motions. A series of example problems in path planning and geometric design are presented showing the effectiveness of the overall approach. Dynamic interference checking and distance calculation using swept volume segments are discussed at the end of the chapter.

12.1.1 Significance

The need for automated machine motion programming has significantly increased in recent years. It is widely understood that the economic efficiency of production lines can be improved with a fast and smooth machine programming facility. For example, initial off-line programming can save considerable time in the on-line programming of robots (which expends the valuable operation time of production facilities). Automated machines are often operated in an environment where interference with fixtures, parts, and other machines is a major concern. Programming the motions of industrial machine parts in concert with other objects is time consuming and there is substantial risk for damaging equipment or work pieces. Reliable and effective computerized path planning, with the capability of providing obstacle clearance, can be a useful tool for increasing productivity and flexibility.

In addition to motion planning for production, the concepts presented here also extend to mechanical and geometric design in the presence of mechanical motion. The integration of geometric design with motion planning can be important in the design of parts, fixtures, robots, or tools for the manufacturing process. Combined geometry and motion considerations may also play an important role in product design, especially in complex and compact assemblies. In such cases, avoiding interference is the most basic requirement, and clearance specifications must be satisfied in many engineering design projects. In the early stages of a design, geometry and configurations are continually changing in response to system trade-offs and manufacturing and assembly requirements. In this critical period, design support may be provided on the basis of "best it can be by the end of the day." The advantage of engineering optimization with automated clearance checking could be very helpful in the design of complicated assemblies on a timely basis and in the continuing effort to reduce development cycle time.

12.1.2 Current Knowledge

Automated-motion planning with collision detection has been considered by a number of researchers over the past decade as a general problem in automation. These efforts can be divided into two categories: algorithmic approaches and optimization-based methods. The algorithmic approaches consider workspace geometry including the robot and surrounding objects and establish a free space for robot motion. This space is characterized by discretized subdivisions, and in a transformed "configuration space" the robot appears as a point with regions of collision excluded from the space. A search algorithm is followed to maneuver the point within the configuration space to effectively move the robot from an initial position to a desired final position without collision. The configuration space strategies of [4, 25, 32] are examples of this type of approach. In contrast to the algorithmic approaches, the optimization-based methods work with the continuous robot joint space instead of an approximating subspace. In general, the optimization methods represent a more natural design procedure. Some of the early work in this area was directed toward the optimization of robot motion based on kinematic or dynamic criteria without considering the presence of obstacles in the workspace [3, 35]. More recent efforts have included obstacle avoidance strategies using potential function concepts [14], interior penalty function methods [8, 10], and distance-to-contact measures with recursive quadratic programming [28] to avoid contact.

Robot path planning by optimization methods, including obstacle avoidance, requires repeated evaluations of candidate robot paths and calculations of distance-to-contact to evaluate possible obstacle collisions. The optimization problem can easily become quite large with many constraints. The function and constraint evaluations within the problem naturally require intensive computation, and an accurate optimization of a specific robot motion has the real possibility of requiring a prohibitive amount of computation time. The use of gradient-based methods has been found (e.g., [13]) to be especially attractive for efficient optimization of such computationally intensive problems. Sandgren and Ragsdell demonstrated that gradient-based constrained optimization was clearly superior to penalty function optimizations [33] while Schittkowski [34] established the superiority of RQP methods for efficient optimization. The development of RQP algorithms was recognized as a "great breakthrough" by Arora [1]. More recent developments in optimization such as the simulated annealing method may offer useful characteristics from the viewpoint of global optimization but are simply not comparable in terms of efficiency for computationally intensive constrained optimization. A major portion of the discussion in this chapter is devoted to formulating the motion planning problem so it can be handled conveniently by a recursive quadratic programming approach.

Geometric design with obstacle avoidance has now been considered by a few people (e.g., [16, 22, 37]). In [37], the optimal design of robot link length is discussed by using a two-link robot in a two-dimensional workspace. Kim and Gossard [16] present an assembly packaging problem approach by using a penalty optimization method. A study of general geometric design problems coupled with obstacle avoidance path planning is described in [22] and is one focus of this chapter.

Distance calculations are important in obstacle avoidance by optimization-based approaches. Efficient distance calculations can improve computational efficiency and enhance the overall performance of the path planning process. A particular problem for RQP optimization is that RQP does not require design variables to stay in the feasible region during an optimization. Interference with obstacles may take place during the iterative computations and part of the efficiency of recursive quadratic programming rests on the requirement that violated constraints be fully evaluated. Therefore, "interference" and derivatives must be handled accurately. Based on trajectory planning experience [22], distance calculations take most of the computation time for path planning problems. A general calculation strategy for handling interference cases is necessary to streamline the motion planning process.

Existing distance calculation methods can be characterized as vertex-based calculations [23, 28], surface-to-surface methods [32], and convex body approaches [5, 6, 10]. The vertex-based calculations and surface-to-surface methods suffer from the basic disadvantage that the computation time may increase proportionally to the number of "elements"—vertices and surfaces. Recent research has shown that distances between bodies can be efficiently calculated by considering them on a convex body basis. In this work, the distance-to-contact between bodies is obtained from an individual quadratic programming problem based on the assumption of convex objects. For any interference situation, the distance-to-contact is zero and this creates ambiguous constraint information for the RQP path planning optimization. The concept of "*distance-past-contact*" is presented here to define the amount of interference and the numerical value of distance-past-contact is calculated based on two basic strategies—the "move-back" and "shrink-object" approaches—as described in [24].

The most common way of detecting collisions during a trajectory is to perform static interference checking at a series of discretized positions along the trajectory of a motion. In general, the results obtained by this approach depend on obstacle shape and on the step size between two adjacent checking positions. Another method of interference checking is referred to as *dynamic checking* [9] and utilizes the swept volume of moving objects for collision detections. The advantage of this method is that collision detection is performed throughout the entire motion. The representation of volumes swept out by moving solids and applications of this approach can be found in the literature [9, 15, 20, 27, 36]. This chapter closes with a discussion of swept volume ideas for path planning by RQP.

This chapter is divided into five additional sections. The formulation of the path planning problem in an optimization format is presented in Section 12.2. This is followed by a rather detailed discussion of distance calculations including the quadratic programming approach. Section 12.4 develops algebraic derivatives to obtain accurate gradient information and Section 12.5 provides several 2-D and 3-D path planning and geometric design examples. Section 12.6 summarizes an enhanced distance calculation strategy based on swept volumes and presents a 2-D path planning example.

12.2 Problem Formulation

The trajectory problem considered here is to plan a motion from an initial robot configuration to a final position. The planning specifically focuses on reaching the final end effect or position and the desire for interference-free motion relative to the robot links and obstacles in the workspace. The first step in the process is to represent the robot trajectory.

12.2.1 Robot Trajectory Representation

During a robot trajectory, each robot degree of freedom is a function of time between the starting and the ending positions. There are two ways to represent the robot joint motions: either by closely discretizing the motion curves into a sequence of points or by an interpolation approach with a small number of controlling variables. The former method suffers from the large number of points which would need to be defined to describe the trajectory especially with added degrees of freedom. Interpolation techniques are more appropriate (e.g., [2, 10]) and the attractive feature is that the infinite dimensional problem can be represented by a finite dimensional approximation of modest size. Motion interpolations may take place at the end of the planning process to obtain the trajectory in as much detail as necessary.

Interpolation methods were initially studied in approximation theory and now are heavily used in the field of continuous computational geometry. As a result, numerous interpolation functions can be found. The interpolation here focuses on a typical Bezier interpolation described in [12] as

$$\hat{x}(t,\bar{\mathrm{x}}) = \sum_{k=0}^{n_s} x_k \Phi_{k,n_s}(t)\ ; \qquad 0 \le t \le 1 \tag{12.1}$$

where

$$\Phi_{k,n_s}(t) = \frac{n_s!}{k!\,(n_s-1)!}\,(t)^j\,(1-t)^{n_s-k} \tag{12.2}$$

The function $\hat{x}(t, \bar{\mathbf{x}})$ is the interpolated curve on the interval $0 \le t \le 1$ with the Bernstein polynomial of degree n_s, while the n_{s+1} dimensional vector $\bar{\mathbf{x}} = [x_0, x_1, \ldots, x_{n_s}]^T$ contains the control points. These control points become the design variables in the motion planning optimization. The shape of $\hat{x}(t, \bar{\mathbf{x}})$ can be directly changed by adjusting the control points and a convenient equation can be obtained for the derivative relations between the curve and the control points.

As an example, Figure 12.1 shows one of the robot's rotational degrees of freedom during a trajectory. The solid line presenting the curve for the ith joint motion is defined by the control points, $x_{i,0}, \ldots, x_{i,3}$, of the Bezier interpolation. The control point $x_{i,0}$ is the starting position of θ_i. The remaining control points $x_{i,1}, x_{i,2}, x_{i,3}$ may be used as design variables in the motion planning optimization. This value of $n_s = 3$ yields a cubic Bezier interpolation, and has been used in most of the examples shown in Section 12.6.

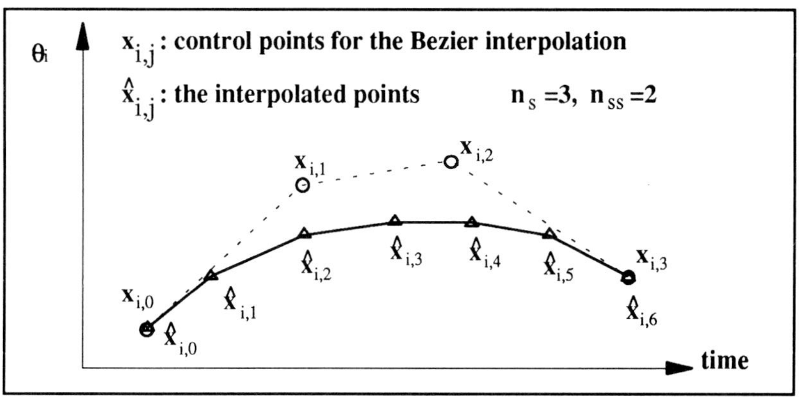

Figure 12.1: Trajectory of One Degree of Freedom with Bezier Interpolation

12.2.2 RQP Optimization

The motion planning and geometric design optimization problem in a nonlinear programming format is

$$\begin{aligned}
\text{minimize} \quad & f(\bar{\mathbf{X}}) \\
\text{subject to} \quad & \bar{\mathbf{X}}_u \ge \bar{\mathbf{X}} \ge \bar{\mathbf{X}}_l \\
& h_q(\bar{\mathbf{X}}) = 0, \qquad q = 1, \ldots, m_q \\
& g_r(\bar{\mathbf{X}}) \ge 0, \qquad r = 1, \ldots, m_r
\end{aligned} \tag{12.3}$$

The design vector $\bar{\mathbf{X}}$, which is bounded by the lower bound $\bar{\mathbf{X}}_l$ and the upper bound $\bar{\mathbf{X}}_u$, may contain two sets of variables: the Bezier control points $\bar{\mathbf{x}}_i$ of each degree of freedom and the shape variables $\bar{\mathbf{y}}$ determining the geometry of robot links and/or workspace features:

$$\bar{\mathbf{X}}^T = [\bar{\mathbf{x}}_1^T, \ldots, \bar{\mathbf{x}}_{n_d}^T, \bar{\mathbf{y}}^T], \tag{12.4}$$

where n_d is the number of degrees of freedom in the robot motions. Each subvector $\bar{\mathbf{x}}_i$ corresponding to the degree of freedom i has n_s elements $x_{i,1}$, $x_{i,2}$, $\ldots$, x_{i,n_s}, and $x_{i,j}$ is the jth Bezier control point for the ith degree of freedom. The shape vector $\bar{\mathbf{y}}$ depends on the object features being designed. It may contain the length of a robot link, the coordinates of a workspace vertex, and so on.

Objective Function

The objective function used here is a combination of two cost functions, $f^p(\bar{\mathbf{x}}_i)$ and $f^s(\bar{\mathbf{y}})$.

$$f(\bar{\mathbf{X}}) = f\Big(f^p(\bar{\mathbf{x}}_i), f^s(\bar{\mathbf{y}})\Big) \tag{12.5}$$

The cost function $f^p(\bar{\mathbf{x}}_i)$ can be formulated according to different performance indices such as minimum time, shortest end-effector path, or minimum energy consumption. Several possible design criteria may involve dynamic considerations such as the minimum time requirement. However, since the concentration of this chapter is on obstacle avoidance and maintaining clearance during a trajectory, the cost function f^p has been selected as a simple summation of weighted robot motions to define a kinematically efficient trajectory

$$f^p(\bar{\mathbf{x}}_i) = \sum_{i=1}^{n_d}\Big(w_i \sum_{k=0}^{n_t-1} (\hat{x}_{i,k+1} - \hat{x}_{i,k})^2\Big) \tag{12.6}$$

Here the w_i are the weighting factors for movement of the ith degree of freedom and the summation describes the overall squared movement of the robot joints in the global coordinate system. The total number of discrete robot motion intervals considered in the objective is $n_t = (n_s \times n_{ss})$ where $(n_s + 1)$ is the number of control points in the Bezier interpolations for each degree of freedom and n_{ss} is the number of subsegments defining intermediate robot positions. In the inner summation of Eq. (12.6), $\hat{x}_{i,k}$ corresponds to the interpolated values resulting from the control points $\bar{\mathbf{x}}_i^T = (x_{i,1}, x_{i,2}, \ldots, x_{i,n_s})$ and the starting point $x_{i,0}$ as shown in Figure 12.1. The interpolated values describe the actual positions which make up the trajectory. The other cost function f^s related to the shape variables $\bar{\mathbf{y}}$ varies on a case-by-case basis.

The Equality Constraint

Robot motions may have some specific requirements during a trajectory. These various requirements, such as velocity profiles, specified angular orientation of the robot end effector, or the desired clearance between objects, can be described in the equality constraints of general optimization methods. In this chapter, an equality constraint is used to bring the

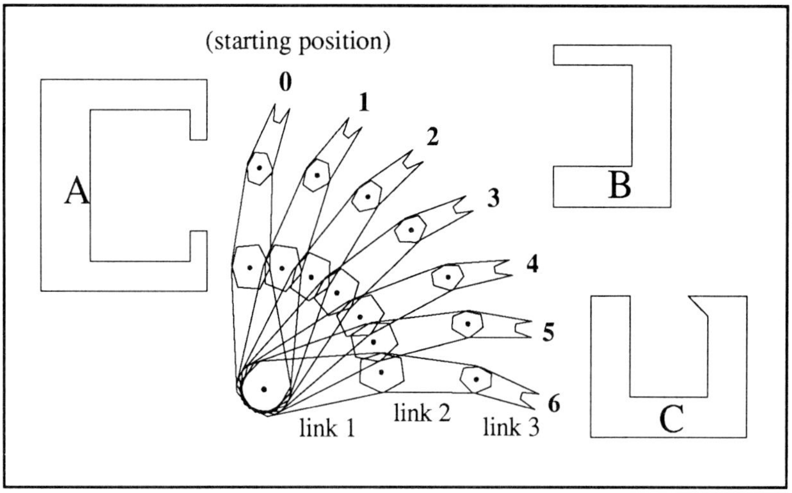

Figure 12.2: Distance-to-Contact Calculations

end effector to its desired final position by evaluating the square of the distance between the robot end effector and its desired location. The equality constraint on the final position of the robot end effector can be formulated as

$$h(\bar{\mathbf{X}}) = |\, (\bar{\mathbf{j}}_e - \bar{\mathbf{p}}_d)\, |^2 \tag{12.7}$$

where $\bar{\mathbf{j}}_e$ represents the coordinates of the final position of the robot end effector and $\bar{\mathbf{p}}_d$ is its desired destination.

Inequality Constraints

Inequality constraints can provide many general restrictions on robot motions. One of the most common considerations for robot dynamics is actuator saturation [8]. Other considerations could be the velocity, acceleration, and jerk constraints for obtaining smoothness of the trajectory [18]. Inequality constraints

$$g_r(\bar{\mathbf{X}}) = \delta_r(\bar{\mathbf{X}}) - \epsilon \geq 0 \ , \quad r = 1, \ldots, m_r \tag{12.8}$$

have been used here to maintain a desired clearance ϵ between robot links and workspace obstacles and prevent possible interference. Measures of distance-to-contact $\delta_r(\bar{\mathbf{X}})$ have been the focus of this work and are the basis of the inequality contact constraints. They have been computed by evaluating distances for each robot link relative to each obstacle at intervals defined by n_s and n_{ss} throughout a trajectory. The number of distances for a robot trajectory can be extremely large. As an illustration, the three-link robot trajectory shown in Figure 12.2 corresponds to $n_{ss} = 2$ and $n_s = 3$ with the number of distance checking positions of $n_{ss} \times n_s = 6$, not including the starting position. There are three obstacles and three robot links. The total number of distances-to-contact between links and obstacles is 54 $(3 \times 3 \times 6)$.

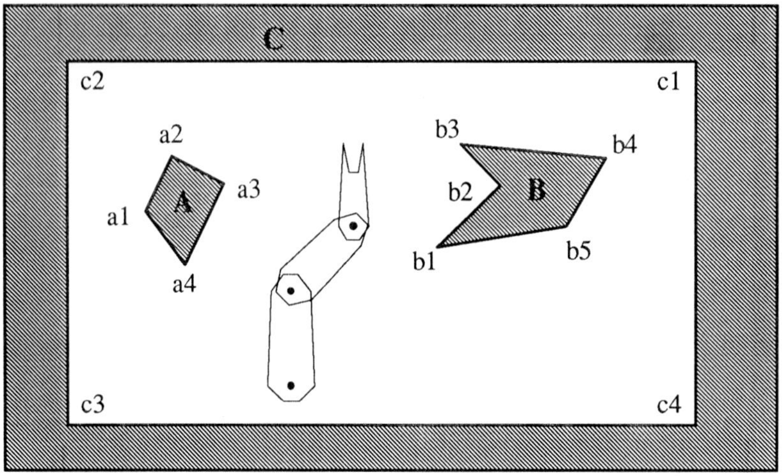

Figure 12.3: Polygons in Two-dimensional Distance Calculations

It is not feasible to use all the distances as the contact constraints even for this rather simple problem. Therefore a set of the smallest distances-to-contact has been used to form the "nearest-contact constraints." These m_r constraints can be a more reasonable number, such as 5. A procedure for finding the δ_r and obtaining the information for the derivatives of the constraints with respect to the variables $\bar{\mathbf{X}}$ is presented in the next two sections.

12.3 Static Distance Calculation

Vertex-based distance computation methods can be found in [19, 23, 28]. Those methods have certain disadvantages in solving three-dimensional problems. However, in two-dimensional cases, they are still very attractive. Before discussing the general convex body formulation, based on quadratic programming (QP) distance calculations and considering three-dimensional problems, we will briefly describe the vertex-based method developed by Mayne [28] and Liu [23].

12.3.1 Vertex-based Approach

Polygons are used to describe objects in two dimensions and a set of n vertices and edges can represent a polygon of arbitrary shape. A workspace can be composed of a polygonal border and obstacles with a free space for robot operation inside the border and outside the obstacles. Figure 12.3 shows a workspace example with bounding polygon C, obstacles A and B, and a robot with polygonal links. The boundary is treated as a polygon complement which is empty in the interior and filled in the exterior. The polygon vertices are ordered consecutively so that "the blocked region is to the right-hand side of the boundary lines." In this way the free space of a confined workspace can be defined by the appropriate sets

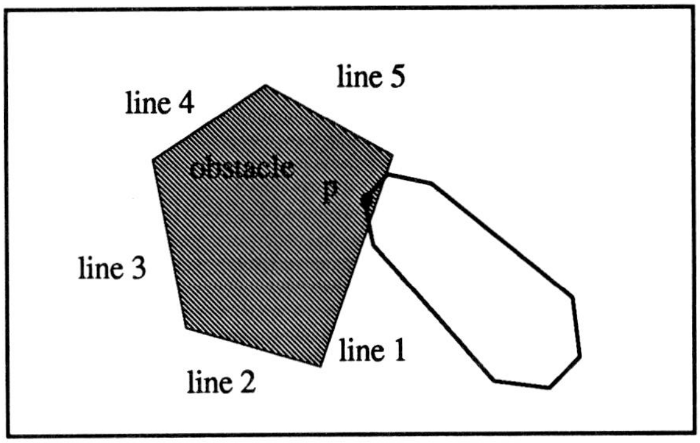

Figure 12.4: Just-Contact Interference

of vertices. For instance, the sets of vertices for the workspace polygons in Figure 12.3 are arranged as A := (a1, a2, a3, a4), B := (b1, b2, b3, b4, b5), C := (c1, c2, c3, c4). The robot link vertices are arranged in a similar manner. The ordered vertices of the robot and workspace polygons form the basis for the 2-D polygon distance calculations required for trajectory planning.

The minimum distance δ_{p-l} between a point p and a line segment may be found as the length of the perpendicular line connecting the point and the line segment or as the distance from point p to either end of the line segment. This distance δ_{p-l} can be directly used in the inequality constraint equations if there is no interference between robot links and the workspace. Unlike interior penalty function methods, distance calculations used in the RQP algorithm-based optimizations must consider interference cases which correspond to constraint violations. The assumption that contact has "just occurred" [28] is used for the case of polygon intersection (interference). This simplifies the difficulties with "negative distance-to-contact" and permits modest contact violations during trajectory optimization. Interference checking takes place at the "object" level, where each polygon body is taken into consideration. Basically, the distance between a vertex point p and polygon A is positive when p is not inside polygon A; otherwise it is negative. Inside checking for a point and polygon can be done by the angle summation method.

For the shortest distance between a point p and a polygon A, where p is not inside A, the outward normal of the polygon boundary line which is closest to the point p always points toward p. The point-to-line distances for boundary lines which are not facing the considered point may be ignored. This is somewhat similar to a hidden line calculation. For p inside A, there is ambiguity in defining how "deep" the vertex p penetrates polygon A. The number of edges of a polygon is at least three and a way to specify the active edge in depth penetration should be described to make the measurement of penetration unique. Considering polygon A slightly overlapping a corner of polygon B with the vertex p located at that corner of polygon B (Figure 12.4), the shortest distance between the polygon boundary lines and the point p can be defined as the penetration distance. This means that the point p can move out from the polygon A through line 1 with minimum effort and is equivalent to assuming that interference has just occurred.

According to the above, (1) the positive or negative distance, (2) the "near" hidden line problem, or (3) the smallest penetration distance may be necessary to evaluate distance-to-contact for a given point and a given polygon boundary line. To save computation time, the three measures may be considered in a single computational process achieved by using the ordered polygon vertices and presetting the point-to-line distances positive or negative. Each line formed by the extension of a polygon boundary line divides the two dimensional space into two half-spaces. We can distinguish these two half-spaces as the right-hand half-space and the left-hand half-space. If the point p falls in the right-hand half-space, the point-to-line distance is negative; otherwise it is positive. If p is not inside polygon A, only positive point-to-line distances and polygon boundary line segments where point p is in the left half-space need be considered. The distance between point p and polygon A is the smallest of those positive vertex-to-line distances. For p inside A, only negative distances are considered and the least negative point-to-line distance is the point-to-polygon distance. In this way, a vertex-polygon distance becomes the distance between a vertex and a typical polygon line.

12.3.2 Quadratic Programming Optimization Method

By using a convex body as the elementary form for workspace obstacles and links, efficient optimization methods can be applied to the distance calculations in both the two-dimensional and three-dimensional spaces. The approach described in this section concentrates on the well-developed quadratic programming (QP) optimization procedure. The concept is that the shortest distance between two convex bodies can be found if the squared distance between two points $\mathbf{p}_1$ and $\mathbf{p}_2$ is minimized while constraints to hold $\mathbf{p}_1$ inside object 1 and $\mathbf{p}_2$ inside object 2 are satisfied. Object 1 is described by a matrix $[A_1]$ and a vector d_1 while $[A_2]$ and d_2 are used for object 2.

A general quadratic programming problem is of the form

$$\begin{aligned} \text{minimize} \quad & f(\bar{\mathbf{z}}) = \bar{\mathbf{g}}^T\bar{\mathbf{z}} + \tfrac{1}{2}\bar{\mathbf{z}}^T H\bar{\mathbf{z}} \\ \text{subject to} \quad & A_\circ\bar{\mathbf{z}} = \bar{\mathbf{d}}_\circ \\ & A\bar{\mathbf{z}} \geq \bar{\mathbf{d}} \end{aligned} \tag{12.9}$$

where the vectors $\bar{\mathbf{g}}$, $\bar{\mathbf{d}}_\circ$, and $\bar{\mathbf{d}}$ and the matrices H, $A_\circ$, and A are constants. Using the QP method to find the distance between two convex bodies is to minimize the distance between two points p_1 and p_2, subject to point 1 being in body 1, $p_1 \in \mathbf{B}_1$, and point 2 being in body 2, $p_2 \in \mathbf{B}_2$. Equality constraints are not used, $A_\circ = [0]$.

The objects of interest here are convex polyhedra which are represented by a set of linear inequality equations. An n_j-surface convex polyhedron $\mathbf{B}_j$ is the intersection of n_j half-spaces $\mathbf{S}_1, \mathbf{S}_2, \ldots, \mathbf{S}_{n_j}$, $\mathbf{B}_j = \{\mathbf{S}_1 \cap \mathbf{S}_2 \cap \ldots \cap \mathbf{S}_{n_j}\}$. Each half-space $\mathbf{S}_i \subset \mathbf{R}^3$ can be expressed by a linear inequality $a_i\ x + b_i\ y + c_i\ z \geq d_i$, where the a_i, b_i, c_i, and d_i are constants. Let the matrix A_j, the vector $\bar{\mathbf{d}}_j$, and the coordinate vector $\bar{\mathbf{p}}_j$ be

$$[A_j]_{n_j \times 3} = \begin{bmatrix} a_1 & b_1 & c_1 \\ a_2 & b_2 & c_2 \\ \vdots & \vdots & \vdots \\ a_n & b_n & c_n \end{bmatrix} \qquad \bar{\mathbf{d}}_j = \begin{bmatrix} d_1 \\ d_2 \\ \vdots \\ d_n \end{bmatrix} \qquad \bar{\mathbf{p}}_j = \begin{bmatrix} x_j \\ y_j \\ z_j \end{bmatrix} \tag{12.10}$$

then the statement of $p_j \in \mathbf{B}_j$ can be written as $A_j \, \bar{\mathbf{p}}_j \geq \bar{\mathbf{d}}_j$. The constraint equations for the two-body distance problem in QP form are accordingly

$$\begin{bmatrix} A_1 & 0 \\ 0 & A_2 \end{bmatrix}_{(n_1+n_2)\times 6} \begin{bmatrix} \bar{\mathbf{p}}_1 \\ \bar{\mathbf{p}}_2 \end{bmatrix}_{6\times 1} \geq \begin{bmatrix} \bar{\mathbf{d}}_1 \\ \bar{\mathbf{d}}_2 \end{bmatrix}_{(n_1+n_2)\times 1} \tag{12.11}$$

and the objective function in the minimum distance problem may be written as

$$f = |(\bar{\mathbf{p}}_1 - \bar{\mathbf{p}}_2)|^2 \tag{12.12}$$

for the two points $\mathbf{p}_1$ and $\mathbf{p}_2$. This becomes

$$f = \frac{1}{2} \begin{bmatrix} \bar{\mathbf{p}}_1 \\ \bar{\mathbf{p}}_2 \end{bmatrix}^T \mathbf{H} \begin{bmatrix} \bar{\mathbf{p}}_1 \\ \bar{\mathbf{p}}_2 \end{bmatrix}; \qquad \mathbf{H} = \begin{bmatrix} 2 & 0 & 0 & -2 & 0 & 0 \\ 0 & 2 & 0 & 0 & -2 & 0 \\ 0 & 0 & 2 & 0 & 0 & -2 \\ -2 & 0 & 0 & 2 & 0 & 0 \\ 0 & -2 & 0 & 0 & 2 & 0 \\ 0 & 0 & -2 & 0 & 0 & 2 \end{bmatrix} \tag{12.13}$$

By letting A, $\bar{\mathbf{z}}$, and $\bar{\mathbf{d}}$ be the $(n_1 + n_2) \times 6$ matrix, 6×1 vector, and $(n_1 + n_2) \times 1$ vector shown in Eq. (12.11), the final QP distance optimization becomes

$$\begin{aligned} &\text{minimize} \quad f(\bar{\mathbf{z}}) = \tfrac{1}{2}\bar{\mathbf{z}}^T H \bar{\mathbf{z}} \\ &\text{subject to} \qquad A\bar{\mathbf{z}} \geq \bar{\mathbf{d}} \end{aligned} \tag{12.14}$$

and is performed with respect to the x, y, and z coordinates of the two points.

12.3.3 Distance-Past-Contact

Direct solution of the distance QP problem above is meaningful when the objects are separated from each other and the result is the "separation distance." When the two convex bodies have interference, zero separation distance is obtained and a point located in both bodies is returned from the QP solver. The returned point might be anywhere in the overlapped region of the two objects and the locations of that point within each body are not sufficient to define or evaluate the amount of interference (i.e., the distance-past-contact). In this section, two distance-past-contact measures are described based on "move-back" and "shrink-robot" approaches and an initial study of interference calculations is presented.

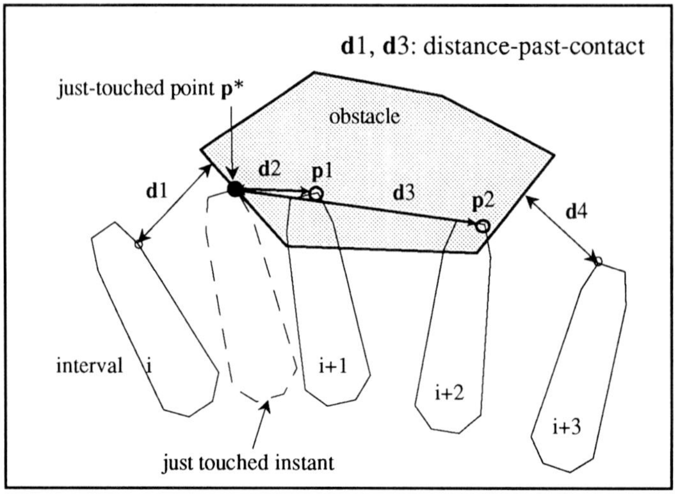

Figure 12.5: Distance-Past-Contact: The Move Back Approach

The move-back approach

In planning a robot trajectory, the initial position of the robot may be assumed to be in free space. If a collision takes place, it will occur as the robot moves from the free space into an obstacle. Therefore, when a collision occurs over a series of discrete robot positions, the position where the robot and obstacle "just touched" can be determined by tracing back through the robot path. Based on this, a distance-past-contact can be defined as "the distance between the contact point at the just-touched instant and the same point on the robot link in the interfering position." In Figure 12.5, a robot link is moving from left to right and collision occurs along the path. Assuming that distance calculations are performed at positions i, $i+1$, $i+2$ and $i+3$, the just-touched instant can be found between positions i and $i+1$ along the robot link path, and the just-contact point is p^*. According to the definition above, the distance-past-contact of the robot link can be found as d_2 at position $i+1$ and as d_3 at position $i+2$. The distance-to-contact is then considered negative for these two cases (indicating a constraint violation) and can be expressed as $-d_2$ and $-d_3$. An iterative strategy may be readily implemented to find the just-touched position to the desired accuracy. The A_j matrix and the $\bar{\mathbf{d}}_j$ vector in Eq. (12.10) which are used to describe the convex robot link must be reconstructed for each of the iterations. However, from our observations, a reasonable estimate of the just-touched instant can be obtained with derivative information and a few linearized approximations. By assuming that the robot link motion can be linearly interpolated between each interval, updating the A_j matrix and the $\bar{\mathbf{d}}_j$ vector is simplified. For example in the interval between time $t_\circ$ and t_1 $A_j^\circ\ \bar{\mathbf{p}}_j \geq \bar{\mathbf{d}}_j^\circ$ at time $t_\circ$ and $A_j^1\ \bar{\mathbf{p}}_j \geq \bar{\mathbf{d}}_j^1$ at time t_1, the A_j matrix and the $\bar{\mathbf{d}}_j$ vector in the equation of $A_j\ \bar{\mathbf{p}}_j \geq \bar{\mathbf{d}}_j$ for $t_\circ < t < t_1$ may be represented as

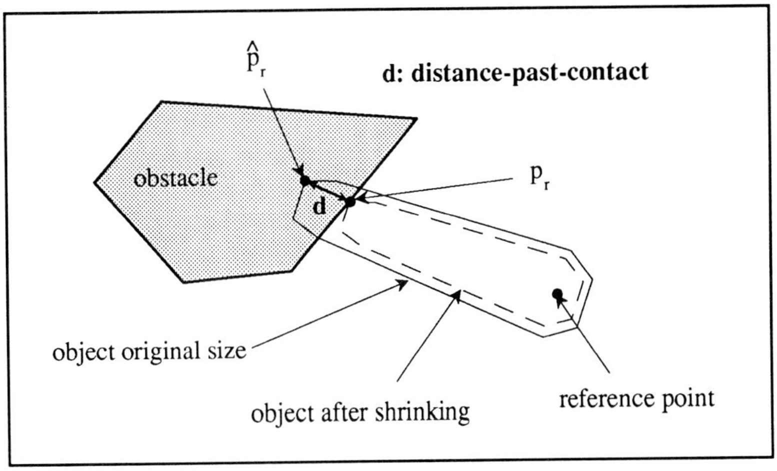

Figure 12.6: Distance-Past-Contact: The Shrink-Robot Approach

$$A_j = A_j^{\circ} + \frac{t - t_{\circ}}{t_1 - t_{\circ}}(A_j^1 - A_j^{\circ}) \tag{12.15}$$

$$\bar{\mathbf{d}}_j = \bar{\mathbf{d}}_j^{\circ} + \frac{t - t_{\circ}}{t_1 - t_{\circ}}(\bar{\mathbf{d}}_j^1 - \bar{\mathbf{d}}_j^{\circ}) \tag{12.16}$$

The shrink-robot approach

An alternate method for defining and finding the distance-past-contact is to computationally shrink the robot. The obstacle growing technique has similarly been used to allow the representation of moving objects or robot links as motion of a point and has been applied in path planning problems (e.g., [31, 26]). With interference between a robot link and an obstacle, consider the process of shrinking the robot link to a point. The link may be scaled so that it just touches the obstacle. In this situation, a contact point p_r may be found as shown in Figure 12.6. Let the point $\hat{p}_r$ on the unshrunk robot link correspond to the point p_r with shrinking. The distance-past-contact may then be calculated as the distance d between the two points p_r and $\hat{p}_r$. The robot shrinking is a linear process and the global coordinates, $\mathbf{p}$, of the scaled robot link point are

$$\bar{\mathbf{p}} - \bar{\mathbf{c}} = r \cdot (\bar{\mathbf{p}}_{\circ} - \bar{\mathbf{c}}) \;, \qquad 0.0 \leq r \leq 1.0 \tag{12.17}$$

where $\bar{\mathbf{p}}_{\circ}$ represents the global coordinates of the unshrunk robot link points, $\bar{\mathbf{c}}$ is the center of the shrinking process, and r is the shrinking ratio.

The robot joints are the focus of link actuation and motion and are a natural choice as centers for a link-shrinking process in the distance-past-contact calculations. Recalling

Eqs. (12.10) and (12.11), a convex boundary can be expressed by a set of linear equations $A\bar{\mathbf{p}} \geq \bar{\mathbf{d}}$. In this equation, the vector $\bar{\mathbf{d}}$ will be affected by the object-shrinking process of Eq. (12.17), and the A matrix remains the same. This may be seen in the following. The point $\bar{\mathbf{p}}_o$ in the unshrunk convex robot link satisfies the equation

$$A\bar{\mathbf{p}}_o \geq \bar{\mathbf{d}} \tag{12.18}$$

For shrinking about robot joint $\bar{\mathbf{j}}$, we have

$$\bar{\mathbf{p}} - \bar{\mathbf{j}} = r(\bar{\mathbf{p}}_o - \bar{\mathbf{j}}) \tag{12.19}$$

from Eq. (12.17). This can be rearranged to

$$\bar{\mathbf{p}}_o = \frac{\bar{\mathbf{p}} - \bar{\mathbf{j}}}{r} + \bar{\mathbf{j}} \tag{12.20}$$

Using Eq. (12.20) for $\bar{\mathbf{p}}_o$ in Eq. (12.18), yields

$$A(\frac{\bar{\mathbf{p}} - \bar{\mathbf{j}}}{r} + \bar{\mathbf{j}}) \geq \bar{\mathbf{d}} \tag{12.21}$$

Eq. (12.21) can be rewritten in the following form

$$\begin{aligned} A\bar{\mathbf{p}} &\geq r(\bar{\mathbf{d}} - A\bar{\mathbf{j}}) + A\bar{\mathbf{j}} \\ &\geq r\bar{\mathbf{d}} + (1 - r)A\bar{\mathbf{j}} \end{aligned} \tag{12.22}$$

and letting $\bar{\mathbf{d}}' = r\bar{\mathbf{d}} + (1 - r)A\bar{\mathbf{j}}$, we have

$$A\bar{\mathbf{p}} \geq \bar{\mathbf{d}}' \tag{12.23}$$

Thus, the A matrix will not be changed in the shrinking process but the $\bar{\mathbf{d}}$ vector will. Eqs. (12.22) and (12.23) provide a convenient mathematical formulation for shrinking an object. The computation procedure for finding the just-touched point by shrinking may be by linear approximation or by a section-cutting approach.

Combining the Two Approaches

Each of the described approaches for distance-past-contact measurement may be attractive for particular situations. For an overall improved measurement of distance-past-contact, these two strategies may be combined. One of the approaches to combine the two strategies may be described as follows: (1) Use the move-back scheme in calculating distance-past-contact for robot link 1 if robot link 1 has only rotational degrees of freedom; otherwise use the shrink-robot scheme; (2) use the shrink-robot method directly for the robot links 2, 3, ..., and so on if their joints are in free space; (3) when a link joint falls within an obstacle, the shrinking process is applied to that link and the preceding link simultaneously.

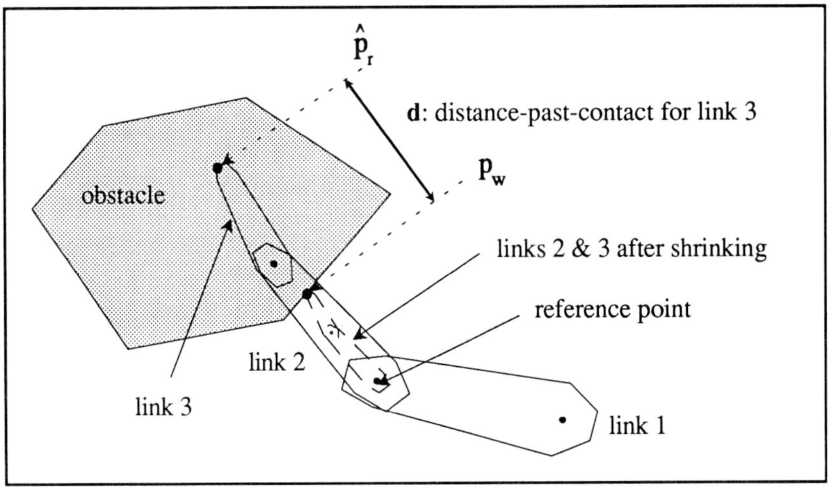

Figure 12.7: Distance-Past-Contact: The Combined Method

In the example of Figure 12.7, this procedure can be described as follows. For the robot link 3 the shrinking process is considered as two separate subprocesses. The first is to shrink link 3 with respect to joint 3:

$$\bar{\mathbf{p}} = r(\bar{\mathbf{p}}_o - \bar{\mathbf{j}}_3) + \bar{\mathbf{j}}_3 \tag{12.24}$$

where $\bar{\mathbf{p}}$ represents the global coordinates of a point in the shrunk robot link 3 and $\bar{\mathbf{p}}_o$ represents the global coordinates of a point in the unshrunk link. The similar shrinking of robot link 2 results in a linear translation for robot link 3. The amount of translation will be $(1-r)(\bar{\mathbf{j}}_2 - \bar{\mathbf{j}}_3)$, where $\bar{\mathbf{j}}_2$ represents the link joint 2. The final equation for $\bar{\mathbf{p}}$ is then obtained by adding the amount of linear transformation to the right-hand side of Eq. (12.24)

$$\bar{\mathbf{p}} = r(\bar{\mathbf{p}}_o - \bar{\mathbf{j}}_3) + \bar{\mathbf{j}}_3 + (1-r)(\bar{\mathbf{j}}_2 - \bar{\mathbf{j}}_3) \tag{12.25}$$

When the separation distance is calculated, the matrix A and the vector $\bar{\mathbf{d}}$ are used in the inequality constraint

$$A\ \bar{\mathbf{p}}_o \geq \bar{\mathbf{d}} \tag{12.26}$$

From Eq. (12.26), we find the equation of $\bar{\mathbf{p}}_o(t_i)$ to be

$$\bar{\mathbf{p}}_o = \frac{\bar{\mathbf{p}} - (1-r)(\bar{\mathbf{j}}_2 - \bar{\mathbf{j}}_3)}{r} + (1 - \frac{1}{r})\bar{\mathbf{j}}_3 \tag{12.27}$$

To show how the matrix A and the vector $\bar{\mathbf{d}}$ for link 3 are affected by the successive shrinking process, replace $\bar{\mathbf{p}}_o$ in Eq. (12.27) with Eq. (12.26), to find

$$A\Big[\frac{\bar{\mathbf{p}} - (1-r)(\bar{\mathbf{j}}_2 - \bar{\mathbf{j}}_3)}{r} + (1 - \frac{1}{r})\bar{\mathbf{j}}_3\Big] \geq \bar{\mathbf{d}} \tag{12.28}$$

which may be rewritten as

$$A\bar{\mathbf{p}} \geq \bar{\mathbf{d}}' \tag{12.29}$$

with

$$\bar{\mathbf{d}}' = rA\Big[\frac{(1-r)(\bar{\mathbf{j}}_2 - \bar{\mathbf{j}}_3)}{r} + (1 - \frac{1}{r})\bar{\mathbf{j}}_3\Big] + r\bar{\mathbf{d}} \tag{12.30}$$

These show that the matrix A is unchanged and the vector $\bar{\mathbf{d}}$ becomes a function of r. A 3-D example of the distance-time curves for a moving three link robot and an obstacle are shown in Figure 12.8. The robot trajectory starts at time zero and ends at a time of one unit. An obstacle lies within the robot path so that links 2 and 3 go through the obstacle as well as joint 3. The resulting distance-to-contact curves for links 2 and 3 are shown in Figure 12.9. Interference between link 3 and the obstacle begins at $t = 0.34$ and ends at $t = 0.87$. The interference for link 2 follows a similar pattern. In both cases the distance-to-contact curves are reasonably smooth and a negative distance-to-contact defines the amount of interference. Computing interference by the shrink-robot approach associates the amount of interference with an appropriate direction. An ideal distance function should have a smooth and convex-shaped curve in the interference regions and should be as continuous as possible, particularly at the zero distance points. Of course, the distance-to-contact is zero when collision just occurs and becomes more negative as the robot link moves further into the obstacle. It reaches a maximum value and then returns to zero as the robot link moves out of the obstacle.

12.4 Gradient Information

The first derivatives of the objective function and the constraint functions are required by the RQP algorithm and other efficient optimization methods. The accuracy of the first derivatives is important because the second derivative estimates, which are also involved in the search direction evaluations, are approximated and updated by the first derivative information at each stage. Generally speaking, finite difference derivative calculations are less attractive because of two factors: first, results of questionable accuracy and second, CPU time intensive computations. In this case, the path planning and geometric design problem formulation based on global reference angles permit the isolation of each degree of freedom of robot motion. This facilitates the derivative computations of the constraints, particularly the more difficult nearest-contact constraints, with respect to the trajectory control points forming the design vector $\bar{\mathbf{X}}$ in the optimization. In this section, several key factors are presented for determining the various gradient equations. The objective function derivatives and the strategy for assembling derivative equations for the contact constraints are discussed.

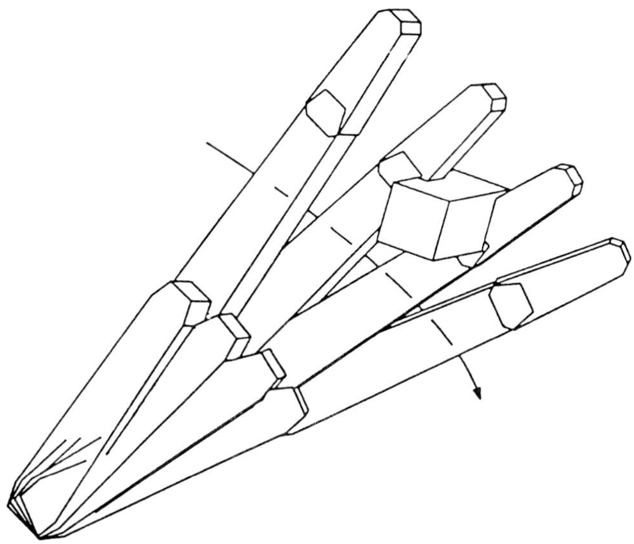

Figure 12.8: Example of QP Distance Problem with the Combined Method

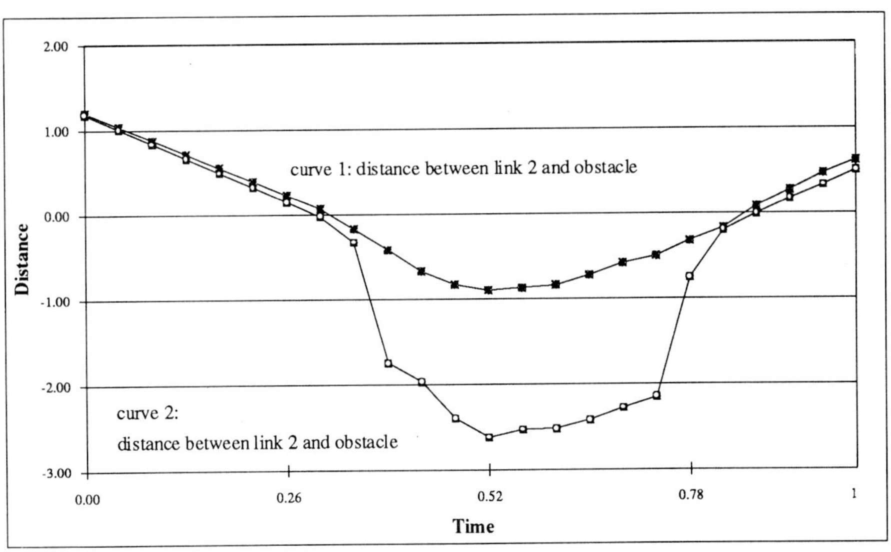

Figure 12.9: Results of QP Distance Problem with the Combined Method

12.4.1 Objective Function Gradient

The objective function of Eq. (12.6) is composed of two functions based on robot motions and the geometry of robot and/or workspace. Because of the variable separation of robot motion and object geometry, the gradient of $f(\bar{\mathbf{X}})$ is a function of f^p, f^s, $\frac{d\, f^p}{d\,\bar{\mathbf{x}}_i}$, and $\frac{d\, f^s}{d\,\bar{\mathbf{y}}}$. The derivative of f^s with respect to the shape variables will be presented in Section 12.5. We will discuss only the term $\frac{d\, f^p(\bar{\mathbf{x}}_i)}{d\,\bar{\mathbf{x}}_i}$ now.

Defining $f_i^p(\bar{\mathbf{x}}_i)$as the part of $f^p(\bar{\mathbf{X}})$ containing the motions of the ith degree of freedom corresponds to

$$f_i^p(\bar{\mathbf{x}}_i) = \sum_{k=0}^{n_t-1} (\hat{x}_{i,k+1} - \hat{x}_{i,k})^2; \qquad i = 1, 2, \ldots, n_d \tag{12.31}$$

and the motion objective function can be expressed as

$$f^p(\bar{\mathbf{X}}) = \sum_{i=1}^{n_d} w_i\; f_i^p(\bar{\mathbf{x}}_i) \tag{12.32}$$

Correspondingly, $\bigtriangledown f^p(\bar{\mathbf{x}}_i)$ is

$$\bigtriangledown\, f^p = \Big[(\frac{\partial f^p}{\partial \bar{\mathbf{x}}_1})^T,\; (\frac{\partial f^p}{\partial \bar{\mathbf{x}}_2})^T,\; \ldots,\; (\frac{\partial f^p}{\partial \bar{\mathbf{x}}_{n_d}})^T \Big]^T \tag{12.33}$$

This equation shows $\bigtriangledown f^p(\bar{\mathbf{x}}_i)$ as a vector which contains the subvectors of $\frac{\partial f^p}{\partial \bar{\mathbf{x}}_i}$. With the robot degrees of freedom separated in the objective function, f_i^p is only a function of $\bar{\mathbf{x}}_i$, the Bezier control vector for degree of freedom i. Each $\frac{\partial f^p}{\partial \bar{\mathbf{x}}_i}$ can be formulated from the derivatives of $f_i^p(\bar{\mathbf{x}}_i)$ with respect to $\bar{\mathbf{x}}_i$ as

$$\frac{\partial f^p}{\partial \bar{\mathbf{x}}_i} = w_i (\frac{d\; f_i^p(\bar{\mathbf{x}}_i)}{d\; \bar{\mathbf{x}}_i}) \tag{12.34}$$

and the subvector may be extended as

$$\frac{d\; f_i^p(\bar{\mathbf{x}}_i)}{d\; \bar{\mathbf{x}}_i} = \Big[\frac{\partial f_i^p(\bar{\mathbf{x}}_i)}{\partial x_{i,1}},\; \frac{\partial f_i^p(\bar{\mathbf{x}}_i)}{\partial x_{i,2}},\; \ldots,\; \frac{\partial f_i^p(\bar{\mathbf{x}}_i)}{\partial x_{i,n_s}}\Big]^T \tag{12.35}$$

The elements $\frac{\partial f_i^p(\bar{\mathbf{x}}_i)}{\partial x_{i,j}}$ in the above equation are the partial derivatives of f_i^p with respect to the Bezier control point j for the degree of freedom i. Taking the derivative of Eq. (12.31) with respect to $x_{i,j}$, yields the jth element of the vector of Eq. (12.35)

$$\frac{\partial f_i^p(\bar{\mathbf{x}}_i)}{\partial x_{i,j}} \;=\; \sum_{k=0}^{n_t-1} \frac{\partial (\hat{x}_{i,k+1} - \hat{x}_{i,k})^2}{\partial x_{i,j}}$$

$$= 2\sum_{k=0}^{n_t-1}\Big(\big(\frac{\partial \hat{x}_{i,k+1}}{\partial x_{i,j}} - \frac{\partial \hat{x}_{i,k}}{\partial x_{i,j}}\big)(\hat{x}_{i,k+1} - \hat{x}_{i,k})\Big) \quad ; \; j = 1, 2, \ldots, n_s \tag{12.36}$$

Based on the Bezier formulation of Giloi [12] as shown in Eq. (12.2),

$$\frac{\partial \hat{x}_{i,k}}{\partial x_{i,j}} = \Phi_{j,n_s}\big(\frac{k}{n_t}\big) = \frac{n_s!}{j!\,(n_s-1)!}\,\big(\frac{k}{n_t}\big)^j\,\big(1-\frac{k}{n_t}\big)^{n_s-j} \tag{12.37}$$

Substituting $\Phi_{j,n_s}(\frac{k}{n_t})$ and $\Phi_{j,n_s}(\frac{(k+1)}{n_t})$ for $\frac{\partial \hat{x}_{i,k}}{\partial x_{i,j}}$ and $\frac{\partial \hat{x}_{i,k+1}}{\partial x_{i,j}}$, Eq. (12.36) yields

$$\frac{\partial f_i^p(\bar{\mathbf{x}}_i)}{\partial x_{i,j}} = 2\sum_{k=0}^{n_t-1}\Big(\big(\Phi_{j,n_s}\big(\frac{(k+1)}{n_t}\big) - \Phi_{j,n_s}\big(\frac{k}{n_t}\big)\big)(\hat{x}_{i,k+1} - \hat{x}_{i,k})\Big) \quad ; \; j = 1, 2, \ldots, n_s \tag{12.38}$$

The combination of Eqs. (12.33) through (12.35) and Eq. (12.38) now defines the objective function gradient for the globally defined motion variables.

12.4.2 Derivative of Distance-to-Contact

The design vector $\bar{\mathbf{X}}$ in the path planning and geometric design optimization contains n motion subvectors, $\bar{\mathbf{x}}_1, \ldots, \bar{\mathbf{x}}_n$, according to the number n of robot degrees of freedom.

The derivative of distance constraints contains two portions: $\frac{\partial g(\bar{\mathbf{X}})}{\partial \bar{\mathbf{x}}_i}$ and $\frac{\partial g(\bar{\mathbf{X}})}{\partial \bar{\mathbf{y}}}$. Again, we will focus on the first term $\frac{\partial g(\bar{\mathbf{X}})}{\partial \bar{\mathbf{x}}_i}$ and leave the second term for Section 12.5. Each subvector of the $\bar{\mathbf{x}}_i$ controls the time history of a degree of freedom (e.g., a joint angle). The gradient of the distance constraint g with respect to the design subvector $\bar{\mathbf{x}}_i$ of the generalized robot degrees of freedom $\bar{\lambda} = \{\lambda_1, \ldots, \lambda_n\}$ is

$$\frac{\partial g}{\partial \bar{\mathbf{x}}_i} = \sum_{j=1}^{n}\Big(\sum_{k=1}^{m}\frac{\partial g}{\partial \bar{\mathbf{v}}_{\hat{i},k}}\frac{\partial \bar{\mathbf{v}}_{\hat{i},k}}{\partial \lambda_j}\Big)\frac{\partial \lambda_j}{\partial \bar{\mathbf{x}}_i} \; ; \qquad i = 1, \ldots, n \tag{12.39}$$

where the $\hat{i}$ indicates the robot link associated with the considered distance-to-contact and $\bar{\mathbf{v}}_{\hat{i},k}$ represents the global coordinates of the vertex k on link $\hat{i}$. Since the robot degrees of freedom are independent of each other, the term $\frac{\partial \lambda_j}{\partial \bar{\mathbf{x}}_i}$ is zero when $i \neq j$. Eq. (12.39) becomes

$$\frac{\partial g}{\partial \bar{\mathbf{x}}_i} = \Big(\sum_{k=1}^{m}\frac{\partial g}{\partial \bar{\mathbf{v}}_{\hat{i},k}}\frac{\partial \bar{\mathbf{v}}_{\hat{i},k}}{\partial \lambda_j}\Big)\frac{\partial \lambda_j}{\partial \bar{\mathbf{x}}_i} \; ; \qquad i = j \tag{12.40}$$

The upper limit m of the summation in Eq. (12.40) is the number of vertices involved in the calculation of the distance constraint g. The derivative $\frac{d\,\lambda_i}{d\,\bar{\mathbf{x}}_i}$ is dependent on the Bezier

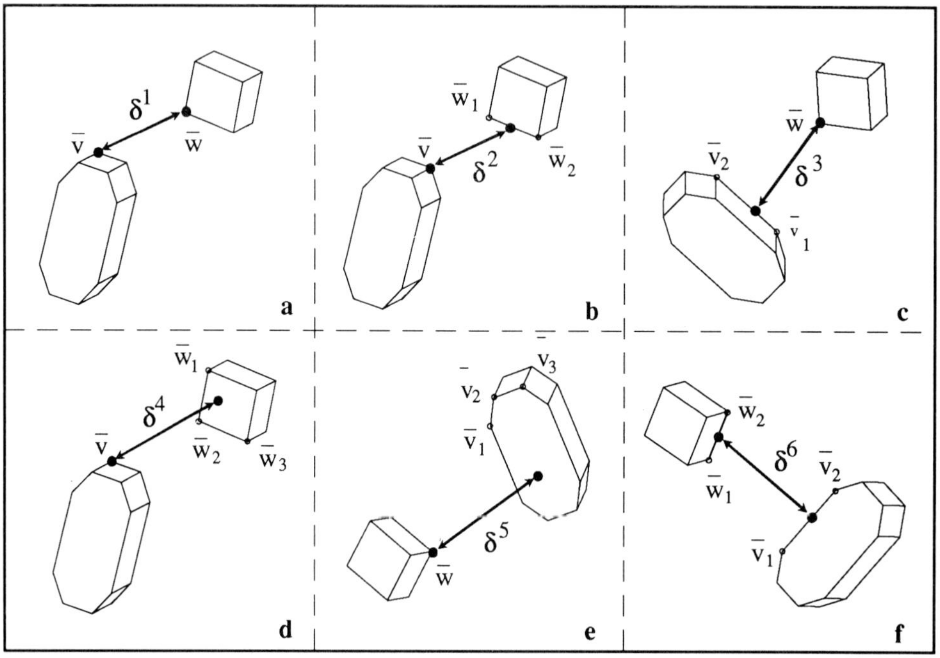

Figure 12.10: Distances in 3-D Space

interpolation function used to define the robot joint and the robot base motions [23, 24]. The details of the $\frac{\partial \bar{\mathbf{v}}_{\hat{i},k}}{\partial \lambda_j}$ operation may be found in [22].

The term $\frac{\partial g}{\partial \bar{\mathbf{v}}_{\hat{i},j}}$ relates a distance-to-contact constraint to the motion of a particular vertex. There are nine types of distance-to-contact situations between two 3-D bodies. The distance from robot links to obstacles can be vertex-to-vertex, vertex-to-line, line-to-vertex, vertex-to-plane, plane-to-vertex, line-to-line, line-to-plane, plane-to-line, or plane-to-plane. The line-to-plane (or plane-to-line) and the plane-to-plane cases rarely happen, especially in the computation of digital computers. The discussion below considers the six common cases illustrated in Figure 12.10 where δ represents the distance-to-contact and the superscript identifies the situations of interest

(a) **vertex-to-vertex**

Let $\bar{\mathbf{v}}$ and $\bar{\mathbf{w}}$ represent the coordinates of the robot link vertex and the obstacle vertex (Figure 12.10a). Then, the vertex-to-vertex distance-to-contact is

$$\delta^1 = |\bar{\mathbf{v}} - \bar{\mathbf{w}}| \tag{12.41}$$

There is only one robot link vertex in Eq. (12.41), therefore $m = 1$ in Eq. (12.40).

(b) **robot link vertex-to-obstacle boundary line**

Let $\bar{\mathbf{w}}_1$ and $\bar{\mathbf{w}}_2$ represent the coordinates of two end points of the obstacle boundary line and $\bar{\mathbf{v}}$ represent the coordinates of the robot link vertex (Figure 12.10b). Then, the distance between the robot link vertex and the obstacle boundary line is

$$\delta^2 = \frac{|(\bar{\mathbf{v}} - \bar{\mathbf{w}}_1) \times (\bar{\mathbf{w}}_2 - \bar{\mathbf{w}}_1)|}{|\bar{\mathbf{w}}_2 - \bar{\mathbf{w}}_1|} \tag{12.42}$$

Again there is one robot link vertex $\bar{\mathbf{v}}$ in Eq. (12.42) and $m = 1$ in Eq. (12.40).

(c) **robot link line-to-obstacle vertex**

Let $\bar{\mathbf{v}}_1$ and $\bar{\mathbf{v}}_2$ represent the coordinates of the two end points of the robot link boundary line and $\bar{\mathbf{w}}$ represent the coordinates of the obstacle polyhedron vertex (Figure 12.10c). Then, the distance-to-contact in this case is

$$\delta^3 = \frac{|(\bar{\mathbf{w}} - \bar{\mathbf{v}}_1) \times (\bar{\mathbf{v}}_2 - \bar{\mathbf{v}}_1)|}{|\bar{\mathbf{v}}_2 - \bar{\mathbf{v}}_1|} \tag{12.43}$$

Now, $m = 2$ since two vertices are necessary to define the link line.

(d) **robot vertex-to-obstacle plane**

For the three vectors $\bar{\mathbf{a}} = \bar{\mathbf{v}} - \bar{\mathbf{w}}_2$, $\bar{\mathbf{b}} = \bar{\mathbf{w}}_1 - \bar{\mathbf{w}}_2$, and $\bar{\mathbf{c}} = \bar{\mathbf{w}}_3 - \bar{\mathbf{w}}_2$ (Figure 12.10d), where $\bar{\mathbf{w}}_1$, $\bar{\mathbf{w}}_2$, and $\bar{\mathbf{w}}_3$ are three sequentially adjacent points on the obstacle plane and $\bar{\mathbf{v}}$ represents the coordinates of the robot link vertex, the distance-to-contact is

$$\delta^4 = \frac{|\bar{\mathbf{a}} \cdot (\bar{\mathbf{b}} \times \bar{\mathbf{c}})|}{|\bar{\mathbf{b}} \times \bar{\mathbf{c}}|} \tag{12.44}$$

The vector $\bar{\mathbf{a}}$ has one robot link vertex $\bar{\mathbf{v}}$; consequently $m = 1$.

(e) **obstacle vertex-to-robot plane**

As shown in Figure 12.10e, the vectors $\bar{\mathbf{a}} = \bar{\mathbf{w}} - \bar{\mathbf{v}}_2$, $\bar{\mathbf{b}} = \bar{\mathbf{v}}_1 - \bar{\mathbf{v}}_2$, and $\bar{\mathbf{c}} = \bar{\mathbf{v}}_3 - \bar{\mathbf{v}}_2$, where $\bar{\mathbf{v}}_1$, $\bar{\mathbf{v}}_2$, and $\bar{\mathbf{v}}_3$ are three sequentially adjacent points on the plane of the robot link and $\bar{\mathbf{w}}$ represents the coordinates of the obstacle vertex. The distance-to-contact is then

$$\delta^5 = \frac{|\bar{\mathbf{a}} \cdot (\bar{\mathbf{b}} \times \bar{\mathbf{c}})|}{|\bar{\mathbf{b}} \times \bar{\mathbf{c}}|} \tag{12.45}$$

and $m = 3$ to define the plane.

(f) **robot line-to-obstacle line**

Let $\bar{\mathbf{a}} = \bar{\mathbf{v}}_2 - \bar{\mathbf{v}}_1$ and $\bar{\mathbf{b}} = \bar{\mathbf{w}}_2 - \bar{\mathbf{w}}_1$ be the vectors of the robot link boundary line and obstacle boundary line respectively, where $\bar{\mathbf{v}}_1$ and $\bar{\mathbf{v}}_2$ are the coordinates of two end points of the link line and $\bar{\mathbf{w}}_1$ and $\bar{\mathbf{w}}_2$ are the coordinates of two end points of the obstacle line (Figure 12.10f). Defining $\bar{\mathbf{c}} = \bar{\mathbf{w}}_1 - \bar{\mathbf{v}}_1$, the distance-to-contact is

$$\delta^6 = \frac{|\bar{\mathbf{c}} \cdot (\bar{\mathbf{a}} \times \bar{\mathbf{b}})|}{|\bar{\mathbf{a}} \times \bar{\mathbf{b}}|} \tag{12.46}$$

with $m = 2$.

The gradient of δ^i with respect to $\bar{\mathbf{v}}_{\hat{i},k}$, can be derived directly from the above distance equations. Since our definition for the negative distance-to-contact δ^n identifies a "just-contact-point," the derivatives of δ with interference can be treated in the same manner as the vertex-to-vertex case. For instance, let the coordinates of the just-contact-point on the obstacle be $\bar{\mathbf{w}}$ (p_r in Figure 12.6) and let $\bar{\mathbf{v}}$ represent the coordinates of the just-contact-point on the full-size robot link ($\hat{p}_r$ in Figure 12.6). The measurement of interference is then equal to $-\delta^1$ and its derivatives may be obtained accordingly.

12.5 Results

This section presents some of the results of robot trajectory planning and geometric design by a recursive quadratic programming approach. The procedure described in Sections 12.2 to 12.4 is directed toward off-line planning and design applications. It has been programmed in conjunction with a FORTRAN subroutine CVM01 [38] which is a RQP algorithm developed for general purpose nonlinear optimization problems. Users provide an objective function and constraint equation calculations in a subroutines and required derivative information can be evaluated by finite difference calculation strategies built into CVM01 or by user-provided exact calculations (in a subroutine form).

In two dimensional examples, vertex-based distance calculations are used and therefore the robot links and the workspace can be arbitrary shapes and need not be convex. Since the QP distance methods are used in three-dimensional examples, we will use convex polyhedra as primitive shapes in constructing the 3-D geometry of both robot and workspace.

12.5.1 2-D Motion Planning Examples

In the example of Figure 12.11, only a rotational motion is permitted for each robot link. The initial configuration shown as H defines the beginning position of the trajectory and the specified end-effector position P indicates its desired location at the end of the trajectory. The number n_s of Bezier control points for the robot joint motions is three, the number n_{ss} of the distance checking positions between two adjacent control points is two, the number of active distance constraints is five, and the initial guess of the trajectory is no motion. The clearance between the workspace polygons and the robot links along the optimized trajectory is achieved by setting a positive value for ϵ in the distance constraints. In the resulting trajectory, the first robot link, hinged at the base, rotates counterclockwise initially in order to move the second and the third links to the other side of obstacle B. Meanwhile, links 2 and 3 move in a clockwise direction so as not to hit the obstacle B. When there is enough room for link 2 and link 3 to clear the obstacle, the first robot link starts to rotate clockwise and links 2 and 3 move toward the desired position.

A wide variety of two-dimensional trajectory planning problems have been used in evaluating trajectory optimization and particularly in studying the behavior of the distance-to-contact constraints in the process of motion optimization. CPU times for 2-D trajectory

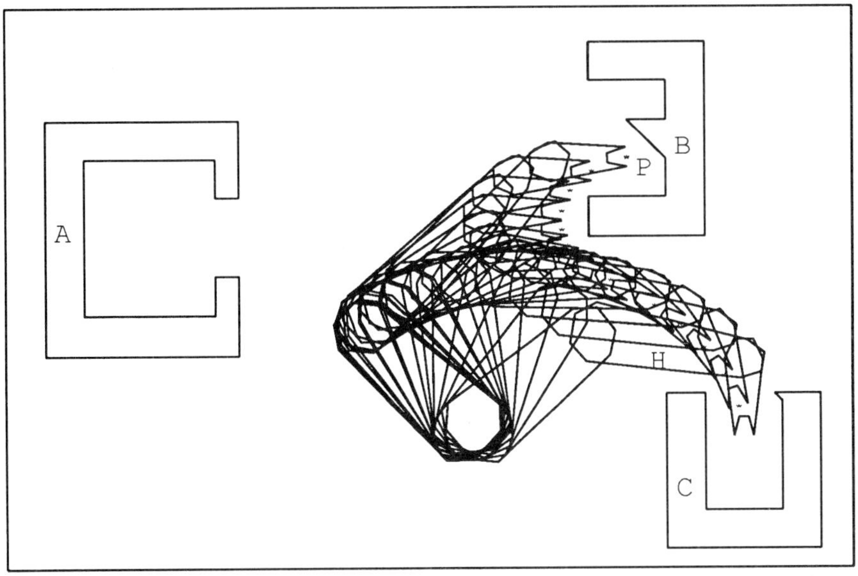

Figure 12.11: 2-D Example of Robot Trajectory with 3 Degree of Freedom

optimizations on a VAX 8700 machine have typically ranged from 30 to 90 seconds.

12.5.2 2-D Geometric Design

The integration of path planning with geometric design has been motivated by situations where movements associated with machine operation or part assembly become important factors in the design process. This might be the case in the design of a machine tool to be used in the assembly of a particular production part where not only are the motions of the machine to be specified but also the shape of its components. On the other hand, if the production tools are fixed, it may be necessary to design a desired part to be compatible with the available tools and their possible movements. The discussion below focuses on two simple examples of these situations.

Reshaping of Robot Links

Figure 12.12 shows a robot configuration that must perform a desired operation at point P in object A from a fixed base position. The end effector dimensions are specified but the lengths of links 1 and 2 are to be selected to provide access to point P and a reasonably efficient robot trajectory. The design vector for this optimization problem is now augmented to include the lengths of link 1 and link 2, ℓ_1 and ℓ_2, as well as the Bezier control points for the three-link joint rotations. The lengths are naturally held constant for trial trajectories during the optimization and are varied only at each iteration of the optimization procedure. In general, the size of the robot link, which is to be designed, is proportional to its link length. In other words, the local coordinates of the vertices on the robot link i may be functions of ℓ_i

$$\bar{\mathbf{u}}_{i,j} = \bar{\mathbf{u}}_{i,j}(\ell_i) \tag{12.47}$$

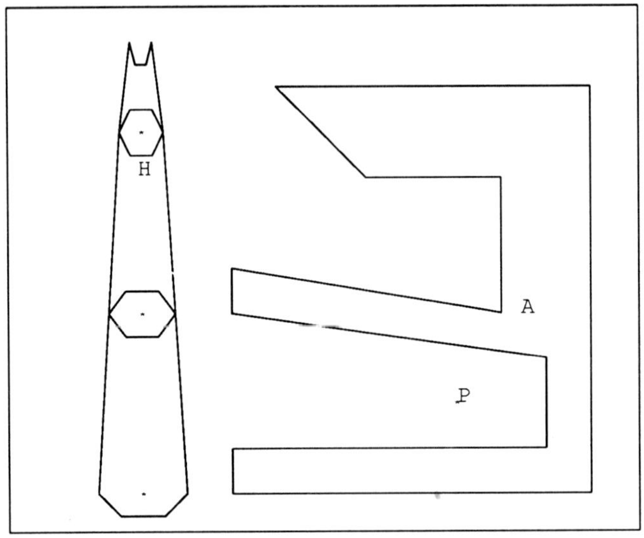

Figure 12.12: Original Geometry of Robot Link Length Design

In this example, $\bar{\mathbf{u}}_{i,j}$ is a linear function of ℓ_i

$$\bar{\mathbf{u}}_{i,j} = \bar{\mathbf{u}}^{\circ}_{i,j}\left(\frac{\ell_i}{\ell^{\circ}_i}\right) , \quad i = 1, 2 \tag{12.48}$$

where $\bar{\mathbf{u}}^{\circ}_{i,j}$ represents the vertex local coordinates when the link length is equal to ℓ°_i. The objective function is modified, to penalize the overall robot length in conjunction with the kinematic objective equations, by adding the link lengths to the objective function with weighting factor w_i.

$$f(\bar{\mathbf{X}}) = \sum_{i=1}^{3}\left(\left(w_i \sum_{k=i}^{3} \ell_k\right) \sum_{k=0}^{n_t-1} (\hat{x}_{i,k+1} - \hat{x}_{i,k})^2\right) \tag{12.49}$$

In revising the gradient computations for this new problem, the derivatives of the objective function are only slightly changed. The gradient entries for the equality constraint 12.7 with respect to ℓ_1 and ℓ_2 can be derived from the robot kinematic transformation equations without difficulty. The partial derivatives of the distance constraints are

$$\frac{\partial g}{\partial \ell_i} = \frac{\partial g}{\partial \bar{\mathbf{v}}_{\hat{i},k}} \frac{\partial \bar{\mathbf{v}}_{\hat{i},k}}{\partial \ell_i} \tag{12.50}$$

where the $\dfrac{\partial g}{\partial \bar{\mathbf{v}}_{\hat{i},k}}$ follow the pattern discussed in Section 12.4. Again, the partial derivatives $\dfrac{\partial \bar{\mathbf{v}}_{\hat{i},k}}{\partial \ell_i}$ can be found from the robot kinematic transformation equations.

Figure 12.12 shows the original robot configuration and link lengths. The optimized link lengths and trajectory are shown in Figure 12.13. The resulting design has shorter link lengths which allow access to point P while maintaining the desired obstacle clearance.

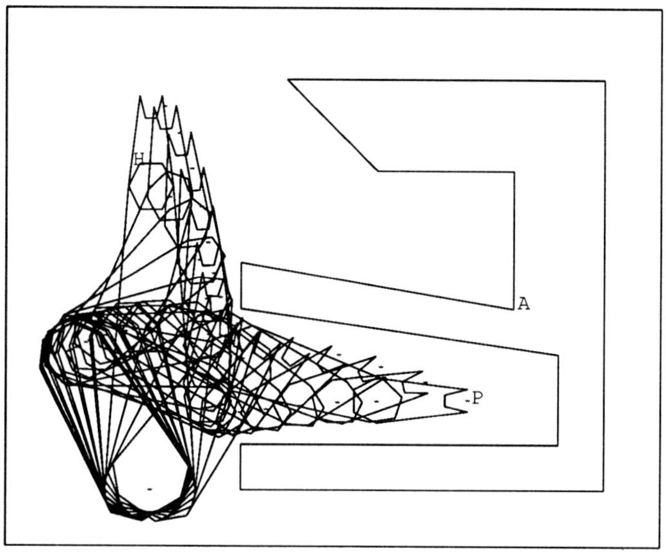

Figure 12.13: Result of Motion Planning and Robot Link Length Design

Reshaping a Workspace Element

Figure 12.14 is an example of a part design problem where an end-effector wrench must have access to the bolt head B. The horizontal positions of vertices d (the vertex 5 of workplace polygon 1) and e (the vertex 3 of workplace polygon 1) are made adjustable on object A and their locations become part of the design vector. The objective function has been modified to make the section widths c-d and e-f as wide as possible while maintaining a reasonable kinematic trajectory and satisfying clearance requirements.

$$f(\bar{\mathbf{X}}) = \sum_{i=1}^{3}\Big(w_i \sum_{k=0}^{n_t-1} (\hat{x}_{i,k+1} - \hat{x}_{i,k})^2\Big) - w_c\Big((x_d - x_c) + (x_f - x_e)\Big) \tag{12.51}$$

Here x_c, x_d, x_e, and x_f are the x coordinates of points c, d, e, and f respectively, and w_c is a positive weighting constant for the geometry changes. There is no difficulty in finding

$$\frac{\partial f}{\partial x_d} = -w_c, \quad \frac{\partial f}{\partial x_e} = w_c, \quad \frac{\partial h}{\partial x_d} = \frac{\partial h}{\partial x_e} = 0 \tag{12.52}$$

As in the previous discussion, the inequality constraints for distance-to-contact are the primary concern. In this case, the derivatives of the distance constraints with respect to the position of a moving vertex x_w, either x_d or x_e, on workspace polygon $\hat{k}$ are based on

$$\frac{\partial g}{\partial x_w} = \sum_{l=1}^{n_w} \left(\frac{\partial \bar{\mathbf{w}}_{\hat{k},l}}{\partial x_w} \frac{\partial g}{\partial \bar{\mathbf{w}}_{\hat{k},l}} \right) \tag{12.53}$$

where n_w is the number of vertices on polygon $\hat{k}$. Since, in the example of Figure 12.14, the moving vertices x_d and x_e are not affecting the positions of other vertices, the expressions for the derivatives of the distance constraints with respect to x_d (or x_e) become

$$\frac{\partial g}{\partial x_d} = \frac{\partial g}{\partial \bar{\mathbf{w}}_{1,5}} \frac{\partial \bar{\mathbf{w}}_{1,5}}{\partial x_d}; \quad \text{and} \quad \frac{\partial \bar{\mathbf{w}}_{1,5}}{\partial x_d} = \begin{bmatrix} 1 \\ 0 \end{bmatrix} \tag{12.54}$$

The $\dfrac{\partial g}{\partial \bar{\mathbf{w}}_{1,5}}$ can be derived by taking partial derivatives of δ^i as discussed in Section 12.4 for different cases. The $\dfrac{\partial\, g}{\partial\, x_e}$ can be derived in the same way as $\dfrac{\partial\, g}{\partial\, x_d}$.

The original configuration of Figure 12.14 did not allow access to the bolt. The optimization result in Figure 12.15 changes the geometry as little as possible and permits the wrench to reach the bolt while satisfying the desired clearance requirements relative to object A.

12.5.3 3-D Examples

The workspace in this example is more like a machine assembly. As shown in Figure 12.16, several cylinders and columns rest on a square-shaped base. The three-link robot is located at the back of this assembly and is to reach into the opening between the long cylinder and the column to approach the small cylinder at the front. The result is shown in Figure 12.17. For this example, a computation time of two to five minutes (on a VAX 8700) may be expected depending the assumed starting trajectory.

The "workspace" presented in Figure 12.16 has also been used to demonstrate a three dimensional geometric design problem. This example is to maximize the height h of the vertical part in front of the robot and behind the tall column while maintaining a sufficient opening for the robot to reach the small cylinder from the other side of the workspace. The reciprocal of height "$1/h$" is used as a weighted component of the optimization objective function and as a variable in the design vector. The exact derivatives of distance-to-contact constraints with respect to the variable h were also derived and added to the set of derivative equations. The necessary modifications of the objective function and the derivatives in this problem follow the pattern of those described in the two-dimensional examples. By applying the same approach and techniques to this 3-D problem, the robot trajectory shown in Figure

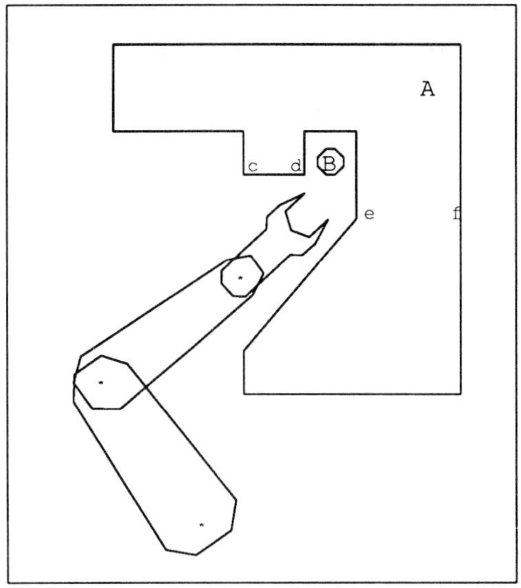

Figure 12.14: Original Geometry of Workspace Geometric Design

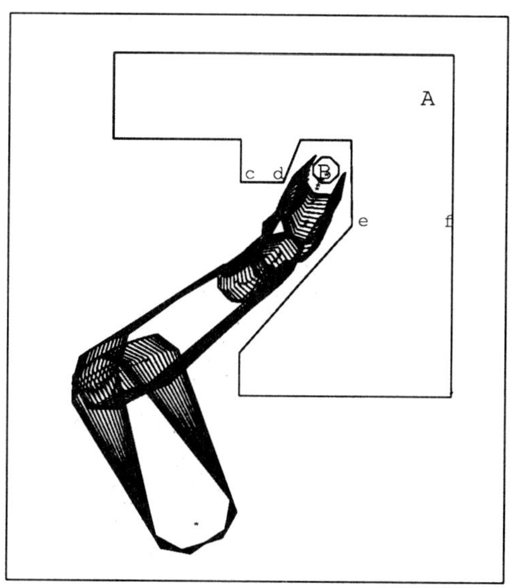

Figure 12.15: Result of Path Planning and Workspace Geometric Design

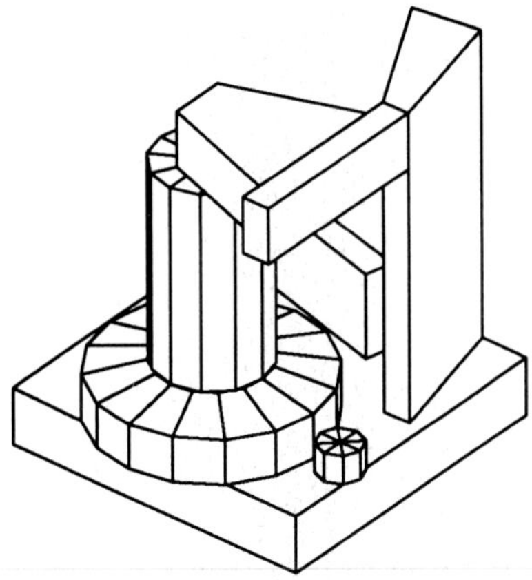

Figure 12.16: 3-D Workspace of Robot Motion Planning Example

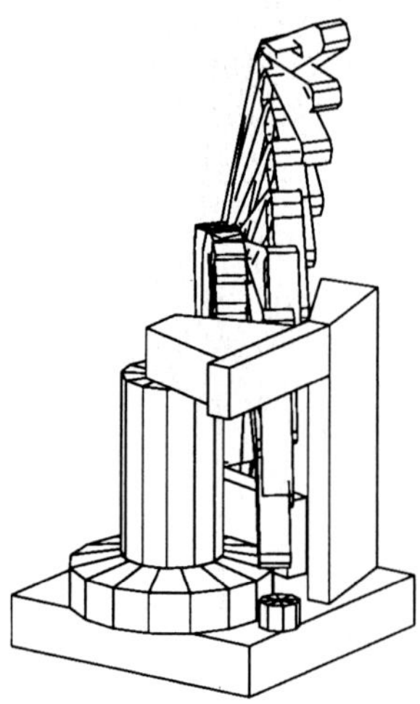

Figure 12.17: Result of 3-D Path Planning Example

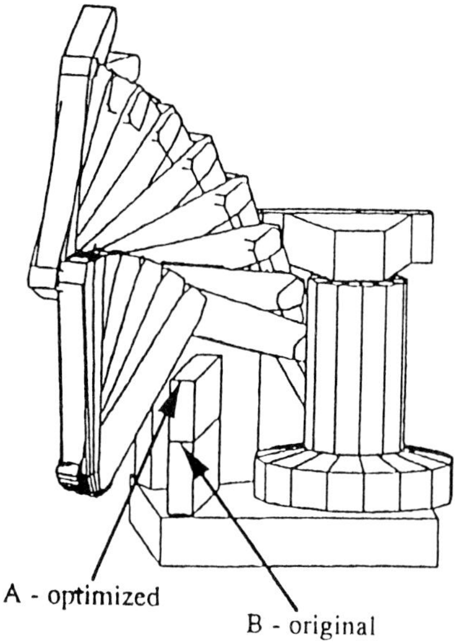

Figure 12.18: Result of Workspace Design and Robot Motion Planning

12.18 results in the maximized height of the vertical part. The middle line shown at B in the figure indicates the original height of the part.

12.6 Dynamic Distance Checking

In this section, an enhanced strategy for distance calculation based on quadratic programming and swept volumes is discussed considering both interference and noninterference cases. The quadratic programming method can handle distance-to-contact problems efficiently by describing convex polyhedra through sets of linear equations as shown in Section 12.3. This method may be adapted to dynamic distance calculations for a trajectory by representing the path of a moving object as a sequence of approximate swept volume segments which are convex. The distance problem becomes converted into the question of how to efficiently construct convex volumes representing the real swept volumes without losing the desired accuracy. Two-dimensional cases are studied here in order to understand the nature of the complexity and three-dimensional problems will then be considered as this chapter proceeds.

12.6.1 QP Distance Problems for Swept Volume Segments

In a two-dimensional space, the boundary of a swept volume consists of three different sets of curves (surfaces in three-dimensional space): the back enclosure, the front enclosure, and the connecting envelope between the two enclosures. The swept volume segment starts from the back enclosure, ends at the front enclosure, and is bounded by the connecting envelope. The back and front enclosures can be computed by checking the outward normal and velocity directions of the moving object boundary. Let $\mathbf{n}_o$ be the outward boundary normal and $\vec{v}$

be the boundary velocity vector.

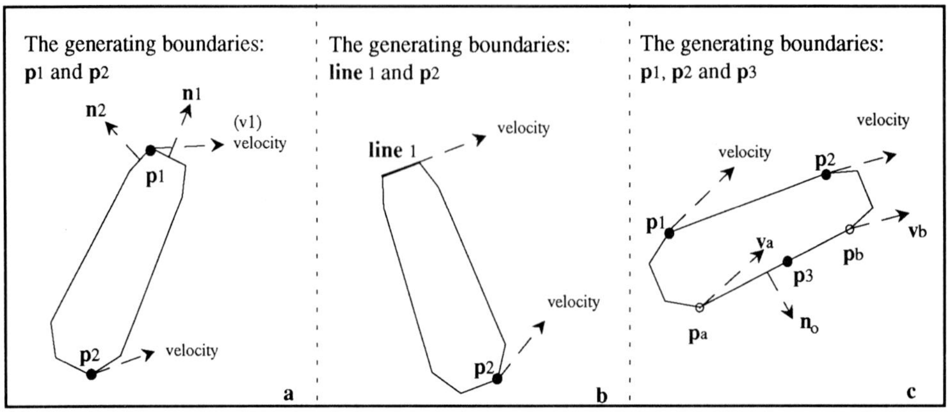

Figure 12.19: Cases of Two-dimensional Polygon Motions

Along the boundaries of the back enclosure, the equation

$$\vec{\mathbf{n}}_o \cdot \vec{v} \leq 0 \tag{12.55}$$

is satisfied and along the boundaries of the front enclosure

$$\vec{\mathbf{n}}_o \cdot \vec{v} \geq 0 \tag{12.56}$$

is satisfied. The curves of the connecting envelope are formed by a certain portion of the object which is called the generating boundary. A necessary condition to be satisfied by the generating boundary is that the outward normal vector be perpendicular to its own velocity vector,

$$\vec{\mathbf{n}}_o \cdot \vec{v} = 0 \tag{12.57}$$

This is applied to moving objects with smooth boundaries. When working with polygons, Eq. (12.57) is difficult to use to find the generating boundary. As illustrated for two dimensions in Figure 12.19, the generating boundary for a polygon may consist of polygon vertices, a polygon boundary line, or a point along a polygon boundary line. For the first case, as shown in Figure 12.19a, the generating boundaries are two polygon vertices $\mathbf{p}_1$ and $\mathbf{p}_2$. There is no unique outward normal at point $\mathbf{p}_1$ or point $\mathbf{p}_2$ and Eq. (12.57) does not apply directly. However, as pointed out by [9], this situation can be efficiently handled by the vector inner product where the vertex velocity and the outward normal of two adjacent polygon lines are compared. The generating boundary $\mathbf{p}_1$ satisfies the condition

$$(\vec{\mathbf{n}}_1 \cdot \vec{v}_1) \times (\vec{\mathbf{n}}_2 \cdot \vec{v}_1) < 0 \tag{12.58}$$

where $\vec{\mathbf{n}}_1$ and $\vec{\mathbf{n}}_2$ are the outward normals of the adjacent lines and $\vec{v}_1$ is the velocity vector of $\mathbf{p}_1$ as shown in Figure 12.19a. The second case, in which an entire polygon line (line 1 in Figure 12.19b) is the generating boundary, rarely happens and may be considered as a special variation of the first case. Figure 12.19c illustrates the third case. This occurs since Eq. (12.57) holds for a certain point, $\mathbf{p}_3$ in Figure 12.19c, along the polygon boundary line. It can be determined by comparing the outward boundary normal and velocity directions of the two adjacent vertices. As shown in Figure 12.19c, let $\mathbf{p}_a$ and $\mathbf{p}_b$ be the vertices of the polygon line with the outward normal $\vec{\mathbf{n}}_o$, and $\vec{v}_a$ and $\vec{v}_b$ be the velocity vectors of $\mathbf{p}_a$ and $\mathbf{p}_b$ respectively. The point $\mathbf{p}_3$ is a generating boundary if

$$(\vec{\mathbf{n}}_o \cdot \vec{v}_a) \times (\vec{\mathbf{n}}_o \cdot \vec{v}_b) < 0 \tag{12.59}$$

and $\mathbf{p}_3$ can be calculated from the following equation

$$\mathbf{p}_3 = \mathbf{p}_a + (\mathbf{p}_b - \mathbf{p}_a)\Big(\frac{|\vec{\mathbf{n}}_o \cdot \vec{v}_a|}{|\vec{\mathbf{n}}_o \cdot \vec{v}_a| + |\vec{\mathbf{n}}_o \cdot \vec{v}_b|}\Big) \tag{12.60}$$

In this case, the swept volume of a moving polygon becomes a complicated nonconvex shape. It is questionable whether computational efficiency can be achieved by the application of quadratic programming to distance computations in such a situation. However, some simplification is possible by using average vertex velocity instead of the individual vertex velocities. By this simplification, the situation can resemble Figure 12.19a. The velocity lines become parallel to each other and the generating boundaries again become two vertices, in this case $\mathbf{p}_2$ and $\mathbf{p}_a$ in Figure 12.19c.

Modeling an entire swept volume for a moving polygon can be approached by using a series of overlapped swept volume segments [21]. Each swept volume segment is developed from polygons representing two neighboring static positions along the trajectory. At each static position, the generating boundary of the swept volume is identified by using vertex velocity vectors and outward normal vectors of the polygon boundary line. Since the velocity of a generating boundary is tangent to the swept volume surface, an approximate envelope for a swept volume segment can be constructed from the velocity vectors of the generating boundary points. As an illustration, the swept volume of a polygon in linear translational motion which starts from the left polygon i and ends at the right polygon $i+1$ is shown in Figure 12.20a. In this case there are two generating vertices 1 and 2 at the starting polygon of this swept volume segment and the back enclosure falls between these two vertices. The generating vertices 3 and 4 belong to the ending polygon and form the end points of the front enclosure. In addition to these eight line segments found in the two end enclosures, two other lines, $\vec{v}_1 - \vec{v}_3$ and $\vec{v}_2 - \vec{v}_4$, can be found from the velocity vectors of the generating vertices. These ten lines form the boundary of the swept volume segment from position i to position $i+1$. This means that a swept volume can be developed from the description of the enclosures and the generating velocity vectors when linear translation takes place.

For a polygon in more general motion, the swept volume needs to be carefully examined. In Figure 12.20b, the extensions of $\vec{v}_1$ and $\vec{v}_{10}$ cross at point 11 and form a reasonable boundary for the conservative approximation of a swept volume segment. This does not happen to the lower portion of the swept volume while looking at the extension of $\vec{v}_5$ and $\vec{v}_6$. The extension of $\vec{v}_5$ passes through the front enclosure and divides the swept

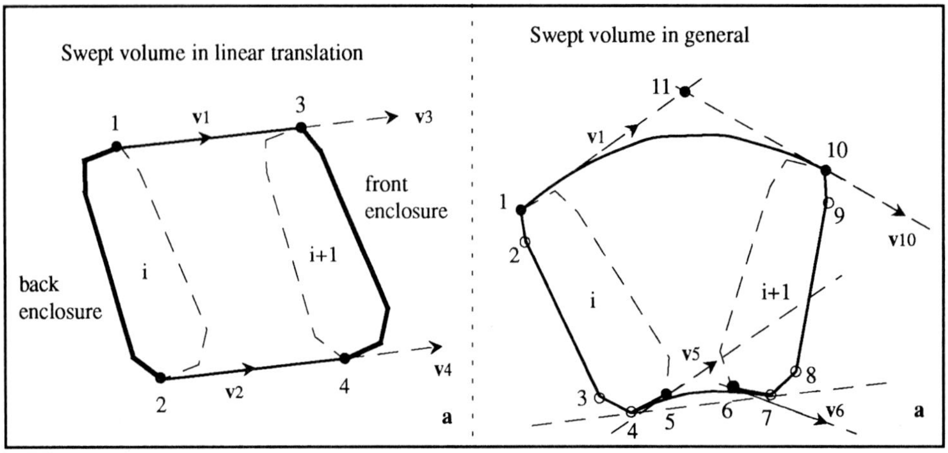

Figure 12.20: Swept Volume Segment Generation

volume into two portions. The back enclosure also intersects with the extension of $\vec{v}_6$. The intersection of velocity lines and enclosures can be detected computationally by using parameter representations for the line segments of the end enclosures. Take Figure 12.20b as an example. The extension of $\vec{v}_5$ can be represented as a function of t_1

$$f_1(t_1) = \bar{\mathbf{p}}_5 + t_1\vec{v}_5 \tag{12.61}$$

and line 8-9 of the front enclosure can be described as a function of t_2

$$f_2(t_2) = \bar{\mathbf{p}}_8 + t_2(\bar{\mathbf{p}}_9 - \bar{\mathbf{p}}_8) \tag{12.62}$$

Unless these lines are parallel to each other, a unique solution of t_1 and t_2 can be derived for their intersection point. If the solution of t_2 is within the range of [0, 1.0], an intersection occurs. When this happens, a simple combination of the two enclosures and the velocity lines does not provide proper boundary information for the swept volume segment. In motion planning problems, a conservative convex swept volume segment can be more useful for obstacle avoidance purposes. So a line tangent to both polygons between vertices 4 and 7 can be used to replace the two enclosure-intersecting velocity lines. A convex polygon of nine boundary lines 1-2-3-4-7-8-9-10-11-1 is then constructed as the swept volume segment. The boundary of such a swept volume segment is composed of straight lines for the end enclosures, velocity lines, and the co-tangent line so that the approximate swept volume segment is a convex polygon and can be described as

$$A\,\bar{\mathbf{p}} \geq \bar{\mathbf{d}} \tag{12.63}$$

consistent with the form of Eq. (12.14). Again $\bar{\mathbf{p}}$ is an interior point, the matrix [A] is composed of the normal vectors of the line segments, and the vector $\bar{\mathbf{d}}$ contains the distance parameters. This convex swept volume representation can then be used directly in QP distance calculations as described in Eq. (12.14).

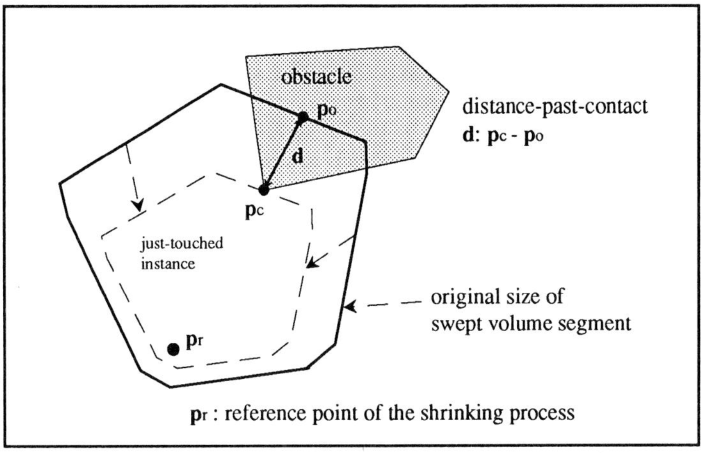

Figure 12.21: Interference of Swept Volume Segments

A shrink-link strategy has been developed in the static distance calculations [21, 24] for use in interference situations and has yielded promising results. In the dynamic distance calculations with a swept volume approach, a shrink-volume scheme may well be appropriate. If interference occurs between a swept volume segment and an obstacle, a shrinking process takes place for the swept volume segment until the just-touched instance is found. A point of contact p_c, as shown in Figure 12.21, may be located at the just-touched instance on the boundary of the shrunken swept volume segment. As the shrunken swept volume segment grows back to its original size, a point p_o corresponding to the contact point p_c can be identified. The distance between these two points p_o-p_c is calculated as the distance-past-contact for the two interfering objects. The shrinking process with a reference point $\bar{\mathbf{p}}_r$ can be described in linear equations as [24]

$$A\ \bar{\mathbf{p}} \geq r\bar{\mathbf{d}} + (1-r)\ A\ \bar{\mathbf{p}}_r \tag{12.64}$$

where the parameter r is the shrinking ratio within the range of [0, 1.0] while the matrix A and the vector $\bar{\mathbf{d}}$ are as in Eq. (12.63).

12.6.2 Example of Distance-to-Contact Calculations

As shown in Figure 12.22, the three-link robot starts from the initial position I and moves to the final position F where the end-effector tip position is specified. From static distance calculations using 30 checking positions between I and F, a distance-to-contact curve is obtained as in Figure 12.23. The most severe interference occurs at $t \approx 0.53$ with a negative distance-to-contact of -0.323. Figure 12.24 shows the robot motion and the approximate swept volume of the robot link 3 with only three swept volume segments (the first segment is emphasized for illustration). Interference is identified for the second swept volume segment and the negative distance-to-contact is −0.379. This is a reasonable estimate with only three segments. Of course, a more accurate swept volume can be obtained by using more swept volume segments, however, this will increase the required CPU time.

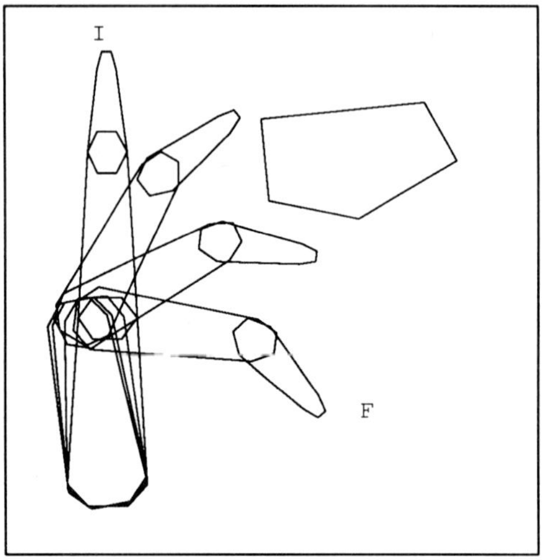

Figure 12.22: Robot Motion in Dynamic Distance Calculation Example

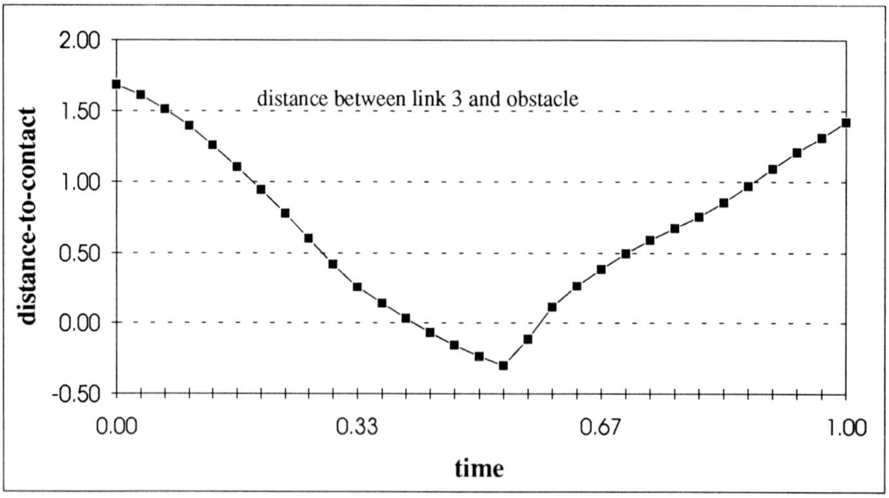

Figure 12.23: Distance Curve from Intensive Static Distance Calculations

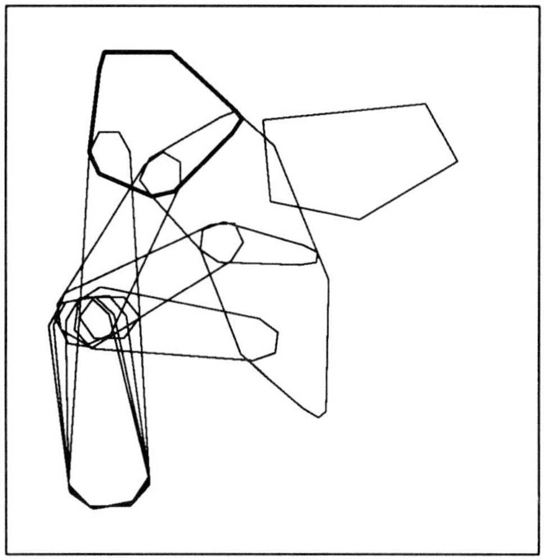

Figure 12.24: Distance Calculation in Three Swept Volume Segments

Preliminary study shows that the total CPU time, including the computation and program initiation time, of a static distance computation procedure with 30 static checking positions is more than three times as expensive as the swept volume approach with three swept volume segments. It can be expected that the computation time is proportional to the number of swept volume segments in the dynamic distance calculations. Therefore, more accurate swept volumes using six, nine, and twelve swept volume segments have been studied. The results from a VAX 8700 are shown in Table 12.1. The comparison shows the possibility of obtaining reasonably accurate distance-to-contact information with reduced computational effort. Dynamic distance calculations by a swept volume approach show promise for yielding distance calculations and distance-past-contact information within a desired accuracy range using less computation time.

Table 12.1: Computation Time and Interference Measurement

	Static method 30 positions	S.V. 3 segments	S.V. 6 segments	S.V. 9 segments	S.V. 12 segments
CPU time	0.64 s	0.18 s	0.21 s	0.28 s	0.30 s
Penetration	-0.323	-0.379	-0.328	-0.352	-0.346

12.6.3 Example of Trajectory Planning

Several two-dimensional path planning examples have been tested in [7]. A typical result is shown in Figure 12.25. The three-link robot starts from the vertical orientation and moves to reach the destination point P inside the C-shaped object. A desired clearance between the robot path and the C-shaped object is maintained through the distance-to-contact constraints. Figure 12.25 shows several discrete robot positions which describe the trajectory. The swept volume representation for the optimal trajectory is shown in Figure 12.26 for links 2 and 3.

12.7 Remarks

This chapter has presented a strategy for planning the motions of automated machines and for integrating motion considerations with geometric design. By using optimization methods, the approach requires that the planning problem be cast in a very specific form to make use of the intelligence offered by nonlinear programming algorithms. The form is not really very restrictive, however, and it is quite easy to maintain the optimization format while applying it to variations in the motion planning problem. Of course, the extension of this approach to the inclusion of geometric design is a good example of its flexibility.

Throughout the examples presented here, the objective function based on kinematic motion has been used to obtain reasonably smooth and effective robot trajectories. Of course, this can be changed as desired while retaining the treatment of constraints and the RQP solution process. An objective function based on speed of response could be applied to obtain fast robot motions. Constraints could also be added to include actuator torque or force capabilities as limitations on the response time. In our development of the optimization approach, we have tried to focus on the avoidance of obstacles through inequality constraints as the critical portion of the process. The QP strategy for distance calculation seems to be quite effective and the discrete position approach can be quite satisfactory. At this point, however, the adaptation of the QP distance calculation to a swept volume representation of the trajectory appears to offer an accurate evaluation of clearance suited for the optimization setting as well as improved computational efficiency.

References

[1] Arora, J.S. (1989). *Introduction to Optimum Design*, McGraw-Hill, New York.

[2] Bobrow, J.E., Dubowsky, S., and Gibson, J.S. (1985). "Time-Optimal Control of Robotic Manipulators along Specified Paths," *The International Journal of Robotics Research*, Vol. 4, No. 3.

[3] Bobrow, J.E. (1989). "A Direct Minimization Approach for Obtaining the Distance between Convex Polyhedrons," *International Journal of Robotics Research*, Vol. 8, No. 3, pp. 65-76.

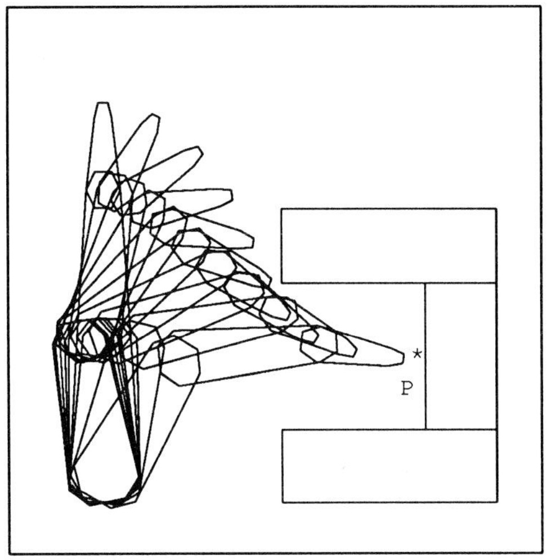

Figure 12.25: Path Planning with Dynamic Distance Computations

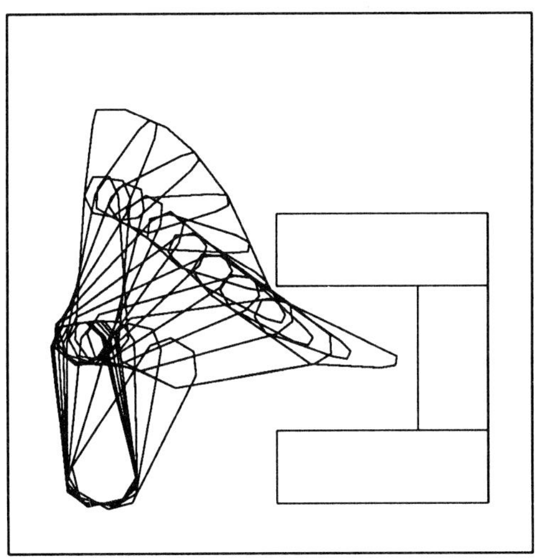

Figure 12.26: Swept Volume Segments in the Robot Trajectory

[4] Brooks, R.A., and Lozano-Perez, T. (1985). " A Subdivision Algorithm in Configuration Space for Findpath with Rotation," *IEEE Transaction on System, Man and Cybernetics*, Vol. SMC15, No. 2, pp. 224-233.

[5] Buchal, R.O., Cherchas, D.B., Sassani, F., Duncan, J.P. (1989). "Simulated Off-line Programming of Welding Robots," *International Journal of Robotics Research*, Vol. 8, No. 3, pp. 31-43.

[6] Buckley, C.E. (1989). "A Foundation for The 'Flexible-Trajectory' Approach to Numeric Path Planning," *International Journal of Robotics Research*, Vol. 8, No. 3, pp. 44-64.

[7] Chen, W.R. (1991). "Optimization in Motion Planning Using Dynamic Distance Calculations," M.S. Thesis, State University of New York at Buffalo, Department of Mechanical & Aerospace Engineering.

[8] Dubowsky, S., Norria, M.A., and Shiler, Z. (1986). "Time Optimal Trajectory Planning for Robotic Manipulators with Obstacle Avoidance: A CAD Approach," *IEEE International Conference on Robotics and Automation*, San Francisco, pp.1906-1912.

[9] Ganter, M.A., and Ucker, J.J., Jr. (1986). "Dynamic Collision Detection Using Swept Solids," *J.of Mechanisms, Transmissions, and Automation in Design*, Vol. 108, pp. 549-555.

[10] Gilbert, E.G., and Johnson, D.W. (1985). "Distance Functions and Their Application to Robot Path Planning in the Presence of Obstacles," *IEEE Journal of Robotics and Automation*, Vol. RA-1, No. 1, pp. 21-30.

[11] Gilbert, E.G., Johnson, D.W., and Keerthi, S.S. (1988). "A Fast Procedure for Computing the Distance between Complex Objects in Three-Dimensional Space," *IEEE Journal of Robotics and Automation*, Vol. 4, No. 2.

[12] Giloi, W.K. (1978). *Interactive Computer Graphics*, Prentice Hall, Inc., Englewood Cliffs, NJ.

[13] Haug, E.J., Arora, J.S. (1979). *Applied Optimal Design*, Wiley Interscience, New York.

[14] Khatib, O. (1986). "Real-Time Obstacle Avoidance for Manipulators and Mobil Robots," *The International Journal of Robotics Research*, Vol. 5, No. 1.

[15] Kieffer, J., and Litvin, F.L. (1990). "Swept Volume Determination and Interference Detection for Moving 3-D Solids," in: *The 1990 ASME Design Technology Conferences - 16th Design Automation Conference*, Chicago, II.

[16] Kim, J.J., and Gossard, D.C. (1989). "Reasoning on the Location of Components for Assembly Packaging," *ASME Design Automation Conference*, Canada.

[17] Lai, M.J. (1988). "Interactive Robot Motion Planning in a Two-Dimensional Workspace by Optimization Methods," M.S. Thesis, State University of New York at Buffalo, Department of Mechanical & Aerospace Engineering.

[18] Lee, B.H., Chien, Y.P. (1987). "Time-varying Obstacle Avoidance for Robot Manipulators: Approaches and Difficulties," in: *IEEE International Conference on Robotics and Automation*, Vol. 3.

[19] Lee, C.K. (1988). *Contact Detaction in the Motion-Planning Problem*, M.S. Project, State University of New York at Buffalo, Department of Mechanical & Aerospace Engineering.

[20] Leu, M.C., Park, S.H., and Wang, K.K. (1986). "Geometric Representation of Translational Swept Volumes and Its Applications," *Journal of Engineering of Industry*, Vol. 108, pp. 113-119.

[21] Li, S.J. (1989). "A Swept Volume Strategy for Contact Distance in Robot Motion Planning," M.S. Project, State University of New York at Buffalo, Department of Mechanical & Aerospace Engineering.

[22] Liu, C.Y. (1989). "Optimization in Motion Planning and Geometric Design Including Obstacle Avoidance," Ph.D. Dissertation, State University of New York at Buffalo.

[23] Liu, C.Y., and Mayne, R.W. (1990). "Nonlinear Optimization in Collision Free Trajectory Planning and Geometric Design," *SME Transactions on Robotics Research*, Vol. 1, pp. 6.11-6.48.

[24] Liu, C.Y., and Mayne, R.W. (1990). "Distance Calculations in Motion Planning Problems with Interference Situations," in: *Advances in Design Automation, Proceedings of 1990 ASME Design Technical Conference*, Chicago, II, DE-Vol. 23-1, ASME, New York, pp. 145-152.

[25] Lozano-Perez, T. (1983). "Spatial Planning: A Configuration Space Approach," *IEEE Transaction on Computers*, Vol. C-32, No. 2, pp. 108-120.

[26] Lozano-Perez, T. (1987) "A Simple Motion-Planning Algorithm for General Robot Manipulators", *IEEE J. of Robotics & Automation*, Vol. RA-3, No. 3.

[27] Martin, R.R., and Stephenson, P.C. (1990). "Sweeping of Three-Dimensional Objects," *Computer-Aided Design*, Vol. 22, No. 4, pp. 223-234.

[28] Mayne, R.W. (1987). "Position and Motion Planning in a Confined Region via Nonlinear Programming," in: *Advances in Design Automation, Vol. 1, The 1987 ASME Design Technology Conferences - The Design Automation Conference*, Boston, MA, September 27-30.

[29] Paul, R.P. (1981). *Robot Manipulators: Mathematics, Programming, and Control*, MIT Press, Cambridge, MA.

[30] Red, W.E. (1983). "Minimum Distance for Robot Task Simulation," *Robotica*, Vol. 1, pp. 231-238.

[31] Red, W. E, and Truong-Cao, H. V. (1984) "The Configuration Space Approach to Robot Path Planning," in: *1984 American Control Conference.*

[32] Red, R.E., and Truong-Cao, H.V. (1985). "Configuration Maps for Robot Path Planning in Two Dimensions," *Transactions of the ASME*, Vol. 107, pp. 292-298.

[33] Sandgren, E., and Ragsdell, K.M. (1980). "The Utility of Nonlinear Programming Algorithm: A Comparative Study," *ASME Journal of Mechanical Design*, Vol. 102, No. 3, pp. 540-551.

[34] Schittkowski, K. (1980). "Nonlinear Programming Codes: Information, Tests, Performance," *Lecture Notes in Economics and Mathematical Systems*, No. 183, Springer-Verlag, New York.

[35] Shin, K.G., and McKay, N.D. (1984). "Open-Loop Minimum-time Control of Mechanical Manipulators and Its Application," *Proc. 1984 American Control Conference*, San Diego, CA, pp. 1231-1236.

[36] Wang, W.P., and Wang, K.K. (1986). "Geometric Modeling for Swept Volume of Moving Solids," *IEEE Computer Graphics & Applications* (December), pp. 8-17.

[37] Yoshimura, M., Yoshikawa, N., and Hitomi, K. (1990). "Design Optimization of Industrial Robots Considering the Working Environment," *International Journal of Production Research*, Vol. 28, No. 5, pp. 805-820.

[38] Zhou, J. (1985) "The Program CVM01," Department of Mechanical Engineering, Huazhong Institute of Technology, Wuhan, China.

Chapter 13

Collision Identification between Convex Objects Using Neural Networks

JING YUAN, YURY STEPANENKO, and ZUOMIN DONG
Department of Mechanical Engineering, University of Victoria

13.1 Introduction

13.1.1 Background

The collision identification and object-to-object distance calculation play an important role in the motion planning for robots, CNC machine tools, coordinate measuring machines, and other manufacturing facilities.

Collision-free motion planning is normally conducted based on a geometric model of the working environment and the computer algorithms for generating a collision-free trajectory of the moving object between given initial and end positions. The geometric model of a working environment presents the surrounding obstacles and workspace of the moving objects. The trajectory of the moving objects is often generated by minimizing the total travel distance (or energy consumption) with the obstacles serving as geometric constraints of the optimization [11, 3, 19]. Compared with the two other major alternative methods for motion planning, this optimization-based approach presents more realistic results than the forbidden region identification method [20], and requires less computation effort than the swept volume method [10, 17, 24] for a complex work environment.

Nevertheless, in the optimization-based motion planning approach, the geometric constraints for collision avoidance must be calculated and inspected in large numbers at each iteration of the optimization. This computation has been recognized as the primary cost of computation effort in motion planning and forms a bottleneck for enhancing motion planning strategies. An efficient algorithm for identifying possible collision and for computing the distance between the moving object and its surrounding obstacles is a major research focus in this field [11, 3, 19].

The present collision identification and distance calculation method forms another distance minimization problem. The distances between the moving object and its surrounding obstacles are calculated by finding the shortest distance between two points. One point is within the boundary of the moving object. The other point is contained in one of the obstacles. The distances from all convex bodies of the moving object to all surrounding obstacles must be calculated and checked at each iteration of the global motion planning optimization. If any one of the calculated distances becomes zero, a collision occurs. The trajectory of the moving object must be adjusted, and the global motion planning optimization starts another iteration until the total travel distance is minimized with all the moving object-to-obstacle distances being non-zero. It would be ideal if the suboptimization problem for object-to-object distance calculation could be replaced by other more computational efficient mechanisms. A large, complex, nested optimization problem can thus be considerably simplified.

Most of the motion planning problems that we are dealing with today are of an off-line type. Planning time is not a major concern. However, for the intelligent robotic and manufacturing systems working in unstructured and time-variant work environments, the on-line motion planning capability is an essential function. The unstructured work environment exists for intelligent robots working under the ocean and in space. The time-variant work environment occurs in small-batch manufacturing where changes of parts to be processed are frequent. In these applications, the work environment is either not known beforehand, or keeps changing constantly. The efficient on-line environment perception/modeling and on-line motion planning methods become critical. With the rapid progress in the development of intelligent robotic and manufacturing systems and the increasing demand for higher productivity, fast on-line motion planning techniques will see more need and applications.

In this chapter, we present a new collision identification and distance calculation method using neural networks. This method can significantly speed up the motion planning of robotic and manufacturing systems. It can be combined with the on-line working environment perception/modeling technique and the optimization-based motion planning method discussed in this book.

13.1.2 Related Work

Research on object-to-object collision identification and distance calculation has gained considerable attention recently. The Euclidean distance has been recognized as the most natural way to measure a collision between two objects [11]. The computation of the Euclidean distance between two objects is a topic studied in computational geometry [14]. If only two convex objects, one moving object and one fixed obstacle, are involved in the motion planning, each of these two objects can be modeled by a polyhedron with M_i vertices ($i = 1, 2$). The object-to-object collision identification and distance calculation can be done by checking the location of each vertex of one object with respect to the other object; the number of operations is a function of $M = M_1 + M_2$. For a two-dimensional (2-D) problem, the asymptotic computational time is about $O(\log^2 M)$ by Schwartz's algorithm [23]. More efficient algorithms were reported by Chin and Wang [6] and Edelsbrunner [9]. When three-dimensional (3-D) polyhedrons are involved, the algorithms by Morlowski [21] and Dobkin and Kirkpatrick [8] require about $O(M \log M)$ and $O(M)$ operations respectively. When the number of involved convex objects increases, the required computational efforts increase significantly. Much more research has been conducted to improve the efficiency of the collision identification and distance calculation algorithm [25, 1, 2, 22, 4, 5, 12]. Recently, the gradient projection method was applied by Bobrow [3], and the quadratic programming method was reported by Liu and Mayne [19]. All these methods are iterative in nature.

The recent development of neural networks brings new light to collision identification and path planning. The well-known Hopfield model was applied by Lee and Bien [15] to collision-free control of robots where the objects and robot arms are approximated by circles and line segments. A more sophisticated approach was recently reported by Lee and Park [16]. They introduce a collision penalty for point-to-polyhedron collision detection. When applied to 3-D trajectory planning, the moving object is represented by a number of via points, which are somewhat similar to the vertices of polyhedrons. Similar work has also been reported by the authors [7]. The neural-network-based collision identification and distance calculation approach utilizes the parallel processing mechanism of the neural networks to accelerate the computation speed of motion planning. It is a promising alternative to the optimization-based method.

In this chapter, a new neural-network-based object-to-object collision identification and distance calculation method is presented. The collision identification neural network (CINN) developed in our research, is designed to measure the distance between two convex polyhedrons A and B. A represents a moving object. B represents an obstacle within the workspace of A. The developed CINN system is based on the Hopfield neural network model. It consists of two subnets. The first subnet is designed for detecting collision between a moving point $\mathbf{p}$ on A and the convex obstacle polyhedron B. The second subnet is used for constraining the point $\mathbf{p}$ within the moving polyhedron A and pushing this moving point away from the center of polyhedron B. The constrained outcome represents the distance between the two polyhedrons, A and B.

In case a collision occurs, the CINN will return the position of the deepest interfering point of the moving object A within the obstacle B, a "negative" distance measure between the two objects, and a direction vector leading to the shortest way out of the collision. The calculated negative distance is used for interference measurement. The obtained direction vector points out the ideal direction for collision removal. A collision can be detected by a CINN within the transient response time of a few nonlinear amplifiers.

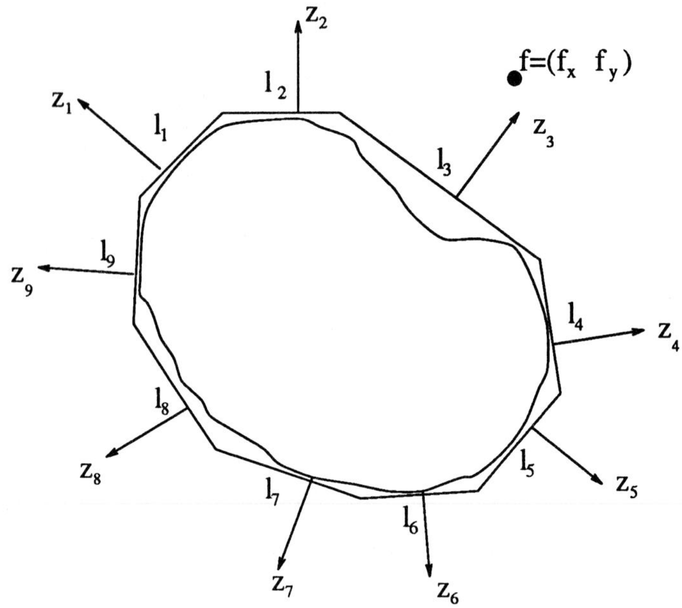

Figure 13.1: Convex N-polygon Approximating an Obstacle

13.2 Point-to-polyhedron Collision Identification

In collision identification and distance calculation, convex polyhedrons are the focus of attention, since a nonconvex object can always be decomposed into, or be approximated by, a number of convex polyhedrons. In general, cylindrical, spherical, and sculptured objects can all be modeled this way. The boundaries of these objects can be represented by many plane surface patches according to the required resolution.

A convex polyhedron is the union of N half-spaces defined by the corresponding boundary planes. Since an arbitrary transverse plane of a convex polyhedron is a convex polygon, it is easier to visualize a convex polyhedron by considering a convex polygon cross section. In this chapter, all illustrations concerning convex polyhedrons are plotted as convex polygons. A typical convex polygon is shown in Figure 13.1. It is the union of nine half-spaces defined by nine boundary lines $\{l_i\}_{i=1}^{9}$. A boundary plane of a convex polyhedron cuts the 3-D space into two halves; the body of the convex polyhedron belongs to only one of them. This feature is reflected in Figure 13.1. For any two half-spaces separated by a boundary line l_i, the convex polygon resides in only one half-space. This is a very useful feature of convex polyhedrons to be explored in this study.

The objective of this work is to develop a neural network for collision identification between multiple convex polyhedrons. As the first step, this section addresses the way to identify a possible collision between a point and a convex polyhedron. In order to identify whether a given point, $\mathbf{p}$, collides with the polyhedron, the following information is of interest:

1. Whether point $\mathbf{p}$ is inside or outside the polyhedron?

2. How far away is point $\mathbf{p}$ from the polyhedron if no collision occurs?

3. How much is the interference and in what direction should point $\mathbf{p}$ be moved out if a collision happens?

One may attach a local Cartesian coordinate frame to each of the N boundary planes of a polyhedron as illustrated in Figure 13.1. The x-y plane of a local coordinate frame coincides with a boundary plane of the polyhedron. Its z axis is directed along the direction of the surface normal (pointing towards the outside of the object). The origin of the coordinate frame can be placed anywhere within the boundary of the plane.

An arbitrary point is represented by a 3-D vector $\mathbf{p} = [p_x, p_y, p_z]^T$. It can be expressed either in the global Cartesian coordinate or in the boundary coordinate frames. In order to distinguish the representation of a point with respect to an appropriate coordinate frame, one may use a subscript "n" to indicate that the vector is expressed in the nth frame (attached to the nth boundary plane). When the point is expressed with respect to the global coordinate frame, no subscript is used. According to the theory of geometric transformation, $\mathbf{p}_n$ is related to $\mathbf{p}$ by

$$\mathbf{p}_n = \mathbf{R}_n \mathbf{p} + \mathbf{t}_n \tag{13.1}$$

where $\mathbf{R}_n \in R^{3\times 3}$ is a rotation matrix; $\mathbf{t}_n \in R^3$ is the offset vector representing the distance between the origins of the two coordinate frames. The three rows of $\mathbf{R}_n$ are unit vectors representing the x-y-z axes of the nth coordinate frame. A detailed matrix equation is given by

$$\begin{bmatrix} p_{nx} \\ p_{ny} \\ p_{nz} \end{bmatrix} = \begin{bmatrix} x_{n1} & x_{n2} & x_{n3} \\ y_{n1} & y_{n2} & y_{n3} \\ z_{n1} & z_{n2} & z_{n3} \end{bmatrix} \begin{bmatrix} p_x \\ p_y \\ p_z \end{bmatrix} + \begin{bmatrix} t_{nx} \\ t_{ny} \\ t_{nz} \end{bmatrix}$$

where the subscript "n" is attached to the entries of a matrix or vector.

As mentioned earlier, the x-y plane of a coordinate frame divides the whole 3-D space into two parts. The first half of the space is characterized by a positive z coordinate. This half-space does not contain the given polyhedron. The other half-space is characterized by a negative z coordinate, and in this half-space the given polyhedron stays. Thus only the z coordinates in the boundary frames are of interest. Equation (13.1) can thus be simplified to

$$p_{nz} = \mathbf{z}_n^T \mathbf{p} + t_{nz} \tag{13.2}$$

where, $\mathbf{z}_n^T$ is the last row of the rotation matrix $\mathbf{R}_n$. The following hints are essential to the new formulation:

1. Suppose the given convex polyhedron has N boundary planes; then p_{nz} are obtained for $1 \le n \le N$. If any one of $\{p_{nz}\}_{n=1}^N$ is positive, then there must exist at least one boundary plane separating point $\mathbf{p}$ from the convex polyhedron. Thus point $\mathbf{p}$ is definitely outside the polyhedron.

2. On the other hand, if point $\mathbf{p}$ is inside the N-surface convex polyhedron, then no matter in which one of the N coordinate frames $\mathbf{p}$ is expressed, its z coordinate will always be negative.

According to these two hints, one may introduce an indicator given by

$$p_{n^*z} = \max_{1 \le n \le N} \{p_{nz}\}. \tag{13.3}$$

The collision between point $\mathbf{p}$ and the N-surface polyhedron can then be identified as

$$\begin{cases} \text{If } p_{n^*z} > 0, & \text{collision free;} \\ \text{If } p_{n^*z} \le 0, & \text{collision occurred, return direction vector } z_{n^*}. \end{cases} \tag{13.4}$$

Since p_{n^*z} has the minimum absolute value when a collision occurs—all $\{p_{nz}\}_{n=1}^{N}$ are negative—the n^*th boundary plane is the nearest to point $\mathbf{p}$. Consequently, the direction allowing point $\mathbf{p}$ to be pushed out of the polyhedron with the shortest distance is the z direction of the n^*th boundary frame. This direction is given by z_{n^*}. The distance in which point $\mathbf{p}$ should be pushed out from the polyhedron to remove collision is given by $|p_{n^*z}|$. It is ideal if the neural networks can also automatically identify the direction of z_{n^*} and provide the value of $|p_{n^*z}|$.

The formulation for collision identification between a point $\mathbf{p}$ and a convex polyhedron, given in Eqs. (13.2), (13.3), and (13.4), forms the basis of this work. One may use Figure 13.1 as a simple 2-D example to digest the working principles of the proposed formulation. In Figure 13.1, the point $\mathbf{p}$ is outside the polygon, its z coordinate is positive when expressed in the second and third boundary frames. The indicator can be found to be $p_{n^*z} = \max_{1 \le n \le 9}\{p_{nz}\} = p_{3z}$ and $n^* = 3$. A point inside the polygon will have a negative z coordinate in any one of the coordinate frames attached to the boundaries of the polygon.

13.3 Programming the Hamming Net

As an alternative form of information processing, neural networks have attracted considerable attention in the last few years. Neural networks are particularly good at solving complex pattern recognition problems, such as speech understanding and handwritten character identification. A major advantage of neural networks over the traditional computational methods is their collective computation power. The highly parallel processing capability of neural networks makes them ideal tools to handle large numbers of simple operations. The Hamming net [18] is one of the well-known neural network structures. An N-neuron Hamming net can be viewed as a Hopfield N-flop [13] plus an additional input layer which performs preliminary inner product processing. A typical application of the Hamming net is pattern recognition in a proper inner product space. This section demonstrates how the collision identification formulations, given by Eqs. (13.2) and (13.3), fit into the mechanism of a Hamming net and how to program a modified Hamming net for a given working environment (obstacles and workspace).

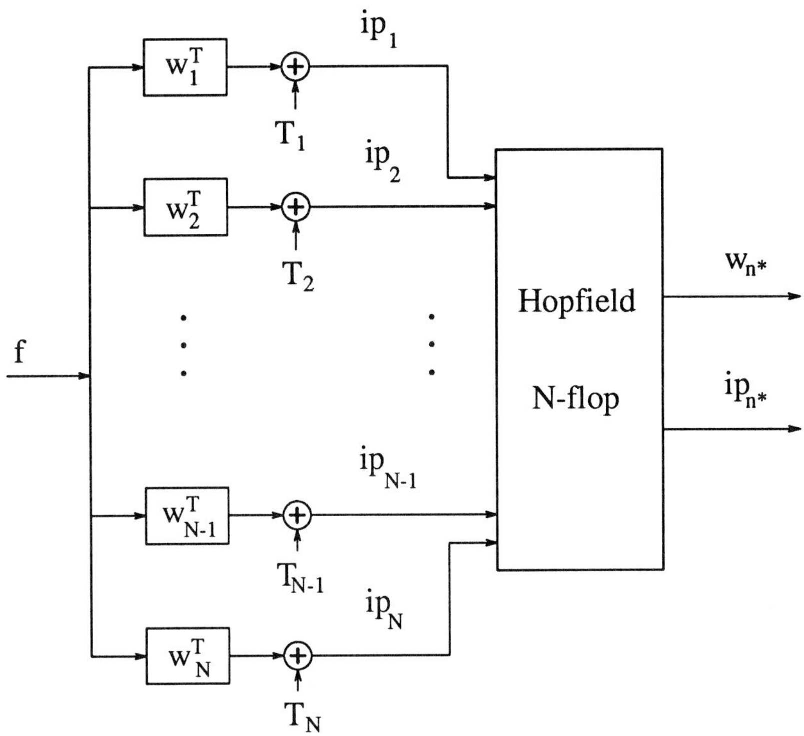

Figure 13.2: N-neuron Hamming Net

A Hamming net for pattern classification in a 3-D Euclidean space is given in Figure 13.2. N patterns are prestored in the Hamming net in the form of N-three element weight vectors, $\{\mathbf{w}_n\}_{n=1}^N$, and N thresholds, $\{T_n\}_{n=1}^N$. Given an input feature vector $\mathbf{f} = [f_x, f_y, f_y]^T$, the first layer (or the input layer) of the Hamming net will match this input with N patterns through N parallel inner product operations:

$$ip_n = \mathbf{w}_n^T \mathbf{f} - T_n \quad (1 \leq n \leq N) \tag{13.5}$$

The resulting N inner products $\{ip_n\}_{n=1}^N$ are then fed into the second layer of the Hamming net (the Hopfield N-flop).

The detailed structure of the Hopfield N-flop is not given in Figure 13.2. Interested readers may refer to textbooks on artificial neural networks or Hopfield's earlier work [13]. The name "N-flop" suggests that this layer is an extension of a flip-flop. As a matter of fact, the function of a Hopfield N-flop is very similar to an R-S flip-flop. An R-S flip-flop consists of two amplifiers. Each tries to shut down the other by feeding its negated output to the input terminal of the competitor. If equal connection weights are used, a fair competition between the two amplifiers is guaranteed. When two external voltage signals are fed to the two interconnected amplifiers, the amplifier with a higher external input will eventually win the competition. In a Hopfield N-flop, N artificial neurons (synthesized by nonlinear amplifiers) are involved in the competition. At the end, only one neuron will emerge as the winner by shutting down all other competitors. Since the internal competition is guaranteed to be fair by equal connection weights, the outcome is determined by only the external inputs which can sometimes make the marginal difference.

The final classification made by the Hopfield N-flop is given by

$$ip_{n^*} = \max_{1 \leq n \leq N} \{ ip_n \} \tag{13.6}$$

The input pattern $\mathbf{f} = [f_x, f_y, f_y]^T$ is thus attributed to the prestored vector $\mathbf{w}_{n^*}$, which has the largest inner product with $\mathbf{f}$.

The resemblance between Eqs. (13.2) and (13.3) and Eqs. (13.5) and (13.6) is evident. The term p_{nz} in Eq. (13.2) has the same mathematic expression as ip_n in Eq. (13.5). The selection of p_{n^*z}, defined by Eq. (13.3), is exactly the same as that of ip_{n^*}, given by Eq. (13.6). To apply the Hamming net formulation for point-to-polyhedron collision identification, one can simply replace the input feature vector $\mathbf{f}$ with the point vector $\mathbf{p}$. The Hamming net can be programmed by storing the vectors from the rotation matrix, $\{\mathbf{z}_n\}_{n=1}^N$, as the weighting vector, $\{\mathbf{w}_n\}_{n=1}^N$. The only additional mechanism needed is some switching circuits for directly identifying a collision according to Eq. (13.4). The indicator p_{n^*z} is returned by the nets as the calculated point-to-object distance when $p_{n^*z} > 0$. Otherwise, the direction vector, $\mathbf{z}_{n^*}$, is an output. This output is the optimum direction along which point $\mathbf{p}$ can be pushed out of the polyhedron with a minimum distance $|p_{n^*z}|$.

13.4 Polyhedron-to-polyhedron Collision Identification

The collision identification between two polyhedrons is more complex than the point-to-polyhedron collision identification, but it is directly applicable to practical applications. An extension to the previously discussed neural network formulation can be used for collision identification between two polyhedrons. An example given in Figure 13.3 illustrates the method. In this figure, two polygons A and B, a point $\mathbf{p}$ within polygon A, two boundary edges of the polygon B, l_1 and l_2, the z axes of the boundary l_1 and l_2, $\mathbf{z}_1^{(B)}$ and $\mathbf{z}_2^{(B)}$, as well as the true distance between these two polygons, $\mathbf{d}_{AB}$, are shown.

The modified Hamming net, discussed previously, allows us to identify a possible collision between a given point and a polygon. This can be applied to the point $\mathbf{p}$ and polygon B shown in Figure 13.3. The point $\mathbf{p}$ is now considered as a free point allowed to move within the polygon A. When point $\mathbf{p}$ moves, the indicator, p_{n^*z}, calculated based on point $\mathbf{p}$ and polygon B using Eq. (13.3), will have different values. The indicator p_{n^*z} is a function of the location of $\mathbf{p}$. It can be represented as $p_{n^*z}(\mathbf{p})$. The extended formulation identifies possible polygon-to-polygon collision using a "mini-maximum":

$$p^*_{n^*z} = \inf_{\mathbf{p} \in A} \{p_{n^*z}(\mathbf{p})\} = \inf_{\mathbf{p} \in A} \{ \max_{1 \leq n \leq N} [p_{nz}(\mathbf{p})]\} \tag{13.7}$$

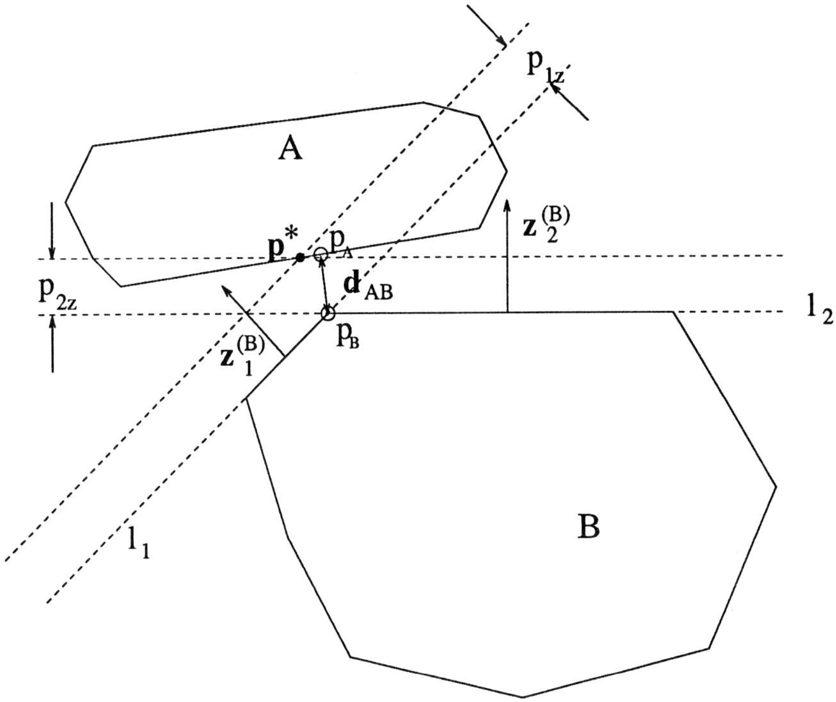

Figure 13.3: Collision Identification between Polygons A and B

where, $p^*_{n^*z}$ is the minimum indicator of point $\mathbf{p} \in A$. In other words, $p^*_{n^*z}$ has the minimum value compared with all calculated indicators of point $\mathbf{p}$ at any other location within polygon A. The location of point $\mathbf{p}$ in polygon A for achieving this minimum indicator is notated as $\mathbf{p}^*$. This point is called the "mini-maximum position" of point $\mathbf{p}$.

For the example given in Figure 13.3, the mini-maximum position is represented by point $\mathbf{p}^*$, which is located on the boundary of polygon A and equally away from the two plane boundaries of object B, l_1 and l_2. Or we may say that point $\mathbf{p}^*$ is at the intersection between the median line of l_1 and l_2 and the edge of A. The calculated mini-maximum value, $p^*_{n^*z}$, equals p_{1z} and p_{2z}. The reason that point $\mathbf{p}^*$ has equal z-coordinates in the two boundary frames attached to l_1 and l_2 can be explained by analyzing the work of the modified Hamming net. Since now the competition of the Hopfield N-flop is between p_{1z} and p_{2z}, a rightward drift of $\mathbf{p}$ along the boundary of A will cause $p_{2z} > p_{1z}$, then $p_{n^*z} = \max\{p_{nz}\} = p_{2z} \geq p^*_{n^*z}$, and a leftward drift of $\mathbf{p}$ along the boundary of A will cause $p_{1z} > p_{2z}$, then $p_{n^*z} = \max\{p_{nz}\} = p_{1z} \geq p^*_{n^*z}$, the only possible result is $p_{n^*z} = p_{1z} = p_{2z}$.

The mini-maximum value $p^*_{n^*z}$ is called a pseudo distance between polygons A and B. Mathematically, $p^*_{n^*z}$ is not the Euclidean distance d_{AB}, as measured between point $\mathbf{p}_a$ and $\mathbf{p}_b$. Nevertheless, the slight difference between $p^*_{n^*z}$ and d_{AB} does not affect the collision identification and the generation of a collision-free path. This fact will be proved later in

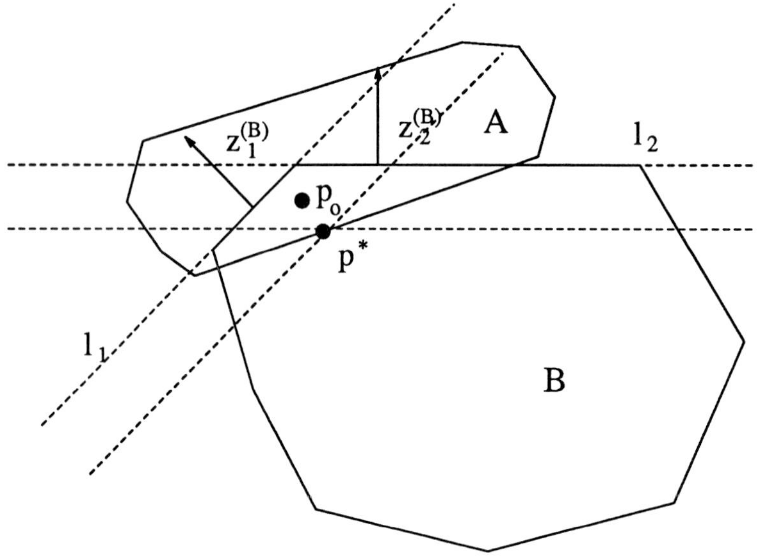

Figure 13.4: Collision between Convex Polygons A and B

this chapter.

A positive mini-maximum value indicates that the two polygons A and B do not contact each other. When one polygon interferes with the other, as illustrated in Figure 13.4, there must exist at least one point $\mathbf{p}_o$ which belongs to both polygon A and B. The z coordinate of point $\mathbf{p}_o$ will be negative, whether the z is expressed in the frames attached to the surfaces of either polygon A or polygon B, that is, $p_{n^*z}(\mathbf{p}_o) < 0$. Furthermore, we have

$$p^*_{n^*z} = \inf_{\mathbf{p} \in A} \{p_{n^*z}(\mathbf{p})\} \leq p_{n^*z}(\mathbf{p}_o) < 0$$

which indicates a collision between the two polygons.

As a general criterion, a collision between any two polygons can be identified by

$$\begin{cases} \text{If } p^*_{n^*z} > 0, & \text{collision free;} \\ \text{If } p^*_{n^*z} \leq 0, & \text{collision occurs.} \end{cases} \tag{13.8}$$

In Figure 13.4, a collision occurs and the mini-maximum position $\mathbf{p}^*$ is on the boundary of polygon A and equally distanced from the boundary l_1 and l_2 of polygon B. To remove this collision, the ideal direction for pushing polygon A away from polygon B is $\mathbf{v}^* = \mathbf{z}_1^{(B)} + \mathbf{z}_2^{(B)}$, where $\mathbf{z}_1^{(B)}$ and $\mathbf{z}_2^{(B)}$ are the z axes of the two boundary frames attached to the boundary l_1 and l_2 of the polygon B. The absolute value of the pseudo distance $|p^*_{n^*z}|$ also provides a good approximation of the required movement along the direction of $\mathbf{v}^*$ for removing the collision. These two z directions and the pseudo distance can be obtained directly from the computation specified by Eq. (13.7). The advantage of using this pseudo distance $p^*_{n^*z}$, rather than the Euclidean distance for collision identification, lies in its capability to provide the extent of the collision and the ideal direction for moving a collided polygon to remove the collision. The Euclidean distance becomes zero and provides no additional information when a collision occurs.

Another advantage of the stated approach is its ability to detect the edge-collisions and surface-collisions quickly and economically. These edge-collisions and surface-collisions can not be detected by the pure vertex-based approach. For instance, the collision illustrated in Figure 13.4 could not possibly be detected by merely checking the vertices of polygon A, since all vertices of A are out of polygon B. This type of edge-collision or surface-collision frequently happens in 2-D and 3-D motion planning. To handle these types of cases, Lee and Park [16] introduced a number of via points on the surfaces of the moving polygon (polyhedron) A. Their approach requires a large number of Hamming nets. Each of these nets detects the collision of one via point with respect to the fixed polygon (polyhedron) B. In the next section, it will be shown that the operation specified in Eq. (13.7) can be implemented by a single Hamming net and some simple neural circuits.

13.5 The Collision-Identification Neural Network

Again, a two-polygon example is used to demonstrate the working principle of the discussed approach. A modified Hamming net is programmed by storing $\{\mathbf{z}_n^{(B)}\}_{n=1}^{N_B}$ and $\{t_n^{(B)}\}_{n=1}^{N_B}$. It identifies a possible collision between an arbitrary point $\mathbf{p}$ (on polygon A) and polygon B. An analog integrator is designed to control the location of point $\mathbf{p}$. The function of this integrator is to force $\mathbf{p}$ to move toward point $\mathbf{p}^*$. A possible straightforward design is given by

$$\dot{\mathbf{p}} = -\mathbf{z}_{n^*}^{(B)} \tag{13.9}$$

where $\mathbf{z}_{n^*}^{B}$ is the direction vector returned by the modified Hamming net. The integration of Eq. (13.9) drives $\mathbf{p}$ along the fastest descendent direction of $p_{n^*z} = \max\{p_{nz}\}$. The main problem with Eq. (13.9) is the absence of the constraint $\mathbf{p} \in A$. As a result, Eq. (13.9) will drive point $\mathbf{p}$ all the way to the center of polygon B, which is certainly not desired.

In order to keep point $\mathbf{p}$ inside polygon A, point $\mathbf{p}$ is fed into a constraint subnet to detect whether it is inside polygon A. This subnet consists of N_A independent neurons. The output of the nth neuron is denoted as o_n, which is obtained by a nonlinear operator

$$o_n = \begin{cases} 1 & \text{if } \mathbf{p}^T\mathbf{z}_n^{(A)} - t_n^{(A)} \geq 0; \\ 0 & \text{otherwise.} \end{cases} \tag{13.10}$$

Let $o = \bigcup_{n=1}^{N_A} o_n$ denote the union of all $\{o_n\}_{n=1}^{N_A}$; then Eq. (13.9) is modified to

$$\dot{\mathbf{p}} = -\sum_{n=1}^{N_A} o_n \mathbf{z}_n^{(A)} - \bar{o}\mathbf{z}_{n^*}^{(B)} \tag{13.11}$$

where $\bar{o}$ is the logic complement of $o = \bigcup_{n=1}^{N_A} o_n$. It satisfies

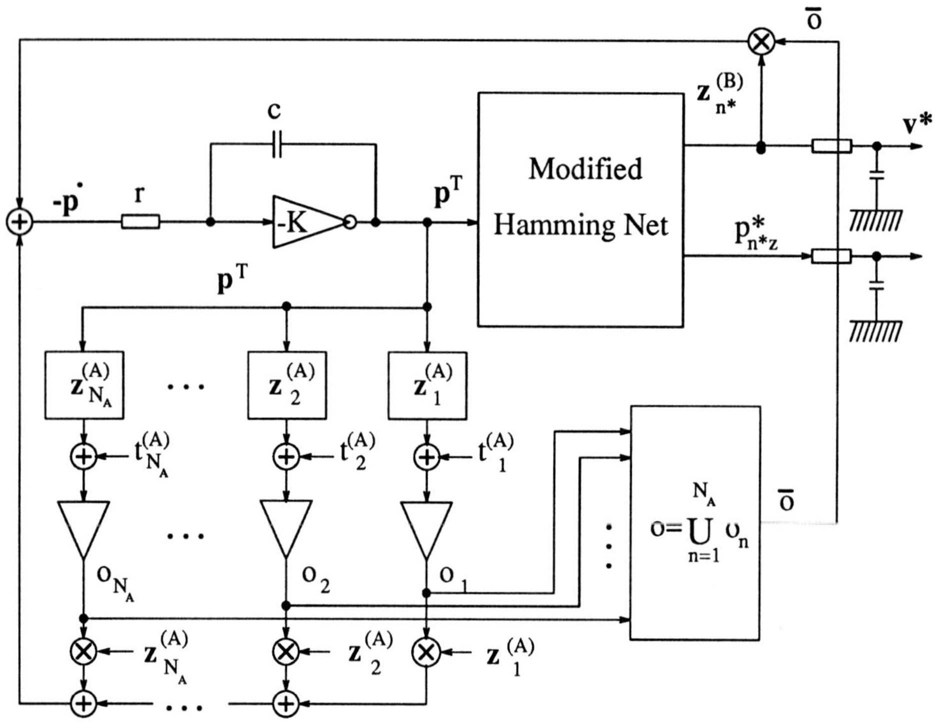

Figure 13.5: Collision-identification Neural Network

$$\bar{o} o_n = 0 \quad \forall \quad 1 \leq n \leq N_A. \tag{13.12}$$

The block diagram of Eq. (13.11) is given in Figure 13.5, where only one analog integrator is plotted for clarity. It represents three integrators needed to integrate the x-y-z components of point $\mathbf{p}$. The nonlinear operation of Eq. (13.10) can be implemented by high-gain switching amplifiers as artificial neurons.

The convergence of $\mathbf{p}$ can be analyzed by considering the following three cases:

1. Point $\mathbf{p}$ is inside polygon A

2. Point $\mathbf{p}$ is outside polygon A

3. Point $\mathbf{p}$ "vibrates" around the boundary of polygon A

In the first case, all $\{o_n\}_{n=1}^{N_A}$ are zero because the z coordinates of $\mathbf{p}$ are negative no matter whether z is expressed in any of the boundary frames of polygon A. As a result, Eq. (13.11) is reduced to Eq. (13.9). It pushes point $\mathbf{p}$ along the fastest descendent direction of p_{n^*z}. This direction is given by $-\mathbf{z}_{n^*}^{(B)}$.

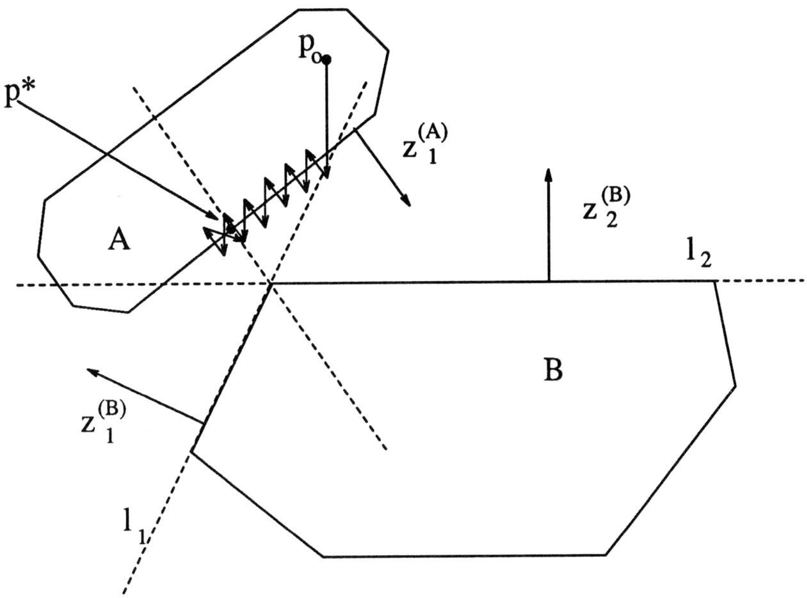

Figure 13.6: Zigzag Convergence Pattern of **p**

In case point **p** is pushed outside polygon A, at least one of the $\{o_n\}_{n=1}^{N_A}$ will be non-zero. Suppose **p** is outside polygon A at the M boundary planes; then one may group the non-zero elements of $\{o_n\}_{n=1}^{N_A}$ as $\{o_{n_m}\}_{m=1}^{M}$. Thus we have $o = \bigcup_{n=1}^{N_A} o_n = 1$ and $\bar{o} = 0$. As a result, Eq. (13.11) is reduced to

$$\dot{\mathbf{p}} = -\sum_{m=1}^{M} \mathbf{z}_{n_m}^{(A)}. \tag{13.13}$$

Consider a positive definite function

$$E(\mathbf{p}) = \sum_{n=1}^{N_A} (\mathbf{p}^T \mathbf{z}_n^{(A)} - t_n^{(A)}) o_n$$

where, according to Eq. (13.10), $o_n = 0$ whenever $\mathbf{p}^T \mathbf{z}_n^{(A)} - t_n^{(A)} \le 0$. Using Eqs. (13.11) and (13.12), one can easily show that

$$\dot{E} = \dot{\mathbf{p}}^T (\sum_{n=1}^{N_A} o_n \mathbf{z}_n^{(A)}) = -\| \sum_{n=1}^{N_A} o_n \mathbf{z}_n^{(A)} \|^2 \le 0.$$

In other words, Eq. (13.13) will push point **p** backward until all $\{o_n\}_{n=1}^{N_A}$ become zero, indicating that point **p** has been moved inside polygon A.

Given the two extreme cases 1 and 2, case 3 is inevitable. When it happens, Eq. (13.11) will switch form between Eqs. (13.9) and (13.13). As a result, point **p** will zigzag along a boundary of polygon A. This situation can be illustrated using the example in Figure 13.6. In this case, point $\mathbf{p}$ is initially at $\mathbf{p}_o$. The modified Hamming net returns $\mathbf{z}_2^{(B)}$ because $p_{2z} = \max\{p_{nz}\}$. The integration equation begins with Eq. (13.9), which pushes $\mathbf{p}$ outside polygon A along the inverse direction of $\mathbf{z}_2^{(B)}$. Once $\mathbf{p}$ is outside polygon A, the constraint subnet signals $\bar{o} = 0$, $o_1 = 1$ and outputs $\mathbf{z}_1^{(A)}$, because $\mathbf{p}^T\mathbf{z}_1^{(A)} - t_1^{(A)} > 0$. The integration equation switches to Eq. (13.13), which pushes $\mathbf{p}$ back into polygon A along the inverse direction of $\mathbf{z}_1^{(A)}$. This switching pattern is repeated, and point $\mathbf{p}$ zigzags toward the median of l_1 and l_2 around the boundary of polygon A from the right side.

Once point $\mathbf{p}$ is pushed across the median line to the left side, the geometric position of $\mathbf{p}$ will cause the indicator to change from $p_{2z} = \max\{p_{nz}\}$ to $p_{1z} = \max\{p_{nz}\}$. Consequently, the Hamming net will return $\mathbf{z}_1^{(B)}$ instead of $\mathbf{z}_2^{(B)}$. When Eq. (13.11) switches to Eq. (13.9), the movement of point $\mathbf{p}$ will be along the opposite direction of $\mathbf{z}_1^{(B)}$, which is different from the original direction $-\mathbf{z}_2^{(B)}$. This direction switch is shown in Figure 13.6 with slight exaggeration. Now the zigzag switch is between $-\mathbf{z}_1^{(B)}$ and $\mathbf{z}_1^{(A)}$. The moving point $\mathbf{p}$ is drawn toward the median line from the left side. Strictly speaking, $\|\mathbf{p} - \mathbf{p}^*\|$ does not converge to zero. Instead, $\mathbf{p}$ will vibrate in the neighborhood of point $\mathbf{p}^*$, as illustrated in Figure 13.6.

When point $\mathbf{p}$ is moved to the neighborhood of the mini-maximum position $\mathbf{p}^*$, the indicator $p_{n^*z}(\mathbf{p})$ returned by the modified Hamming net can be considered an approximation to $p^*_{n^*z}$. Its value can be substituted into Eq. (13.8) for collision identification. When a collision occurs, the modified Hamming net returns a direction vector $\mathbf{z}_{n^*}^{(B)}$, which may be switching between the z-axes of neighboring boundary frames. For the example shown in Figure 13.4, $\mathbf{z}_{n^*}^{(B)}$ will be switching between $\mathbf{z}_1^{(B)}$ and $\mathbf{z}_2^{(B)}$. The DC value outputted by the net is actually $\mathbf{v}^* = \mathbf{z}_1^{(B)} + \mathbf{z}_2^{(B)}$, which is used to direct polygon A to move out of the collision.

The CINN, shown in Figure 13.5, has a simple canonical structure. One can program it by simply calculating and storing two sets of weights and thresholds. This structure requires very simple switching and analog integrations. The complex division operations required by the neurons [15] and the sinusoidal functions required by neurons [16] are not needed.

13.6 Collision-free Motion Planning

In planning the motion of a robot or a manufacturing facility, construction of the geometric models for the moving objects, the obstacles, and the workspace is the first step. The geometry of the moving object is always given, and the moving object model, as one or a group of polyhedrens, can easily be built. The geometric model for the working environment, on the other hand, can be modeled based on either the known geometry of structured obstacles and workspace, or on the computer vision system and on-line modeling schemes for the unstructured and time-variant working environment. In most cases, the model consists of several fixed polyhedrens. Modeling a structured work environment can be easily accomplished using the widely used geometric modeling techniques. The on-line construction of obstacle and workspace models using range-finding devices, sensor fusion, and continuous model updating techniques are discussed in detail in the preceding chapter.

Motion planning is normally conducted by minimizing the total travel distance of the moving objects or the energy consumption caused by the motion, within the modeled workspace and without interfering with the modeled obstacles. The optimization-based motion planning technique is examined in depth in the preceding chapter.

In this section, emphasis will not be given to environment modeling or motion planning. Instead, integration of the new neural-network-based collision identification and distance calculation method with environment modeling and motion planning techniques will be illustrated. The neural-network-based collision identification method can significantly reduce computation time, thus making on-line motion planning possible. The on-line environment sensoring/modeling and on-line motion planning are two basic functions of intelligent robotic and manufacturing systems.

13.6.1 Equations of Motion Planning

To illustrate the use of the neural-network-based collision identification method in motion planning, a 2-D motion planning problem, as shown in Figure 13.7, is used. In this example, a polygon representing a moving object seeks the shortest path from position A_1 to position A_8. The obstacles to be avoided are represented by two rectangles, B and C. The initial trajectory of the moving polygon is represented by eight identical polygons $\{A_i\}_{i=1}^{8}$. The ith location of the moving polygon, A_i, is represented by two fixed (pivot) points on the polygon, $\mathbf{a}_i$ and $\mathbf{b}_i$. $\mathbf{a}_1$ and $\mathbf{b}_8$ are fixed, representing the origin and destination of the path. An imaginary spring connects polygons A_i and A_{i+1} via the pivot points $\mathbf{b}_i$ and $\mathbf{a}_{i+1}$. In the absence of obstacles B and C, these imaginary springs make the trajectory a straight line—the shortest path between the origin and destination as illustrated in Figure 13.7.

To achieve the minimum total travel distance objective, the motion planning problem can be formulated as

$$\min_{w.r.t.\mathbf{x}} \sum_{i=1}^{7} \|\mathbf{a}_{i+1} - \mathbf{b}_i\|^2 \tag{13.14}$$

subject to

$$I_i^B(\mathbf{a}_i, \mathbf{b}_i) \leq 0 \qquad (i = 1, \ldots, 8) \tag{13.15}$$

$$I_i^C(\mathbf{a}_i, \mathbf{b}_i) \leq 0 \qquad (i = 1, \ldots, 8) \tag{13.16}$$

where the design variable $\mathbf{x} = [\mathbf{b}_1, \mathbf{a}_2, \mathbf{b}_2, \ldots, \mathbf{a}_7, \mathbf{b}_7, \mathbf{a}_8]^T$; I_i^K ($K \in B$ or C) represents the interference between the moving polygon at the ith location and the obstacle K. I_i^K is zero or negative when no collision occurs and has a positive value representing the degree of the interference when a collision happens. In the traditional approach, the object-to-object collision identification and distance calculations are conducted by solving a constrained point-to-point distance minimization problem, as discussed in the introduction. The interference calculation is based upon more complex algorithms. When a large number of convex polyhedrons are involved, an extensive computation load results. The neural-network-based object-to-object distance calculation, collision identification, and interference

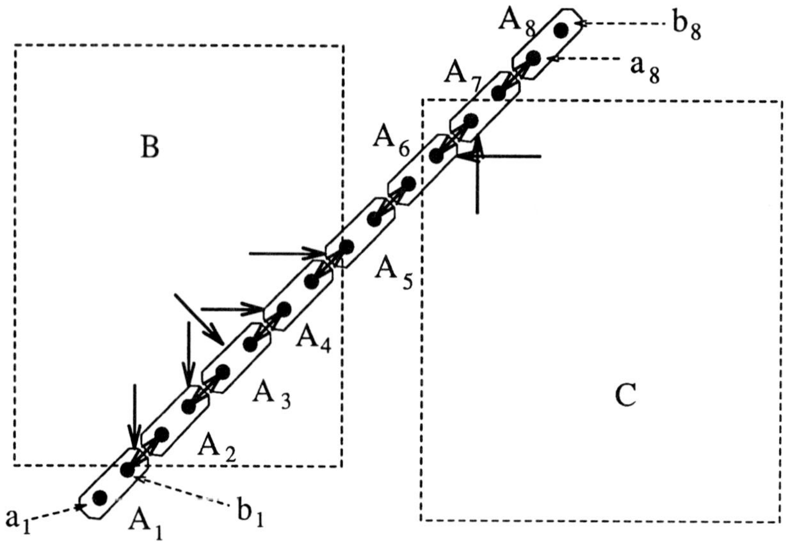

Figure 13.7: Initial Position of the Desired Path

measurement method, discussed in the previous sections, will be used to calculate the value of I_i^K. Specifically, we make

$$I_K^i(\mathbf{a}_i, \mathbf{b}_i) = \begin{cases} 0 & \text{if } {}_K^i p_{n^*z}^* > 0 \\ |{}_K^i p_{n^*z}^*| & \text{otherwise} \end{cases}$$

where, ${}_K^i p_{n^*z}^*$ is the pseudo distance between the moving polygon at the ith location to the obstacle K ($K \in B, C$).

The nice formulation of the neural network problem and the large scale parallel processing conducted by the neural networks makes the calculation of the interfence measurement, I_i^K, an easy task.

The above motion planning formulation forms a constrained nonlinear optimization problem. One way to solve this problem is to convert the original constrained optimization into unconstrained optimization using a transformation with the penalty function. The transformed, unconstrained optimization problem can then be solved using an ordinary unconstrained optimization scheme. The new optimization problem has the following form:

$$\min_{w.r.t.\mathbf{x}} \sum_{i=1}^{7} \|\mathbf{a}_{i+1} - \mathbf{b}_i\|^2 + r\sum_{i=1}^{8} <_B^i p_{n^*z}^* >^2 + r\sum_{i=1}^{8} <_C^i p_{n^*z}^* >^2 \qquad (13.17)$$

$$<_K^i p_{n^*z}^* >= \begin{cases} 0 & \text{if } {}_K^i p_{n^*z}^* > 0 \\ |{}_K^i p_{n^*z}^*| & \text{otherwise} \end{cases}$$

13.6.2 Using CINN in Motion Planning

In collision-free motion, the polygons $\{A_i\}_{i=1}^{8}$ should have no interfence with obstacles B and C. For each polygon position A_i, two CINNs are programmed to detect possible collisions between A_i and obstacles B and C. The z-axes and threshold vectors, $\{\mathbf{z}_i^{(k)}\}_{i=1}^{N_k}$ and $\{\mathbf{t}_i^{(k)}\}_{i=1}^{N_k}$, of the two CINNs, can be directly obtained from the geometric models of obstacles B and C. The main objective of using neural networks is to break a complex numerical problem into a number of simple switching or integration operations.

If A_i collides with an obstacle, the deepest point of A_i involved in the interfence, $\mathbf{p}^*$, the direction vector pointing the ideal direction for removing the collision, $\mathbf{v}^*$, and an approximation to the distance of collision-avoidance movement, $|p^*_{n^*z}|$, are returned by the CINN. As demonstrated by Figure 13.7, A_1, A_2, A_3, A_4, A_6, and A_7 initially collide with either polygon B or polygon C. The deepest interfering points and the suggested collision removal directions are all illustrated in the figure. Based upon the calculated interference and the suggested direction of collision removal, the motion planning optimization will generate a set of new locations for the moving polygon, corresponding to a new search point in the design space:

$$\mathbf{a}_i^{k+1} = \mathbf{a}_i^k + |p^*_{n^*z}| \cdot \mathbf{z}_{n^*} \quad (i \in 2, \ldots, 8)$$

$$\mathbf{b}_i^{k+1} = \mathbf{b}_i^k + |p^*_{n^*z}| \cdot \mathbf{z}_{n^*} \quad (i \in 1, \ldots, 7)$$

and start another iteration of the optimization search.

The collision-free motion planning optimization starts with a relatively small penalty factor, r, to avoid a difficult-to-solve steep valley optimization problem. The value of r gradually increases as the iteration progresses, adding more penalty to the active collision constraints. The optimization ends as r approaches infinity, and all collisions disapper. The final locations of the polygon, $\{A_i\}_{i=1}^{8}$, will present an equilibrium state as shown in Figure 13.8. In this equilibrium state, the eight locations of polygon A look like an elastic string in the presence of the rigid-body obstacles B and C. It is consistent with our physical understanding of the problem.

In the final solution, A_3 still collides with obstacle B, and A_7 collides with obstacle C. This can be avoided by adding a small positive number to the function $<^{i}_{K} p^*_{n^*z} >$.

13.7 Summary

In this chapter, the previous work and advances in object-to-object collision identification and distance calculation methods for motion planning are briefly reviewed. A new collision identification and distance calculation method using neural networks is presented.

The new neural network structure is designed for polyhedron-to-polyhedron collision identification and distance identification. The basic element of this neural network structure is a modified Hamming net for point-to-polyhedron distance calculation. With an additional subnet for constraining the moving point inside a moving object, the formulations are extended to general object-to-object distance calculation and collision identification. A

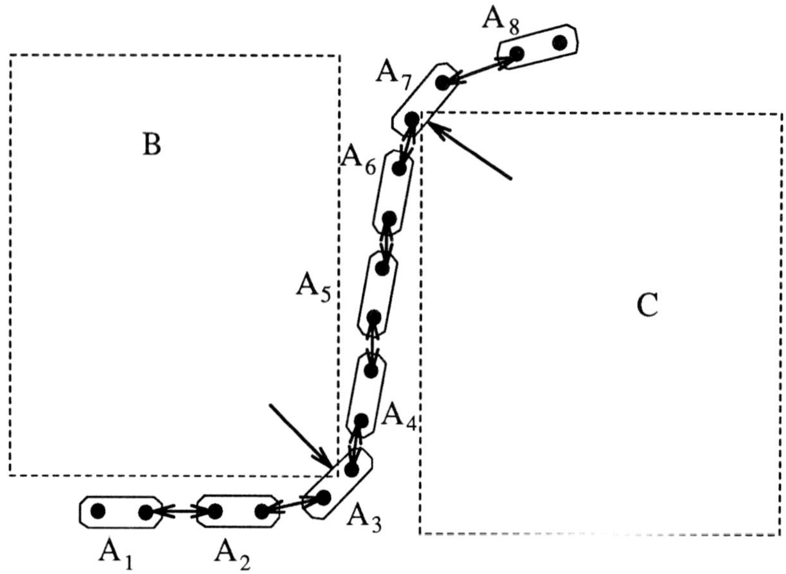

Figure 13.8: Final Position of the Desired Path

collision identification neural network (CINN) is formed. Implementation of the extended formulation using neural networks is illustrated using a neural network diagram.

The motion planning example presented in this chapter demonstrates the basic principle of the CINN, designed for motion planning. The interface of the CINN to geometric models of obstacles and workspace (either structured or unstructured), and the use of the CINN in optimization-based collision-free motion planning are illustrated through the example motion planning problem.

The introduced formulation is derived from the homogeneous geometric transformations of rotation and translation among the local coordinate frames of the obstacle plane boundaries. The method can be used to identify a possible collision between a moving object and obstacles in its workspace. The approach can also be used to calculate the object-to-obstacle distance, measure the interference, and find the optimal direction for removing the collision. The formulated calculation is simple in form and supports large-scale parallel processing. It can considerably decrease required computation time in motion planning and has the potential for being applied to on-line motion planning for intelligent robotic and manufacturing systems.

References

[1] Barr, R.O. (1969). "An Efficient Computational Procedure for a Generalized Quadratic Programming Problem," *SIAM J. Contr.*, Vol. 7, pp. 415-429.

[2] Barr, R.O., and Gilbert, E. G. (1969). "Some Efficient Algorithms for a Class of Abstract Optimization Problems Arising in Optimal Control," *IEEE Trans. Auto. Contr.*, Vol. AC-14, pp. 640-652.

[3] Bobrow, J.E. (1989). "A Direct Minimization Approach for Obtaining the Distance between Convex Polyhedrons," *The International Journal of Robotics Research*, Vol. 8, No. 3 (June), pp. 65-76.

[4] Buckley, C.E., and Leifer, L.J. (1985). "A Proximity Metric for Continuum Path Planning," *Proc. 9th Int. Joint Conf. Artificial Intelligence*, pp. 1096-1102.

[5] Cameron, S.A., and Culley, R.K. (1986). "Determining the Minimum Translation Distance between Two Convex Polyhedrons," *Proc. IEEE Int. Conf. on Robotics and Automation*, pp. 591-596.

[6] Chin, F., and Wang, C.A. (1984). "Optimal Algorithms for the Intersection and Minimum Distance Problems between Planer Polygons," *IEEE Trans. Compute.*, Vol. C-32, pp. 1203-1207.

[7] Dong, Z., and Yuan, J. (1992). "A Formulation for Collision Identification and Distance Calculation in Motion Planning Using Neural Networks," *Int. J. Adv. Manuf. Technol.*, Vol. 7.

[8] Dobkin, D.P., and Kirkpatrick, D.G. (1985). "A Linear Algorithm for Determining the Separation of Convex Polyhedrons," *J. Algorithms*, Vol. 6, pp. 381-392.

[9] Edelsbrunner, H. (1985). "On Computing the Extreme Distances between Two Convex Polygons," *J. Algorithms*, Vol. 6, pp. 515-542.

[10] Ganter, M.A., and Uicker, J.J. (1986). "Dynamic Collision Detection Using Swept Solid," *Journal of Mechanisms, Transmissions, and Automation in Design*, Vol. 108, pp. 549-555.

[11] Gilbert, E.G., and Johnson, D.W. (1985). "Distance Functions and Their Application to Robot Path Planning in the Presence of Obstacles," *IEEE Journal of Robotics and Automation*, Vol. RA-1, No. 1, pp. 21-30.

[12] Gilbert, E.G., Johnson, D.W., and Keerth, S. (1988). "A Fast Procedure for Computing the Distance between Complex Objects in Three Dimensional Space," *IEEE J. Robotics and Auto.*, Vol. 4, pp. 193-203.

[13] Hopfield, J.J. (1982). "Neural Networks and Physical Systems with Emergent Collective Computational Abilities," *Proc. Natl. Acad. Sci. USA*, Vol. 79, pp. 2554-2558.

[14] Lee, D.T., and Preparata, F.P. (1984). "Computational Geometry – A Survey," *IEEE Trans. Compute.*, Vol. C-33, pp. 1072-1101.

[15] Lee, J., and Bien, Z. (1990). "Collision-free Trajectory Control for Multiple Robots Based on Neural Optimization Network," *Robotica*, pp. 185-194.

[16] Lee, S., and Park, J. (1991), "Neural Computation for Collision-free Path Planning," *J. of Intelligent Manufacturing*, Vol. 2, pp. 315-326.

[17] Leu, M.C., Park, S.H., and Wang, K.K. (1986). "Geometric Representation of Translational Swept Volumes and Its Applications," *Journal of Engineering for Industry*, Vol. 108, pp. 113-119.

[18] Lippmann, R. P. (1987). "An Introduction to Computing with Neural Nets," *IEEE ASSP Magazine* (April), pp. 4-22.

[19] Liu, C.Y., and Mayne, R.W. (1990). "Distance Calculations in Motion Planning Problems with Interference Situations," *Proc. of the 1990 ASME Design Tech. Conf.*, DE-Vol. 23-1, pp. 145-152.

[20] Lozano-Perez, T., and Wesley, M.A. (1979). "An Algorithm for Planning Collision-free Paths among Polyhedral Obstacles," *Communications of the ACM*, Vol. 22, No. 10, pp. 560-570.

[21] Morlowski, M. (1985). "The Computation of the Distance between Polyhedrons in 3D-space," presented at *SIAM Conf. on Geometric Modeling and Robotics*, Albany, NY.

[22] Red, W.E. (1983). "Minimum Distance for Robot Task Simulation," *Robotica*, Vol. 1, pp. 231-238.

[23] Schwartz, J.T. (1981). "Finding the Minimum Distance between Two Convex Polygons," *Inform. Processor. Lett.*, Vol. 13, pp. 168-170.

[24] Wang, W.P., and Wang, K.K. (1986). "Geometric Modeling for Swept Volume of Moving Solids," *IEEE Computer Graphics and Applications*, pp. 8-17.

[25] Wolf, P. (1976). "Finding the Nearest Point in a Polytope," *Math. Programming*, Vol. 11, pp. 128-149.

Subject Index

Author Index